The Law and the Dead

The fate of the dead is a compelling and emotive subject, which also raises increasingly complex legal questions. This book focuses on the substantive laws around disposal of the recently deceased and associated issues around their post-mortem fate. It looks primarily at the laws in England and Wales but also offers a comparative approach, drawing heavily on material from other common law jurisdictions including Australia, New Zealand, Canada and the United States.

The book provides an in-depth, contextual and comparative analysis of the substantive laws and policy issues around corpse disposal, exhumation and the posthumous treatment of the dead, including commemoration. Topics covered include: the legal frameworks around burial, cremation and other disposal methods; the hierarchy of persons who have a legal duty to dispose of the dead and who are entitled to possession of the deceased's remains; offences against the dead; family funeral disputes, and the legal status of funeral instructions; the posthumous use of donated bodily material; and the rules around disinterment, and creating an appropriate memorial. A key theme of the book will be to look at the manner in which conflicts involving the dead are becoming more common in increasingly secular, multi-cultural societies where the traditional nuclear family model is no longer the norm, and how such legal contests are resolved by courts.

As the first comprehensive survey of the laws in this area for decades, this book will be of use to academics, lawyers and judges adjudicating on issues around the fate of the dead, as well as the death industry and funeral service providers.

Heather Conway is a Senior Lecturer in Law at Queen's University, Belfast.

The Law and the Dead

Heather Conway

Routledge
Taylor & Francis Group

LONDON AND NEW YORK

First published 2016
by Routledge

2 Park Square, Milton Park, Abingdon, Oxfordshire OX14 4RN
52 Vanderbilt Avenue, New York, NY 10017

Routledge is an imprint of the Taylor & Francis Group, an informa business

First issued in paperback 2020

British Library Cataloguing in Publication Data
A catalogue record for this book is available from the British Library

Library of Congress Cataloging in Publication Data
Names: Conway, Heather (Law teacher)
Title: The law and the dead / Heather Conway.
Description: Abingdon, Oxon ; New York, NY : Routledge, 2016. | Includes
bibliographical references and index.
Identifiers: LCCN 2015045010 | ISBN 9781315867656 (ebk)
Subjects: LCSH: Burial laws--England. | Dead bodies (Law)--England. | Human
body--Law and legislation--England.
Classification: LCC KD3365 .C66 2016 | DDC 344.4204/5--dc23
LC record available at http://lccn.loc.gov/2015045010

ISBN: 978-0-415-70694-0 (hbk)
ISBN: 978-0-367-59684-2 (pbk)

Typeset in Galliard
by Fish Books Ltd.

To Jo and Sarah
The best big sister and 'baby' sister that I never had

Contents

Foreword

Over the years there have been books devoted to the law relating to coroners, inquests and death registration, and books about the law relating to cadaveric organ transplantation, and books about the law relating to the provisions and management of burial grounds and crematoria and to their use, and books about the law relating to exhumation. The common thread to these is, of course, dead bodies. There has never been a book about the law relating to dead bodies as whole. At least in the UK. Until now.

Those of us who have been waiting for such a book will welcome this ambitious first attempt, which Dr Conway's previously published essays show she is well able to write. Her book comes at a most opportune time. Over the past thirty or so years there has been a burgeoning of "death studies", an umbrella term under which the students of many different academic disciplines shelter. There is a peripatetic biennial conference, "The Social Context of Death, Dying and Disposal" (DDD for short), first convened in Sussex in 1995, and since 2011 abroad as well. There is also a related Society, "The Association for the Study of Death and Society". A conference, so far biennial at least, about Death in Scotland was started in 2013. Non-lawyer participants in these ventures often find their researches encounter legal questions which they know not how to answer. There always is an answer, though often not as clear cut as they would like. Dr Conway's book is likely to become their first port of call, and, perhaps, not only for British scholars. Apart from native legal sources it draws on materials from Australia, Canada and America, all of which have multiple jurisdictions generating relevant legislation and case law, and New Zealand.

Likewise for those professionally or occupationally involved in dealing with dead bodies. In her Introduction Dr Conway adverts to two cultural changes which have the capacity for generating disputes which are increasingly resolved by resort to law: the diversification of styles in funerals generated by the growth of multiculturalism, consumerism and individualism and the increase in the dissolution and reformation of marriages/partnerships/families. There are increasingly more ways to deal with a dead body, increasingly more ways to ceremonialise the passing of a life, and increasingly more people connected to a deceased feeling they have some standing to dictate how it should be done. Those caught in the middle of disputes will be able to check their legal ground by consulting Dr Conway's book.

Then, recent scandals have shone a, sometimes unwelcome, light on the law about dead bodies. The 1990s saw disclosures about the unauthorised removal and retention of organs removed at the post-mortems of infants in English and Scottish hospitals. Inquiries following Dr Harold Shipman's conviction for murdering 15 of his patients, revealed that over a twenty-five-year period he had murdered more than 200, undetected by the procedures for investigating and registering deaths and authorising cremations. The Marchioness disaster in 1989 cast a light on procedures for identifying bodies in mass fatalities. And most recently of all, there have been unwelcome disclosures, at first in Scotland, but now in England, about how health care professionals, hospitals and crematoria have been dealing with the cremations of neonates, the stillborn and foetal remains. Many have been touched by these events and there can be few whom they have passed by. One might expect a copy of Dr Conway's book on the shelves of any university, legal or public library to become well thumbed.

Stephen White
Former Senior Lecturer, Cardiff Law School

Preface

One individual's death can affect a large number of people, when we think of the 501,424 deaths that occurred in England and Wales in 2014 alone[1] and the impact on each person's kinship and friendship networks. For those who were close to the deceased, death causes an irreplaceable loss; however, it also triggers a complex series of laws around the fate of the deceased's physical remains and associated decision-making responsibilities. Questions can arise over organ donation, and whether or not an autopsy should be carried out; the method of disposal (usually burial or cremation) and what form the deceased's funeral should take; and, further down the line, what type of memorial should be erected, or whether there are grounds for exhuming an interred corpse or ashes if the family want to relocate the remains. Each of these scenarios raises its own distinct legal issues, which are explored and analysed here.

The Law and the Dead has been two years in the making, and numerous people have helped and supported throughout. Professor Mark Pawlowksi at Greenwich University and Dr Julie Rugg at York University took the time to answer specific questions, as did the Ministry of Justice Coroners, Burials, Cremation and Inquiries Team who kindly assisted with a number of queries on exhumation law. A special word of thanks to Stephen White who has been researching and writing in this area for years, for sharing some of his extensive knowledge and for being kind enough to write an endorsement for this book. I also owe a huge debt of gratitude to colleagues in the School of Law at Queen's University – Professor Norma Dawson, Professor Kieran McEvoy, Dr Anne-Marie McAlinden, Dr John Stannard and Dr David Capper – who were generous with their time and comments in reading draft chapters; any outstanding errors and omissions are my own. On the publishing side, I am extremely grateful to the wonderful team at Routledge who brought *The Law and the Dead* to life, so to speak. To Katie Carpenter, who proposed the book, to Mark Sapwell whom I dealt with initially as the

---

1   Office for National Statistics, *Statistical Bulletin: Deaths Registered in England and Wales, 2014* (see www.ons.gov.uk/ons/rel/vsob1/death-reg-sum-tables/2014/sb-deaths-first-release—2014.html, accessed 30 September 2015).

Editorial Assistant for Law and to his successor, Olivia Manley; Olivia not only kept me on track, but answered my numerous emails about referencing, house-style, format, indexes, etc, with typical patience and professionalism when I was probably driving her to the proverbial point of distraction – her help has been immense. Thanks also to Ashlie Jackman as Production Editor, and to Mark and Karl at Fish Books for copy editing and preparing the final manuscript.

On a more personal level, thanks to Mark, Sinead, Mags and Treena – good friends, who encouraged, supported and kept me sane throughout this entire project. Jo, Sarah, mum and dad have, as ever, been a constant source of love and laughter. And finally, to Peter, my 'once in a lifetime'; there are no words to express my love or my gratitude.

Dr Heather Conway
Belfast
October 2015

Table of cases

Australia

Canada

England and Wales

European Court of Human Rights

France

Israel

New Zealand

Northern Ireland

Scotland

United States of America

Table of statutes

England and Wales

Introduction

A corpse is always a problem – both for the living and for the dead.[1]

Human beings are mortal; all of us will die at some point. Death, as a natural event, touches everyone at some stage during their lives and usually more than once. It also triggers a range of legal consequences, and while thoughts might instinctively turn to distributing assets and wealth, the fate of the deceased's body is the main concern in the immediate post-mortem period.

Once death has been confirmed and legally certified,[2] disposal of the corpse becomes a pressing issue[3] and multiple decisions have to be made by surviving relatives: which funeral director should be appointed; should the body be embalmed; should the deceased be buried or cremated; what form should any funeral ceremony take and who should participate; must the deceased's wishes be upheld; where should the corpse or post-cremation ashes[4] be interred or placed; and, in all these scenarios, who has the right to decide? Subsequent events, such as exhuming the deceased's remains or erecting an appropriate memorial, raise their own distinct issues, as surviving relatives contemplate the available options – for example, can a buried corpse be moved elsewhere, and at whose request; what type of gravestone can be erected, and who has the final say on its design and wording?

These are compelling and emotive subjects, and a substantive body of law exists in England and Wales, dealing with disposal of the dead and related

---

1   B Lovejoy, *Rest in Pieces: The Curious Fate of Famous Corpses* (Simon & Schuster, 2013), p xv.

2   Which generates a range of legal issues in itself, including medico-legal standards for determining when someone has actually died, the investigation of suspicious or sudden deaths by the coroner, and consent requirements for organ donation where this is a viable option.

3   'Disposal' is a somewhat clinical term, but is used deliberately here and throughout the following chapters to denote the process by which human remains are removed from the active realm of the living, eliminating the risk to public health and sensory offence caused by a decaying corpse. In most instances, disposal denotes both burial and cremation as the two main ways of achieving this, though other methods are discussed in Ch 2.

4   Sometimes referred to as 'cremains'.

issues. There is no comprehensive legal framework; the core principles are scattered across a range of sources, from legislation passed from the nineteenth century onwards (remnants of which remain in place today), to statutory instruments, common law and equitable rules laid down in older and more recent cases,[5] and Church of England ecclesiastical law with its unique historical influence and burgeoning volume of case law on areas such as exhumation. The law's treatment of dead bodies has also been influenced by the fact that they are not simply organic material; as a potent "symbol of the pre-mortem person",[6] corpses attract special treatment. Social mores demand respect for the dead (something which is also reflected in the law's attitude towards human remains), while religious and cultural attitudes also play an important role.[7] What legal rights exist in dead bodies has been the subject of intense debate, raising complex issues around the deceased's personhood and autonomy interests,[8] about who has decision-making authority (whether on behalf of the deceased or in their own right),[9] and whether or not human remains can be categorised as property. As physical matter, the answer to the last point might seem self-evident. However, the common law does not regard corpses as property, perpetuating a legal myth that dates back to the seventeenth century[10] and is now firmly entrenched in English law and in common law jurisdictions with derivative legal systems.[11]

5   The vast majority are first instance decisions; there is comparatively little appellate authority.

6   S McGuiness and M Brazier, "Respecting the Living Means Respecting the Dead Too" (2008) 28 *Oxford Journal of Legal Studies* 297, p 305. See also PM Quay, "Utilizing the Bodies of Dead" (1984) 28 *St Louis University Law Journal* 889, p 902 (this disintegrating mass of human material is "still uniquely related to the original person").

7   Different belief systems and cultures have specific views on death and bodily integrity. See generally D Rees, *Death and Bereavement: The Psychological, Religious and Cultural Interfaces* (Whurr, 2nd edn, 2001).

8   See for example, D Sperling, *Posthumous Interests: Legal and Ethical Perspectives* (CUP, 2008) and KR Smolensky, "Rights of the Dead" (2009) 37 *Hofstra Law Review* 763.

9   These are recurring themes throughout Chs 3–8.

10   The decision in *Haynes Case* (1614) 12 Co Rep 113 (that a dead person could not own property) was subsequently misinterpreted as ruling that a dead body was not property in any legal sense. This spawned a line of similar authority – see *Exelby v Handyside* (1749) 2 East PC 652, *R v Lynn* (1788) 2 Term Rep 733, *R v Scott* [1842] 2 QB 248, *Re Sharpe* (1857) Dears and Bell 160, *Williams v Williams* (1882) 20 Ch D 659 and *R v Price* (1884) 12 QBD 247. For an excellent critique, see P Matthews, "Whose Body? People as Property" (1983) 36 *Current Legal Problems* 193, pp 196–221.

11   For example, Australia and New Zealand. Despite misgivings over its origins, the English Court of Appeal endorsed the rule in *Dobson v North Tyneside Area Health Authority* [1996] 4 All ER 474 and *R v Kelly* [1998] 3 All ER 741, the latter case suggesting that it could only be undone by Parliament. See also the discussion in *Yearworth v North Bristol NHS Trust* [2009] EWCA Civ 37.

The consequences are far-reaching,[12] and the 'no property' rule continues to pervade contemporary legal discourse around the post-mortem status of the dead.[13]

Other factors are also exerting pressure on the laws that apply in this area. In recent decades, England and Wales (like other Western societies) has witnessed two major social changes that have shaped attitudes towards death and what happens to the resultant corpse or ashes. The first is increased religious and cultural diversity – not simply a trend towards secularisation (though a decline in religious adherence is part of the overall picture), but what Howarth more accurately describes as "a rejection of orthodox Christianity of the established churches".[14] What we now have is a multi-faith, multi-ethnic society, incorporating a range of different death rites and customs, and creating new tensions around the fate of the dead as the

---

12 For example, corpses cannot be bought and sold in any commercial sense (nor seized by creditors as a means of discharging the deceased's debts – *R v Fox* [1841] 2 QB 246 and *R v Scott* [1842] 2 QB 248), and cannot be stolen because the offence of theft presupposes the existence of property (*Re Sharpe* (1857) Dears and Bell 160, though the position may be different in Scotland as a result of *HM Advocate v Dewar* 1945 SLT 114, which suggested that unburied bodies are capable of being stolen – see the discussion in S White, "The Legal Status of Corpses and Cremains: When and Where Can You Steal a Dead Body?" in S Buckham, PC Jupp and J Rugg (eds), *Death in Scotland 1855–1955: Beliefs, Attitudes and Practices* (Peter Lang, forthcoming)). Funeral instructions are also ineffective, since a person cannot apparently bequeath their body if it is not property – *Williams v Williams* (1882) 20 Ch D 659.

13 Hardly surprising in an area suffused with notions of ownership and control – see for example, TL O'Carroll, "Over My Dead Body: Recognizing Property Rights in Corpses" (1996) 29 *Journal of Health and Hospital Law* 238; PD Skegg, "The 'No Property' Rule and Rights Relating to Dead Bodies" [1997] *Tort Law Review* 222; R Atherton, "Claims on the Deceased: The Corpse as Property" (2000) 7 *Journal of Law and Medicine* 361; and JK Mason and GT Laurie, "Consent or Property? Dealing with the Body and its Parts in the Shadow of Bristol and Alder Hey" (2001) 64 *Modern Law Review* 710. This is part of an ongoing and much broader debate around whether people should own their own bodies. Again, there is a wealth of literature on this topic, but see M Bray, "Personalizing Property: Toward a Property Right in Human Bodies" (1990) 69 *Texas Law Review* 209; A Grubb, "'I, Me, Mine': Bodies, Parts and Property" (1998) 3 *Medical Law International* 299; L Skene, "Arguments Against People Legally 'Owning' Their Bodies, Body Parts and Tissue" (2002) 2 *Macquarie Law Journal* 165; A George, "Is 'Property' Necessary? On Owning the Human Body and Its Parts" (2004) 10 *Res Publica* 15; R Hardcastle, *Law and the Human Body: Property Rights, Ownership and Control* (Hart, 2007); D Dickenson, *Property in the Body: Feminist Perspectives* (CUP, 2007); M Pawlowski, "Property in Human Body Parts and Products of the Human Body" (2009) 30 *Liverpool Law Review* 35; J Wall, "The Legal Status of Body Parts: A Framework" (2011) 31 *Oxford Journal of Legal Studies* 659; M Quigley, "Property in Human Biomaterials: Separating Persons and Things?" (2012) 32 *Oxford Journal of Legal Studies* 659; M Render, "The Law of the Body" (2013) 62 *Emory Law Journal* 549; and the collection of essays in I Goold, K Greasley, J Herring and L Skene (eds), *Persons, Parts and Property: How Should We Regulate Human Tissue in the 21st Century?* (Hart, 2014).

14 G Howarth, "The Rebirth of Death: Continuing Relationships with the Dead" in M Mitchell (ed), *Remember Me: Constructing Immortality* (Routledge, 2007), p 26.

dominant legal paradigm struggles to adapt accordingly. The second major social change has been the decline of the traditional nuclear family model and the emergence of reconstituted or blended families.[15] This re-ordering of kinship networks – alongside an increasingly diverse array of personal relationships and living arrangements – has challenged conventional assumptions about who constitutes the deceased's 'family', and who is legally entitled to decide the funeral arrangements and other death-related issues. Looking beyond these factors, advances in medical science have also raised questions about the use of living material supplied by dead donors, with organ donation and posthumous reproduction raising a host of legal (and ethical) issues;[16] and when it comes to disposal of the dead, environmental concerns are playing an increasingly important role as existing methods become more heavily regulated and new, ecologically sensitive alternatives are introduced.[17] Finally, the digital age is also affecting traditional funerary practices, with live-streaming of funerals via the internet allowing distant relatives and friends to participate in some way[18] and virtual memorials to the dead (often through social media) enabling collective acts of online commemoration regardless of geographical separation.[19] Whether these are socially acceptable mourning rituals is open to debate;[20] from a legal perspective, questions of access and control have yet to be addressed.

15   See for example, D Chambers, *A Sociology of Family Life* (Polity, 2012).
16   See Ch 6.
17   See Ch 2.
18   See for example, G Swerling, "Live Streaming Takes Funerals into Digital Age", *The Times* (London, 27 December 2014) www.thetimes.co.uk/tto/technology/article 4307581.ece (accessed 30 September 2015).
19   See Ch 8.
20   See for example, CM Moreman and AD Lewis (eds), *Digital Death: Mortality and Beyond in the Online Age* (ABC-CLIO, 2014), L Penny, "Mourning in the Digital Age: Selfies at Funerals and Memorial Hashtags", *The New Statesman* (London, 14 April 2014) www.newstatesman.com/culture/2014/04/selfies-funerals-and-memorial-hash tags-mourning-digital-age (accessed 30 September 2015) and J Meese, M Gibbs, M Carter, M Arnold, B Nansen and T Kohn, "Selfies at Funerals: Mourning and Presencing on Social Media Platforms" (2015) 9 *International Journal of Communication* 14.

In recent years, key aspects of death, dying and disposal have been explored from a range of perspectives – most notably by sociologists and anthropologists researching in the area of 'death studies'.[21] From a legal perspective, however, the analysis has been fairly selective, with a disproportionate emphasis on property rights in corpses[22] and medical law issues surrounding bodies and body parts.[23] A handful of recent publications have taken a more holistic view, looking at the legal consequences of death and the post-mortem status of the deceased's remains.[24] However, with the possible exception of Cantor[25] (who, in any event, deals predominantly with the American position), relatively little attention has been paid to disposal of the dead and the various issues identified in the paragraphs above, despite their centrality to both everyday life and modern systems of law.

---

21   The numerous monographs and edited collections include the following: G Howarth and PC Jupp (eds), *Contemporary Issues in the Sociology of Death, Dying and Disposal* (Palgrave Macmillan, 1995); C Seale, *Constructing Death: The Sociology of Dying and Bereavement* (CUP, 1998); E Hallam and J Hockey, *Beyond the Body: Death and Social Identity"* (Routledge, 1999); DJ Davies, *Death, Ritual and Belief: The Rhetoric of Funeral Rites* (Continuum, 2nd edn, 2002); G Howarth, *Death and Dying: A Sociological Introduction* (Polity, 2007); M Mitchell (ed), *Remember Me: Constructing Immortality: Beliefs on Immortality, Life and Death* (Routledge, 2007); J Hockey, C Komaromy and K Woodthorpe (eds), *The Matter of Death: Space, Place and Materiality* (Palgrave Macmillan, 2010); and DJ Davies and H Rumble, *Natural Burial: Traditional-Secular Spiritualities and Funeral Innovation* (Bloomsbury, 2012). A large number of journal articles have also been published in these fields (various examples are referenced in the chapters that follow), and there are several international journals devoted to the interdisciplinary study of death and dying: *OMEGA: Journal of Death and Dying* (published by SAGE publications), *Death Studies* (published by Taylor and Francis) and *Mortality* (published by Taylor and Francis).

22   See the sources cited at n 13.

23   As well as organ donation and the posthumous use of sperm and embryos (see the various sources cited in Ch 6), the legal and ethical issues posed by the use of cadavers and excised body parts in medical research – see D Nelkin and L Andrews, "Do The Dead Have Interests: Policy Interests for Research After Life" (1998) 24 *American Journal of Law and Medicine* 261; M Brazier, "Retained Organs: Ethics and Humanity" (2002) 22 *Legal Studies* 550; MH Klaiman, "Whose Brain Is It Anyway: The Comparative Law of Post-Mortem Organ Retention" (2005) 26 *Journal of Law and Medicine* 475; and RL Walker, ET Juengst, W Whipple and AM Davis, "Genomic Research With the Newly Dead: A Crossroads for Ethics and Policy (2014) 42 *Journal of Law, Medicine and Ethics* 220.

24   Most notably, B Brooks-Gordon, F Ebtehaj, J Herring, MH Johnson and M Richards (eds), *Death Rites and Rights* (Hart, 2007), NL Cantor, *After We Die: The Life and Times of the Human Cadaver* (Georgetown University Press, 2010) and R Madoff, *Immortality and the Law: The Rising Power of the American Dead* (Yale University Press, 2010). Selected aspects are also considered in A Bainhaim, SD Slater and M Richards, *Body Lore and Laws* (Hart, 2002) and in SW Smith and R Deazley, *The Legal, Medical and Cultural Regulation of the Body: Transformation and Transgression* (Ashgate, 2009).

25   *Ibid.*

The Law and the Dead charts a different intellectual path by focusing solely on the fate of the recently dead – looking primarily at disposal, but also the related areas of exhumation and commemoration, as well as the posthumous use of donated bodily material.[26] It begins by examining the legal regime that is triggered by death (for example, registration requirements, coronial investigations and criminal offences against the dead),[27] before analysing the range of contemporary disposal options (not just burial and cremation, but less conventional alternatives) and their regulatory frameworks.[28] The focus then shifts to the duty of disposal and right to possession of the deceased's remains (who determines the funeral arrangements, and the scope of any legal entitlements),[29] and how the law responds when faced with disputes between surviving relatives over the method of disposal and attendant funeral rites – an increasingly common occurrence as the decline of the baby-boom generation coincides with the rise of alternative family forms.[30] Closely aligned with this is the idea of allowing an individual to dictate their own posthumous fate by setting out funeral instructions, which the law would protect and revert to in the event of family disagreements.[31] The same theme is also explored in the law's response to cadaveric organ transplants and posthumous reproduction, alongside the efficacy of current legislative regimes.[32] Finally, *The Law and the Dead* looks at requests to disinter the dead, the strict legal controls that apply here and the various consents that must be secured before this can take

---

26   Since the focus is on the legal issues triggered by death (and not those immediately preceding it), the book does not explore end-of-life issues such as advance directives, or euthanasia and assisted dying. These are analysed elsewhere – see for example, A Maclean, "Advance Directives, Future Selves and Decision-Making" (2006) 14 *Medical Law and Ethics* 291; D Nachman, "Living Wills: Is It Time to Pull the Plug?" (2010) 18 *Elder Law Journal* 289; JT Monahan and EA Lawhorn, "Life-Sustaining Treatment and the Law: The Evolution of Informed Consent, Advance Directives and Surrogate Decision-Making" (2009) 19 *Annals of Health Law* 107; J Keown, *Euthanasia, Ethics and Public Policy: An Argument Against Legalisation* (CUP, 2002); S McLean, *Assisted Dying: Reflections on the Need for Law Reform* (Routledge, 2007); and T Delamothe, R Snow and F Godlee, "Why the Assisted Dying Bill Should Become Law in England and Wales" (2014) 349 *BMJ: British Medical Journal* 4349. Likewise, the focus is on the recently dead, and not on legal contests involving ancient human remains such as museum specimens and preserved corpses. Again, these are explored elsewhere – for a sample, see J Hubert, "Dry Bones or Living Ancestors? Conflicting Perceptions of Life, Death and the Universe" (1992) 1 *International Journal of Cultural Property* 105, PR Afrasiabi, "Property Rights in Ancient Skeletal Remains" (1997) 70 *Southern California Law Review* 805, and A Bernick, "Burying an Injustice: Indigenous Human Remains in Museums and the Evolving Obligations to Return Remains to Indigenous Groups" (2014) 1 *Indonesian Journal of International & Comparative Law* 637.
27   Ch 1.
28   Ch 2.
29   Ch 3.
30   Ch 4.
31   Ch 5.
32   Ch 6.

place,[33] before switching its attention to memorials, what is broadly permissible and how the law deals with competing claims to denote both the deceased's passing and primary relationships – especially in the virtual world.[34]

These are distinct yet intimately connected areas, which have never been addressed in a single source before. Some have been looked at in isolation; for example, there are essential texts on coroners' law and practice,[35] while *Davies' Law of Burial, Cremation and Exhumation*[36] provides a comprehensive overview of those specific topics for professionals working in the field. Other reference works have been out-of-date for years,[37] while the relevant sections of *Halsbury's Law of England*[38] give an encyclopaedic account of the relevant laws with no underlying contextual analysis. In mapping out a complex and highly significant field of study, *The Law and the Dead* differs in a number of key respects – beyond its scope and contemporary resonance. Throughout the various chapters, the book navigates a dense and disparate body of material, drawing together case law, legislation, academic writings, and law reform and policy documents to create a defining set of principles. In providing an in-depth, critical analysis of the issues surrounding disposal and the post-mortem fate of the dead it reveals a number of discrepancies in the applicable legal frameworks, and explores issues that modern-day common law systems have been slow to address – for example, religious and cultural diversity; changing definitions of 'family'; the increased emphasis on rights-based discourse; the greater variation in modern funeral rites and the development of new disposal methods; and the influence of digital technology.

Beyond its consideration of substantive laws and policies, *The Law and the Dead* addresses a number of recurring themes that contribute to its overarching theoretical framework. For example, what are the competing public and private interests affecting the law's treatment of the dead? What social, moral and cultural dictates have influenced the relevant regulatory frameworks, and how will these impact on future reforms? How does the law

---

33   Ch 7.

34   Ch 8.

35   Namely C Dorries, *Coroners' Courts: A Guide to Law and Practice* (OUP, 3rd edn, 2014) and P Matthews, *Jervis on the Office and Duties of Coroners* (Sweet & Maxwell, 13th edn, 2014).

36   The most recent version is DA Smale, *Davies' Law of Burial, Cremation and Exhumation* (Shaw & Sons, 7th edn, 2002). A new edition is planned for late 2017.

37   Other notable contributions to the field are now also out-of-date – in particular CJ Poulson, RP Brittain and TK Marshall, *The Disposal of the Dead* (English Universities Press, 1953) (updated in 1962, and again in 1975), as well as the leading American texts of HY Bernard, *The Law of Death and Disposal of the Dead* (Oceana Publications, 1966) and PE Jackson, *The Law of Cadavers and of Burial and Burial Places* (Prentice Hall, 2nd edn, 1950).

38   See for example, "Cremation and Burial", 24 *Halsbury's Laws of England* (5th edn, 2010).

restrict the choices available to both the living and the dead, and who has decision-making powers? Can an individual determine their own post-mortem fate? And how does the law deal with a clash of rights over the dead – occasionally between the state and surviving relatives, but more usually within families or where the preferences of the living conflict with the wishes of the dead? More generally, the overall contextual analysis is enriched by drawing on literature from a range of disciplines – law, sociology, anthropology, science and medicine – to infuse the discussion and offer alternative perspectives, while media reports offer some insight into public perceptions of the issues around death, dying and disposal.

The Law and the Dead deals with an area of law that has frequently been overlooked, and one that has seen a growing interest in recent years amongst academics, judges and lawyers, as well as the death industry and associated service providers. The subject matter also has immense human interest; treatment of the dead is a central feature of every society, and is hugely important to private citizens faced with a series of decisions when a loved one dies. In terms of jurisdictional remit, the primary focus is on England and Wales, though the same basic laws are applicable in Northern Ireland and (to a lesser extent) Scotland.[39] However, the book also draws heavily on material from other common law jurisdictions – most notably Australia, but also New Zealand, Canada and the United States of America. These jurisdictions have been strongly influenced by English law, and display similar legal (and socio-cultural) attitudes towards dying and what happens to the dead. Looking at these different countries also highlights areas where new approaches are required, and reveals solutions that could be mapped onto our own legal system.

In drawing all of the aforementioned themes and perspectives together in one book, *The Law and the Dead* creates the definitive, modern work in the area – offering an in-depth, contextual and uniquely comparative analysis of the substantive laws and policy issues around disposal of the dead and related topics. With the exception of one or two late additions, the law is stated as at 30 September 2015.

---

39   The same is also true of Jersey, Guernsey, the Isle of Man and the Republic of Ireland.

1 Death and its immediate aftermath

Death is unique ... unlike aught else in its certainty and its incidents.[1]

Introduction

Death not only marks the cessation of human life; it triggers a complex series of laws around the fate of the dead, and disposal of the physical matter that is left behind.[2] Certain legal requirements must be fulfilled before this occurs, involving a number of different institutions and professions. While these are analysed below, defining 'death' is the starting point in any discourse involving the law and the dead. However, this is not as straightforward as it might seem.

I.   The legal definition of 'death'

While life and death are binary concepts, the precise moment at which one ends and the other begins can be hard to determine in a complex multicellular organism like the human body. The transition from living being to corpse is exactly that – and if physical death is a series of discrete yet interlinked biological processes (as opposed to an instantaneous cessation of life), constructing an exact definition is not easy.[3] Medical advances mean

---

1   Lumpkin J in *Louisville & NR Co v Wilson* 51 SE 24, 25 (Ga 1905).
2   The emphasis is on an individual's *physical* demise, as opposed to 'legal death'. The latter invokes the basic rule that a person may be presumed dead if missing or unheard of for seven years, even though there is no corpse or other direct evidence of death.
3   Mims describes it as "like trying to define night as opposed to day, [or] having to pick an arbitrary cut-off point at some stage during that transition period of dusk and dawn" – C Mims, *When We Die: The Science, Culture and Rituals of Death* (Robinson Publishing, 2000), p 116. While most of the discourse here, and in subsequent chapters, focuses on individuals who had some independent existence, the dead can also include stillborn children. Section 1(1) of the Stillbirth (Definition) Act 1992 defines a stillborn child as one born after the 24th week of pregnancy and which did not breathe after birth or show any other sign of life.

that we now know with much greater certainty when someone has died; crude techniques that were once used to detect flickering signs of life – the "legendary feather in front of the nose or mirror in front of the mouth"[4] – have long since been abandoned.[5] However, modern medicine also makes an accurate determination of death imperative in two situations: withdrawal of life-support systems from intensive care patients where certain bodily systems are still functioning, and removal of organs for transplant purposes within a short time of death.[6]

Contemporary death indices are based on identifiable and assessable medical criteria.[7] Like most Western jurisdictions, English law adopts two basic standards.[8] The first relies on the permanent cessation of heart and respiratory function – otherwise known as 'cardiopulmonary death'. The second is known as 'brain stem death' and is the accepted medical definition in the relatively small number of cases where an individual is unconscious, unable to breathe spontaneously and has suffered irreversible damage to parts of the brain stem;[9] that person's heartbeat and breathing are being artificially sustained. In these circumstances, a diagnosis of brain stem death serves two vital purposes: physicians will not face criminal liability if a decision is made to remove life-sustaining care or organs are harvested from someone who is clinically dead, while the living are reassured that medical treatment has only ceased because the individual is incapable of recovery and that organs are only removed from deceased donors.

---

4   NL Cantor, *After We Die: The Life and Times of the Human Cadaver* (Georgetown University Press, 2010), p 12.

5   These and other old methods of determining whether death had occurred were not always reliable, and individuals could be mistakenly buried when merely unconscious. For some of the more macabre tales, see W Hadwen, W Tebb and EP Vollum, *Premature Burial: How It May Be Prevented* (edited by Jonathan Sale, Hesperus, 2012).

6   See Mims (n 3), pp 115–116.

7   While 'death' has numerous meanings, the medico-legal definition is based on biological processes and what happens to the physical body; it may not be compatible with philosophical, theological and societal understandings of death, all of which pose different challenges – see the analysis in PL Chau and J Herring, "The Meaning of Death" in B Brooks-Gordon, F Ebtehaj, J Herring, MH Johnson and M Richards (eds), *Death Rites and Rights* (Hart, 2007).

8   There is no statutory definition. Instead, English courts rely on clinical criteria, developed in *Re A* [1992] 3 Med L Rev 303 and *Airedale NHS Trust v Bland* [1993] AC 789, which are still regarded as the two leading cases. For a more in-depth analysis of this legal definition, see JK Mason and GT Laurie, *Mason & McCall Smith's Law and Medical Ethics* (OUP, 9th edn, 2013), ch 16; J Herring, *Medical Law and Ethics* (OUP, 5th edn, 2014), pp 477–482; and SD Pattinson, *Medical Law and Ethics* (Sweet and Maxwell, 4th edn, 2014), pp 424–428.

9   See Herring, *ibid*, pp 478–479.

The widespread adoption of the brain stem death standard has proved controversial,[10] though Pattinson suggests that this particular definition has a "soothing rhetorical force" by appearing to "facilitate activities that would invite greater challenge if thought to be conducted on the living".[11] Meanwhile shifting and mutable definitions of death are a distinct possibility in the future, as medical science advances further – for example, if dead tissue can be restored by stem cell manipulation,[12] transplanting human heads onto donor bodies becomes a reality,[13] or if cryogenic freezing results in 'dead' subscribers being successfully reanimated in years to come.[14]

II.   Organ donation

Most organ donation occurs within the brain-stem death setting, given the time-sensitive nature of the procedure. However, harvesting the deceased's organs not only invokes strict legal requirements under the Human Tissue Act 2004; it also raises complex issues around respect for autonomy where the deceased pledged to donate their organs, and how the law reconciles conflicting rights and deals with objections from surviving relatives.[15]

III.   Documenting death

Individual deaths trigger compulsory registration requirements, and English law insists that both the fact *and* the cause of death be recorded.[16] As the state's acknowledgment that someone has died, registration paves the way for disposal of the dead to take place, prevents illicit crimes being concealed and (in the longer-term) provides documentary evidence that allows the

---

10   See the discussion in Chau and Herring (n 7). Note that laws enacted in the state of New Jersey allow an attending physician to disapply the brain death standard if he/she has reason to believe this would violate the patient's religious beliefs – see R Madoff, *Immortality and the Law: The Rising Power of the American Dead* (Yale University Press, 2010), p 39.

11   Pattinson (n 8), p 426.

12   See Cantor (n 4), p 21.

13   Surgeons claim that this is a short-term reality – see "Frankenstein-Style Human Head Transplant 'Could Happen in Two Years'", *The Telegraph* (London, 26 February 2015) www.telegraph.co.uk/news/science/science-news/11436319/Frankenstein-style-human-head-transplant-could-happen-in-two-years.html (accessed 30 September 2015).

14   Cryonics is discussed in Ch. 2, Pt V.

15   See Ch 6, Pt I.

16   See generally J Green and M Green, *Dealing with Death: A Handbook of Practices, Procedures and Law* (Jessica Kingsley, 2nd edn, 2008), Chs 2–3 and "Cremation and Burial", 24 *Halsbury's Laws of England* (5th edn, 2010), paras [1130]–[1133] (listing the applicable legislation, though the main provisions can still be found in the Births and Deaths Registration Act 1953, Pt II). While a stillbirth (see n 3) can be registered, a spontaneous delivery (or one resulting from surgical intervention) before the 24-week maturity term is classed as a 'non-viable foetus'; registration is not possible here – Green and Green, *ibid*, pp 118–119.

deceased's estate to be wound up. At a national level, registering deaths allows accurate mortality rates to be compiled; statistical analysis then signposts public health trends and allows the government to plan its longer-term resource allocation.

Every death occurring in England and Wales[17] must be registered in the sub-district in which death occurred (or in which the deceased's body was found) within 5 days.[18] The person who assumes responsibility for this (usually a member of the deceased's immediate family) must fall within the list of qualified informants under the Births and Deaths Registration Act 1953,[19] and must also provide the necessary documentation – including a medical certificate indicating the cause of death.[20] The registrar then issues a certificate of registration,[21] which allows the funeral to go ahead.[22] However, if a suspicious, violent or otherwise unexplained death has been reported to the coroner,[23] registration is dependent on the coroner furnishing subsequent information on the cause of death.[24]

Given the importance attached to recording deaths and the legal consequences that flow from issuing a registrar's certificate, it is hardly surprising that non-compliance attracts criminal law offences. For example, in England and Wales, failure to inform the registrar of a death or to provide the relevant information is an offence under s 36 of the 1953 Act,[25] while s 35 makes it an offence for a registrar to fail to register a death "without reasonable cause".[26]

---

17   And in the rest of the UK, though with slight variances (for example, a death can be registered within 8 days in Scotland).

18   Births and Deaths Registration Act 1953, s 15. However, this period can be extended to 14 days if the registrar is satisfied that the necessary medical certificate (see below) has been issued.

19   1953 Act, s 16 (information concerning death in a house) and s 17 (information concerning other deaths). This is a designated legal responsibility; the qualified informant cannot delegate the function to someone else.

20   1953 Act, s 22.

21   Commonly referred to as the death certificate. While registration is free of charge (1953 Act, s 20) certified copies of the entry on the register incur a fee.

22   1953 Act, s 24 – though the same provision also allows burial (though not cremation) to go ahead without formal registration where the cause of death has been medically certified; this facilitates ethnic minority groups whose religious and/or cultural dictates mandate immediate disposal of the dead. In any event, there is a legal requirement to notify the registrar that disposal has taken place within 96 hours of this occurring; failure to do so within 14 days triggers an inquiry by the registrar, to ensure that the body has been disposed of – Births and Deaths Registration Act 1926, s 3(1) and 1953 Act, s 24(5). See generally 24 *Halsbury's Laws of England* (n 16), para [1132].

23   See Pt III.

24   1953 Act, s 23.

25   As in *R v Wriggleworth* (Leeds Crown Court, 21 January 2005) and discussed in Ch 2, p 43.

26   Losing or damaging the register is also an offence under s 35 (though this is unlikely today, given the use of computerised registers).

There are also specific legal requirements around transporting bodies in and out of England and Wales.[27] Where an individual died abroad, the registrar for the relevant district within England and Wales must issue an appropriate certificate before disposal of the repatriated remains can occur.[28] The coroner may also be involved if death was not due to natural causes.[29] If death occurred in England and Wales but the body is being moved elsewhere, this cannot take place unless four days' notice of the intended removal has been given to the coroner – thus preventing a body being removed before a suspicious or unexplained death can be properly investigated.[30]

IV. Suspicious and unexplained deaths: The coroner's jurisdiction

There are two types of autopsy in England and Wales. The first investigates the deceased's death in a hospital or clinical setting, and is dependent on consent from the next-of-kin.[31] The second occurs under the coronial system, which warrants some form of autopsy for reportable deaths, and symbolises an ongoing connection between the dead and the institutions of the state.[32]

---

27   See generally 24 *Halsbury's Laws of England* (n 16), paras [1130]–[1131]. Multiple air transportation routes and preservation techniques (such as embalming, refrigerated storage) enable repatriation if the deceased died while in another country or the intent is to bury the body elsewhere. While more complex issues arise if the deceased died outside the UK, specific legal requirements also apply to moving bodies between England and Wales, Northern Ireland and Scotland (though not from one county to another, within these individual jurisdictions) – S Wienrich and J Speyer, *The Natural Death Handbook* (Rider Books, 4th edn, 2003), p 98.

28   1953 Act, s 24(2). Where an individual died abroad, his/her death will usually be registered according to the laws of that country.

29   This is because the repatriated body is now within the coroner's area, and (regardless of where death occurred) there is reason to believe that the deceased, for example, died a violent or unnatural death.

30   Births and Deaths Registration Act 1926, s 4 and Removal of Bodies Regulations 1954 SI 1954/448, regs 4 and 6. The only exception is if the coroner indicates, on receiving the notice, that no inquest will be held.

31   The underlying intent is to ascertain or verify the cause of death. However, the number of hospital post-mortems has fallen significantly, as a result of high-profile organ retention scandals at Bristol and Alder Hey hospitals, and elsewhere – see JK Mason and GT Laurie, "Consent or Property? Dealing With The Body and Its Parts in the Shadow of Bristol and Alder Hey" (2001) 64 *Modern Law Review* 710 and M Brazier, "Retained Organs: Ethics and Humanity" (2002) 22 *Legal Studies* 550 as well as the litigation in *AB v Leeds Teaching Hospitals NHS Trust* [2004] EWHC 644 (QB). The legal restrictions (including consent requirements) contained in the Human Tissue Act 2004, Pt 1 and Sch 1, now apply.

32   Here, the availability of a hospital post-mortem is subordinate to the functions of the coroner – Human Tissue Act 2004, s 11.

What follows is an overview of the coroner's functions and remit, and what happens to the deceased's body.[33]

1.   Investigative powers and functions

As in other countries, the coroner's functions in England and Wales are prescribed by statute, and supplemented by case law where individual decisions are challenged by judicial review. Long-overdue reforms have recently been introduced by the Coroners and Justice Act 2009 and associated regulations,[34] the bulk of which came into effect in July 2013.[35]

Under s 1 of the 2009 Act, the relevant coroner has a duty to investigate if he/she has been made aware that "the body of a deceased person is within that coroner's area"[36] and the coroner has "reason to suspect that: (a) the deceased died a violent or unnatural death, (b) the cause of death is unknown, or (c) the deceased died while in custody or otherwise in state detention".[37] Preliminary inquiries can be carried out, to determine whether there is a duty to investigate; if so, the matter proceeds to the investigation stage – either with or without an inquest. The coroner assumes control from the time the death is reported until the investigation is complete,[38] and the medical cause of death established.

An autopsy may be carried out as part of either the preliminary inquiries or the formal investigation. Under s 14, the coroner may decide to order a post-mortem to establish the cause of death or determine whether a duty to investigate arises,[39] and has a statutory power to remove the deceased's body

---

33   Detailed accounts are available in the two leading texts – namely C Dorries, *Coroners' Courts: A Guide to Law and Practice* (OUP, 3rd edn, 2014) and P Matthews, *Jervis on the Office and Duties of Coroners* (Sweet & Maxwell, 13th edn, 2014). The English coronial system has very distinct origins, though these are of historical interest today. More generally, the medico-legal nature of the coroner's investigation has been a source of conflict between the two professions – see for example, B Carpenter and G Tait, "The Autopsy Imperative: Medicine, Law and the Coronial Investigation" (2010) 31 *Journal of Medical Humanities* 205.

34   The main regulations are the Coroners (Investigations) Regulations 2013, SI 2013/1629.

35   For an overview, see Chief Coroner and Ministry of Justice, *The Chief Coroner's Guide to the Coroners and Justice Act 2009* (September 2013), located at www.judiciary. gov.uk/wp-content/uploads/JCO/Documents/coroners/guidance/chief-coroners-guide-to-act-sept2013.pdf (accessed 30 September 2015).These reforms had a lengthy gestation – see T Luce, "Coroners and Death Certification Law Reform: The Coroners and Justice Act 2009 and its Aftermath" (2010) 50 *Medicine, Science and the Law* 171.

36   2009 Act, s 1(1). The presence of a 'body' (see Dorries (n 33), pp 32–36) is not essential; s 1(4) of the 2009 Act authorises a coronial investigation where a coroner has reason to believe a death occurred in his/her area, and the circumstances merit an investigation.

37   2009 Act, s 1(2). For a detailed overview, see Dorries (n 33), pp 40–48.

38   Or discontinued, if the coroner attributes death to natural causes.

39   The post-mortem should be carried out "as soon as reasonably practicable" – 2013 Regulations, reg 11.

under s 15(1).[40] Looking beyond the statute, the coroner also has a common law entitlement to possession of the deceased's body until the inquiry is complete.[41] This is not an unfettered right, but a "limited degree of control over a body for limited purposes".[42] Having been notified of the original decision to investigate the death,[43] the deceased's personal representative or next-of-kin must also be notified of the decision to conduct a post-mortem examination.[44] During this time, any possessory rights that the personal representative has in the corpse (through being legally responsible for its disposal[45]) are temporarily displaced. Section 47(2) of the 2009 Act also lists a number of "interested persons" to any coronial investigation, most notably:

(a) a spouse, civil partner, partner,[46] parent, child, brother, sister, grandparent, grandchild, child of a brother or sister, stepfather, stepmother, half-brother or half-sister;

(b) a personal representative of the deceased;

[...]

(m) any other person whom the coroner thinks has a sufficient interest.[47]

These individuals should be consulted at key stages in the proceedings, and can be legally represented at any inquest.

---

40 Under Sch 5, para 6 of the 2009 Act, the coroner can also exhume a body where a post-mortem is required or as part of any criminal proceedings – see Matthews (n 33), [8-03]–[8-10].

41 *R v Bristol Coroner, ex parte Kerr* [1974] QB 652. Pawlowski suggests an analogous right to possession where police are investigating a crime – M Pawlowski, "Property in Human Body Parts and Products of the Human Body" (2009) 30 *Liverpool Law Review* 35, p 37.

42 *Haydon v Chivell* [1999] SASC 336, [32], discussing the equivalent legislative framework in South Australia. In England and Wales, the coroner also has discretion to release the body at an earlier stage, if it is no longer needed for investigative purposes – 2013 Regulations, reg 21.

43 2013 Regulations, reg 6.

44 2013 Regulations, regs 13(1) and 13(3) – the former creating the duty to notify and the latter listing the persons to be notified (though notification can be dispensed with if it is impracticable or would cause unreasonable delays in carrying out the post-mortem – reg 13(2)). The personal representative or next-of-kin are the first category listed under reg 13(3)(a) which also includes "any other interested person who has notified the coroner in advance of his or her desire to be represented at the post-mortem examination". Note that 'next-of-kin' is not defined under coronial legislation – an omission that Matthews (n 33), [8-24] finds "surprising".

45 See below, and Ch 3.

46 This denotes cohabiting partners. Under s 47(7) a person is the partner of a deceased person "if the two of them (whether of different sexes or the same sex) were living as partners in an enduring relationship at the time of the deceased person's death".

47 What constitutes 'sufficient interest' is discussed in Matthews (n 33), [8-24]–[8-28].

The fact that neither removal of the body nor the decision to order a post-mortem requires the consent of the deceased's personal representative or surviving relatives is the major distinction between coronial autopsies and their hospital equivalents.[48] However, it is also one of the first potential sites of legal conflict involving the dead.

2.   Emotional, religious and cultural objections

The majority of autopsies in England and Wales involve a full internal examination of the deceased's body.[49] Grieving families will often resist such a physically invasive procedure, viewing it as a (further) violation of their loved one's remains.[50] Such 'emotional' objections are noted by the coroner, but will not usually displace the strong public interest in ascertaining the cause of death.[51] However, more complex issues arise where familial opposition is based on religious and cultural values – an increasingly likely scenario in modern societies with a diversity of belief systems. Cutting the corpse can be seen as an act of desecration, which interferes with the deceased's passage into the afterlife; there may also be concerns around delayed burial or cremation where this is time-limited because of the deceased's spiritual beliefs (and those of the bereaved).[52] Given the root of these objections and the core values that they embrace, is the coroner's legislative mandate more susceptible to challenge here?

48   An individual cannot prevent an autopsy by directing (while alive) that one should not be carried out; likewise, the deceased's next-of-kin cannot refuse one if the coroner has jurisdiction – L Skene and B Masters, "What Legal Rights do you Have Over your Body after Your Death?" (2002) 81 *Australian Law Reform Commission Reform Journal* 38, p 38.

49   As opposed to external viewing of the body or partial internal examination.

50   The sudden (and often unexpected) nature of deaths within the coronial jurisdiction induces chronic grief emotions – see Y Neria and BT Litz, "Bereavement by Traumatic Means: The Complex Synergy of Trauma and Grief" (2004) 9 *Journal of Loss and Trauma* 73.

51   See *R v Greater Manchester Coroner, ex parte Worch* [1988] QB 513 (coroner's decision to hold a post-mortem upheld, despite objections from the deceased's widow and a Jewish burial society carrying out the deceased's burial) and *Re Jacobs' Application for Judicial Review* (1999) 53 BMLR 21 (coroner had not acted *ultra vires* in requesting a partial autopsy on the deceased, despite the husband's objections).

52   For example, both Judaism and Islam stress the importance of bodily integrity and non-mutilation of the dead; Islamic law also insists on burial within 24 hours of death – see B Carpenter, G Adkins, M Barnes, C Naylor and N Begum, "Communicating with the Coroner: How Religion, Culture and Family Concerns May Influence Autopsy Decision Making" (2011) 35 *Death Studies* 316, pp 319–320, also noting strong spiritual objections to autopsies amongst Aboriginal peoples in Australia.

Like its predecessor, the 2009 Act is silent on the issue, only granting a right of appeal against a decision *not* to order a post-mortem.[53] Objections based on religious or cultural imperatives have been documented between coroners and members of the Jewish and Muslim communities in England and Wales,[54] with coroners apparently willing to sanction less invasive post-mortem procedures to determine an adequate cause of death.[55] However, if there is a need for information that these alternatives cannot provide, a full internal examination is likely – and where the coroner's decision to order one is challenged, the overriding state interest in investigating suspicious or unexplained deaths tends to prevail.[56]

Similar trends can be seen elsewhere, despite ostensibly stronger legal safeguards. For example, while a small number of US states have legislation that allows individuals to register religious objections to an autopsy, one can still be performed if there are compelling legal or medical reasons to do so.[57] More comprehensive measures exist in Australia where all state legislation contemplates familial objections to a coronial autopsy, with some allowing

---

53   Section 40(1) of the 2009 Act allows an 'interested person' to appeal a specific decision to the Chief Coroner; under s 40(2)(d) this includes a "decision not to request a post-mortem examination under s 14" (though s 40(2)(e) grants a limited right of appeal against a decision to hold a second post-mortem).

54   Luce (n 35), p 174. See also T Elliott, "Religious Belief and Choices Regarding the Human Corpse" (2014) 2 *Journal of Medical Law and Ethics* 89, pp 100–106.

55   The coroner can determine the type of investigation – 2009 Act, s 14(2). This might include an external or more limited internal inspection, though a 'virtual' autopsy (one which uses non-intrusive magnetic resonance imaging (MRI) and other 3-D techniques to create an internal picture of the deceased's body) may be another option – see SP Stawicki, A Aggrawal, AJ Dean, DA Bahner, SM Steinberg, CD Stehly and BA Hoey, "Postmortem Use of Advanced Imaging Techniques: Is Autopsy Going Digital" (2008) 12 *OPUS* 17. Matthews (n 33), [8-68] highlights several difficulties, including the cost and availability of these alternative techniques, and the need to interpret any results accurately (assuming the method employed generates sufficient information). See also R Palmer, "Death and the Coroner: Some Reflections on Current Practice and Proposed Reforms" (2012) 52 *Medicine, Science and the Law* 63, p 68 (alternative techniques "must provide reliable, scientifically validated answers if they are to command authoritative support").

56   "[I]f an autopsy is needed then the law requires that it is carried out" – Dorries (n 33), p 123 citing *R v HM Coroner for Northumberland, ex parte Jacobs* (2000) 53 BMLR 21. Arguments could also be made that forcing coronial autopsies on unwilling groups violates the respective rights to private and family life and to freedom of religion under Articles 8 and 9 of the European Convention on Human Rights (now part of domestic law following the Human Rights Act 1998). There appears to be no direct case law on this to date. However, the coroner's authority would probably still prevail where there are legitimate reasons for carrying out a full post-mortem and alternative methods cannot be used; both Articles 8 and 9 permit interference with these rights in the public interest (in the bodily disposal context, both provisions are discussed in more detail in Chs 4 and 5).

57   AD Renteln "The Rights of the Dead: Autopsies and Corpse Mismanagement in Multicultural Societies" (2001) 100 *South Atlantic Quarterly* 1005.

religious and cultural objections to be lodged as well.[58] These provisions do not confer an absolute power of veto; the coroner can proceed despite objections, though a decision to do so can be challenged by the deceased's next-of-kin.[59] The fact that religious and cultural objections can still be overruled has been identified as particularly traumatic for grieving families, and indicative of a wider disconnect between medico-legal constructs of the body as a mere corpse and the connective attachments that the bereaved have towards their dead[60] – though one could argue that the practice is as much to do with the public interest in establishing an accurate cause of death. Studies also suggest that relational and religious opposition has resulted in less invasive autopsies being ordered in parts of Australia, though conflicting cultural values have had much less impact on the coronial process to date.[61]

3.   Releasing the body

In general, the body must be released for burial or cremation "as soon as is reasonably practicable";[62] the coroner cannot retain it for longer than is necessary to discharge any coronial functions,[63] and the personal representatives or next-of-kin must be notified if this will exceed 28 days.[64] However,

58   See the respective discussions in P Vines, "Objections to Post-mortem Examination: Multiculturalism, Psychology and Legal Decision-Making" (2000) 7 *Journal of Law and Medicine* 422; R Atherton, "Who Owns Your Body" (2003) 77 *Australian Law Journal* 178, pp 189–191; and P Vines and RF Croucher, "Law and Religion: Religion and Death in the Common Law" in P Radan, D Meyerson and RF Croucher, *Law and Religion: God, the State and the Common Law* (Routledge, 2012), pp 305–309.

59   Australian courts have upheld objections where the cause of death is apparent or there is sufficient evidence to indicate natural causes (as in *Green v Johnstone* [1995] 2 VR 175 and *Kantz v Hand* [1999] NSWSC 432), though a post-mortem will probably still be ordered if there is any suggestion of foul play–see *Magdziarz v Heffey*, Supreme Court of Victoria (unreported), 3 December 1995). These and other cases are discussed in Atherton (n 58), pp 190–191, and in Vines and Croucher (n 58), pp 301–309.

60   B Carpenter, G Tait and C Quadrelli, "The Body in Grief: Death Investigations, Objections to Autopsy and the Religious and Cultural 'Other'" (2014) 5 *Religions* 165.

61   See B Carpenter *et al* (n 52). Similar trends have been noted in New Zealand, where traditional autopsy practices often marginalise Maori beliefs – see K Selket, M Glover and S Palmer, "Normalising Post-Mortem: Whose Cultural Imperative? An Indigenous View on New Zealand Post-Mortem Policy" (2012) *New Zealand Journal of Social Sciences Online*.

62   2013 Regulations, reg 20(1). The coroner must also have issued the appropriate burial or cremation order, before disposal can occur – reg 21 and see the detailed discussion in Matthews (n 33), [9-13]–[9-35].

63   See the comments of Lander J in *Haydon v Chivell* [1999] SASC 315, [39], discussing the equivalent legislative framework in South Australia.

64   2013 Regulations, reg 20(2). Decisions on whether or not to release the body are subject to judicial review – Matthews (n 33), [8-12] citing *R v Bristol Coroner, ex parte Kerr* [1974] QB 652 and *R v Bristol County Coroner, ex parte Atkinson*, unreported, 5 May 1983.

the 2009 Act (like its predecessor) is curiously silent on the question of who the body should be released to.

In most cases, the body will be handed over to a funeral director acting on behalf of the person(s) lawfully entitled to possession of the body.[65] For adult testate deaths, this is the deceased's executor; for intestate deaths and those of infants or minors, this is the highest ranking next-of-kin who would be entitled to administer the deceased's estate (regardless of whether or not there is one to administer). After the spouse or civil partner, the law looks to the deceased's children, parents and siblings in descending order of entitlement – though cohabiting partners are excluded, and have no legal right to claim the deceased's remains, despite being "interested persons" under the 2009 Act.[66] From the coroner's perspective, problems will arise where warring factions of the deceased's family – each with different views on the appropriate funeral arrangements – demand that the body is released to them.[67] In *Burrows v HM Coroner for Preston*,[68] the court suggested that coroners should simply apply the designated order of entitlement to decide contested applications, while accepting that there might be circumstances in which this ranking could be varied to allow for human rights arguments and other "special circumstances".[69] However, the source of the coroner's power to make such a ruling is questionable (in the absence of an express provision in the 2009 Act dealing with competing claims to possession),[70] leaving civil courts to resolve the matter if necessary.[71] As with other aspects of the

---

65   This common law entitlement is analysed in Ch 3.

66   Though Dorries (n 33), p 92 queries whether a cohabiting partner could rely on the right to private and family life in Article 8 of the ECHR to claim the deceased's body here.

67   As in *Burrows v HM Coroner for Preston* [2008] EWHC 1387 (Admin) (noted immediately below) and *R v Inner South London Coroner, ex parte Brinsom* (*The Times*, 18 July 1995) (cited in Dorries (n 33), p 92) in which the deceased's wife and mistress both demanded possession of his body.

68   [2008] EWHC 1387 (Admin).

69   [2008] EWHC 1387 (Admin), [29]. The reference to 'special circumstances' relates to s 116(1) of the Senior Courts Act 1981, which allows someone else to be appointed as the deceased's executor or estate administrator – see Ch 4, pp 102–104.

70   Dorries (n 33), p 93 makes the point that *Burrows* ignores the question "whether the coroner actually has a lawful remit to determine such an application" or "feels equipped" to deal with the matter. Although dealing with a different legislative framework, the Australian case of *Gilliott v Woodlands* [2006] VSCA 46 suggests that – in competing claims to possession – the coroner has an implied power to determine the matter, as part of the effective discharge of the coroner's functions.

71   As in *Hartshorne v Gardner* [2008] EWHC B3 (Ch).

coronial process, a decision to release the body to the wrong person is susceptible to judicial review.[72]

In contrast, a small number of coronial statutes in force elsewhere contain specific provisions on who can apply for release of the deceased's body alongside an order of entitlement to deal with competing claims. For example, in the Australian state of Victoria, s 48 of the Coroners Act 2008 favours the executor named in the deceased's will or, failing that, the deceased's senior next-of-kin in descending order (spouse or domestic partner, followed by adult children, parents and then siblings etc).[73]

V.   Facilitating disposal: Funeral directors and the funeral industry

Funeral directors perform a vital role in the ritualised disposal of the dead, and surviving relatives[74] will usually employ one, even though there is no legal requirement to do so. Acting as an "intermediary between the bereaved and the organizations that furnish the mode of disposal",[75] the funeral director is involved in the process from an early stage – advising on the various bodily disposal options, essential documentation, funeral furnishings and types of ceremony. The funeral director will also take custody of the body – albeit temporarily – to prepare it for disposal.[76]

From a legal perspective, the rights and responsibilities of the funeral director are governed by contract – either negotiated between the funeral director and the person who orders the funeral, or contained in a pre-paid funeral plan that the deceased arranged while alive.[77] Basic contractual warranties and liabilities apply if issues subsequently arise about inadequate service provision, or non-payment of agreed funeral costs; however,

---

72   See *R (on the application of Haqq) v HM Coroner for Inner West London* [2003] EWHC 3366 (Admin) (coroner's decision to release deceased's body to his spouse, challenged by 'second' spouse under a polygamous marriage; application for judicial review refused). For similar Australian authority, see *Re Boothman, ex parte Trigg* (Supreme Court of Western Australia, 27 January 1999) (executor granted a writ of *certiori* quashing the coroner's decision to release the deceased's body to his de facto wife) and *Threlfall v Threlfall* [2009] VSC 283 (coroner had erred by releasing the deceased's remains to his brother instead of the deceased's estranged wife as the person entitled to administer her husband's estate).

73   See *Carter v Coroners Court of Victoria* [2012] VSC 561 (coroner's decision that the deceased's domestic partner was her senior next-of-kin under s 48 challenged by the deceased's sister; upheld on appeal).

74   Or anyone else tasked with organising the deceased's funeral.

75   B Parsons, "Conflict in the Context of Care: An Examination of Role Conflict Between the Bereaved and the Funeral Director in the UK" (2003) 8 *Mortality* 67, p 67.

76   For a summary, see Green and Green (n 16), pp 102–103.

77   Liability for funeral costs is discussed in Ch 3, Pt II.

negligent handing of the deceased's remains by the funeral director or similar 'wrongful' acts may result in other forms of civil action.[78]

1.   Industry regulation and consumer protection

Funeral directors are an integral part of the 'death industry' – an umbrella term for those companies, independent businesses and other organisations that provide funeral-related goods and services. Worth an estimated annual £2 billion to the UK economy,[79] there is surprisingly little legal regulation of this lucrative financial sector and individual service providers.[80] Looking specifically at funeral directors, industry organisations play a significant role – establishing and enforcing voluntary codes of conduct, and providing a forum for consumer complaints.[81] However, whether these offer sufficient protection to inexperienced purchasers, faced with making difficult purchases following the death of a loved one and with little time to 'shop around' is questionable.[82]

2.   The modern funeral and the changing role of the funeral industry

The symbolic and ritualistic elements of the funeral have changed dramatically in the twenty-first century. Until fairly recently, the traditional funeral ceremony focused on a particular religious view of death, and was structured in a certain way – with Christianity as the dominant theme. Greater variance in funeral rites is hardly surprising in the multi-faith, ethnically diverse society that is modern Britain; yet, the fact remains that a person's religious (and cultural) background "will typically entail some prescriptive content in regard to funerals".[83]

---

78   See Ch 3, Pt III.

79   IBISWorld, *Funeral Activities in the UK: Market Research Report* (January 2015).

80   While specific methods of bodily disposal are heavily regulated, "funeral directing is not embraced in a legislative framework" and funeral directors can operate without qualification or training – Parsons (n 75), p 68.

81   The two major associations in the UK are the National Association of Funeral Directors and the National Society of Allied and Independent Funeral Directors.

82   Funerals are classic 'distress purchases' and the issue of consumer protection has been a recurring theme in this country and elsewhere. See for example, BD Sher, "Funeral Prearrangement: Mitigating the Undertaker's Bargaining Advantage" (1963) 15 *Stanford Law Review* 414; KE Horton, "Who's Watching the Cryptkeeper: The Need for Regulation and Oversight in the Crematory Industry" (2003) 11 *Elder Law Journal* 425; Office of Fair Trading, "Funerals: A Report of the OFT Inquiry into the Funerals Industry" (2001); and New Zealand Law Commission, *The Legal Framework for Burial and Cremation in New Zealand: A First Principles Review* (Issues Paper 34, October 2013), chs 10–12.

83   T O'Rourke, BH Spitzberg and AF Hannawa, "The Good Funeral: Towards an Understanding of Funeral Participation and Satisfaction" (2011) 35 *Death Studies* 729, p 733.

The decline of organised religion and growing trend towards secularisation within England and Wales (and throughout the rest of the UK) has also had a noticeable impact on funerals. Less attachment to churches and religious ceremonies means that funerals are becoming increasingly de-ritualised and now have more of a 'personal touch'; people want creative and meaningful funeral options that resonate with the life and character of the deceased.[84] This movement away from sombre, highly ritualised funerals to life-celebratory and individualised 'send-offs' has forced funeral directors and the wider death industry to respond to changing consumer needs. An array of goods and services is available, from themed or novelty funerals with more active participation by the deceased's family and friends, to bespoke coffins or urns that are shaped or decorated in a certain way.[85] Technology is also playing a part, with audio-visual systems in crematoria allowing images of the deceased and their favourite music to feature in the ceremony, and live streaming of funerals for those who cannot attend.[86]

VI.   'Mistreatment' of the dead: Criminal offences pending bodily disposal

Cantor notes that the "first line of legal protection [against bodily disturbance] is for the path to the grave".[87] Any improper or indecent interference with corpses attracts a range of circumstantially defined common law and (to a lesser extent) statutory offences.[88] Whether the law is protecting the rights

---

84 See J Kelly, "Happy Funerals: A Celebration of Life?", *BBC News Magazine* (14 June 2015) www.bbc.co.uk/news/magazine-31940529 (accessed 30 September 2015). While some advocate the solemn dignity of traditional funerals, others argue that contemporary funeral symbols and participant behaviours fulfil the same functions as traditional religious symbolism – see S Adamson and M Holloway, "Symbols and Symbolism in the Funeral Today: What Do They Tell Us About Contemporary Spirituality" (2013) 3 *Journal for the Study of Spirituality* 140.

85 See generally, G Cook and T Walter, "Rewritten Rites: Language and Social Relations in Traditional and Contemporary Funerals" (2005) 16 *Discourse & Society* 365 (noting the trend towards more personalised funerals, with diminished rituals, more euphemistic references to death, and increased scope for interaction between celebrant and mourners) and G Sanders, "Themed Death: Novelty in the Funeral Industry" (2008) 10 *Consumers, Commodities and Consumption*. Bespoke coffins are increasingly popular – see Ch 5, n 199 for an illustration.

86 T Walter, R Hourizi, W Moncur and S Pitsillides, "Does the Internet Change How We Die and Mourn? Overview and Analysis" (2011) 64 *OMEGA: Journal of Death & Dying* 275.

87 Cantor (n 4), p 240.

88 For an excellent account, see J Herring, "Crimes Against the Dead" in Brooks-Gordon *et al* (n 7), ch 13 though he accepts that criminalising specific acts raises complex philosophical and moral issues, including the basic question of whether the dead can suffer 'harm' (see pp 231–236 and the various commentaries cited therein).

of the dead here is open to debate;[89] a more likely explanation is that it is reflecting ingrained social and moral norms around treating the dead with dignity and respect,[90] while circumventing some of the problems caused by the fact that corpses are not classed as property (and therefore not protected by specific property-based offences).[91]

Developed on a seemingly *ad hoc* basis over the centuries and sustained by antiquated case law, the common law offences in particular have been repeatedly revived to deal with modern exigencies – creating all sorts of interpretive problems.[92] For example, while there is no specific offence of maltreatment of corpses,[93] there is probably a common law offence of mutilating a dead body[94] – though this would not encompass embalming or other preparatory acts for burial or cremation,[95] and dissection or removal of organs with lawful authority would also be exempt.[96] In contrast, what might be described as disrespectful treatment of corpses would fall under the generic criminal law offence of outraging public decency,[97] and includes

89   There are conflicting views on whether the living retain any interest in their own bodies after they die – see Ch 5, pp 145–147.
90   I Jones and M Quigley, "Preventing a Lawful and Decent Burial: Resurrecting Dead Offences" (2016) *Legal Studies* (forthcoming).
91   As a general rule, corpses cannot be stolen (see ATH Smith, "Stealing the Body and its Parts" [1976] *Criminal Law Review* 623) and do not attract the offence of criminal damage.
92   For example, G McBain, "Modernising the Law on the Unlawful Treatment of Dead Bodies" (2014) 7 *Journal of Politics and Law* 89 argues that the common law offences promote uncertainty and confusion, and should be replaced by a new statutory offence of unlawful treatment of a corpse.
93   Herring (n 88), p 222. This is unsurprising, given the definitional problems and potential scope of a broad offence of mistreatment.
94   S White, "The Law Relating to Dead Bodies" (2000) 4 *Medical Law International* 145, p 154 – though the author also queries whether this offence might simply be an example of the broader offence of outraging public decency (see below).
95   In TW Price, "Legal Rights and Duties in Regard to Dead Bodies, Post-Mortems and Dissections" (1951) 68 *South African Law Journal* 403, p 408 the author argues that if "*any* cutting into a dead body" is regarded as improper, then "every undertaker and embalmer would be offering criminal indignities to dead bodies daily". Of course, the position would be different where, for example, the deceased's family or representatives had expressly prohibited embalming.
96   See for example, *R v Price* (1884) 12 QBD 247, 252 ("[t]he practice of anatomy is lawful and useful …"). However, the Human Tissue Act 2004 creates a number of related offences – these are mentioned briefly in Ch 6, p 164.
97   See A Hooper and DC Omerod (eds), *Blackstone's Criminal Practice* (OUP, 2009), [B3.288]–[B3.292]. The key elements are a lewd, obscene or disgusting act, which outrages minimum standards of public decency – *ibid*, citing *Knuller (Publishing, Printing and Promotions) Ltd v DPP* [1973] AC 435.

things like leaving human remains in public view[98] or using them in an obscene manner.[99] The visual element is important here, alongside the feelings of upset and outrage that the act generates – and, in this respect, there is an intuitively emotional aspect to the offence of outraging public decency. In contrast, the common law offence of causing a public nuisance focuses more on the physical consequences of what has taken place.[100] Typically invoked where someone is accused of burning human remains outdoors, the issue is whether the resultant smoke and smell threaten the welfare and safety of society, though endangering public morals can feature as well.[101] Causing a public nuisance is often (though not exclusively) used in conjunction with the separate common law offence of concealing or destroying a body where an inquest might have been warranted.[102]

Other crimes against corpses that might seem obvious to both lawyer and layperson are less clear-cut. For example, necrophilia has only been a criminal

---

98   *R v Clark* (1883) 15 Cox CC 171 (exposing the naked body of a dead infant on or near a public highway would shock public decency). Presumably the result would be the same, irrespective of whether or not the corpse was clothed.

99   As in *R v Gibson* [1990] 2 QB 619 where the defendants exhibited 'human earrings' made from freeze-dried human foetuses in a commercial art gallery, and were found guilty of outraging public decency. Contrast this with the case mentioned in White (n 94), p 159 where removing a corpse from a morgue in Barnsley (apparently as part of a botched ransom plot), transporting it in a car along a public highway and then dumping it in a car park did not result in a conviction for this particular offence. There was no evidence that the defendants' behaviour had actually seen by anyone (and the fact that members of the public were subsequently outraged was irrelevant).

100   Again, this is a generic criminal law offence where an individual is guilty of an unwarranted act or omission that damages, inconveniences or endangers the rights of the community. For an overview, see *Blackstone's Criminal Practice* (n 97), [B11.76]–[B11.86].

101   The legality or otherwise of 'DIY cremation' is discussed in Ch 2, along with specific offences that arise in the bodily disposal context.

102   As in *R v Stephenson* (1884) 13 QBD 331 where the defendants deliberately burnt the body of an illegitimate child the night before the coroner's enquiry was due to take place. See also *R v Price* (1884) 12 QBD 247 (discussed in Ch 2, p 40) and *R v Purcy* (1933) 24 Cr App R 70 (corpse of a dead baby hidden in a hedge with intent of obstructing an inquest). For more recent examples, see *R v Godward* [1998] 1 Cr App R (S) 385 and *R v Whiteley* [2001] 2 Cr App R (S) 119, as well as *R v Butterworth* [2004] 1 Cr App R 40.

offence since 2003,[103] while cannibalism is probably not unlawful in the absence of specific laws against the non-consensual consumption of another human being's bodily material.[104] However, cannibalism could attract the common law offence of mutilating a dead body[105] or possibly that of outraging public decency.[106]

Conclusion

The state's involvement in the death of individual citizens manifests itself in various ways, from determining when this actually occurs, to mandatory registration requirements and the coronial investigation of sudden and suspicious deaths.[107] Once basic legal requirements have been met, the emphasis shifts to disposal of the corpse – usually facilitated by a funeral director. The following chapter explores the various bodily disposal options that are currently available in this jurisdiction and elsewhere, and the laws surrounding each.

---

103   Under s 70 of the Sexual Offences Act 2003, which created the offence of sexual penetration of a corpse. Before the 2003 Act, doubts persisted as to whether necrophilia was unlawful in this country – see for example, DS Cook and DS James, "Necrophilia: Case Report and Consideration of Legal Aspects" (2002) 5 *Medical Law International* 199. Similar legal discrepancies have been noted elsewhere (see for example, J Troyer, "Abuse of a Corpse: A Brief History and Re-Theorization of Necrophilia Laws in the USA" (2008) 13 *Mortality* 132) though some common law jurisdictions have a specific statutory offence of improperly or indecently interfering with human remains and have prosecuted alleged cases of necrophilia accordingly (see for example, s 182(b) of the Canadian Criminal Code 1985 and *R v Ladue* (1965) 51 WWR 175 where the same charge was brought against a man who had attempted sexual intercourse with a dead woman, claiming he was drunk and unaware that his victim was dead).

104   White (n 94), p 154 citing, *inter alia*, comments made by Hutchinson B in the famous case of *R v Dudley and Stephens* (1884) 14 QBD 273. For an overview of the latter and its legal significance, see AC Hutchinson, *Is Eating People Wrong? Great Legal Cases and How They Shaped the World* (CUP, 2010), ch 2. Of course, killing a victim with the intent of ingesting their flesh would constitute murder.

105   White (n 94), p 154.

106   Assuming that the constituent elements of this particular offence were satisfied.

107   As well as criminal offences for mistreatment of the dead.

2 Bodily disposal laws

Disposal of the dead is a fundamental aspect of our existence; it is an inevitable activity, which cannot be avoided.[1]

Introduction

From time immemorial, removing the dead from the active realm of the living has been an urgent task when someone dies. All societies engage in this basic activity, and despite religious and cultural variations, there is a striking degree of commonality in corpse disposal practices worldwide, with burial and cremation remaining the two most popular.[2] Certain aspects are heavily regulated – for example, English law has detailed legislation on the provision, siting and ongoing management of burial grounds and crematoria.[3] In contrast, there are comparatively few laws governing actual bodily disposal.[4] Common law rules and isolated statutory provisions do exist,[5] while public health requirements alongside the basic premise of respect for the dead also

---

1   L Canning and I Szmigin, "Death and Disposal: The Universal, Environmental Dilemma" (2010) 26 *Journal of Marking Management* 1129, p 1129.
2   While this has always been the case, the popularity of each method has fluctuated over time – see DJ Davies, *A Brief History of Death* (Blackwell, 2005), p 49. Other methods of corpse disposal are permissible – see for example, s 41 of the Births and Deaths Registration Act 1953 ('disposal' of the dead, for the purposes of the Act, "means disposal by burial, cremation or any other means") and s 47(1) of the Public Health (Control of Diseases) Act 1984 (Secretary of State can introduce regulations imposing conditions or restrictions "with respect to means of disposal of dead bodies otherwise than by burial or cremation" if desirable in the interests of public health or safety).
3   These are scattered across a range of statutes (some of which are mentioned below). For an overview see the relevant sections of "Cremation and Burial", 24 *Halsbury's Laws of England* (5th edn, 2010), and DA Smale, *Davies' Law of Burial, Cremation and Exhumation* (Shaw & Sons, 7th edn, 2002), chs 5–7 (burial) and chs 9–10 (cremation).
4   S White, "The Law Relating to Dead Bodies" (2000) 4 *Medical Law International* 145. For example, there are no prescribed time limits for disposing of the dead.
5   Some of which are identified below.

exert a strong influence.[6] Aligned to the latter concept is the idea that a particular disposal method must be socially acceptable to be a viable option – something which depends on contemporary mores, as well as the death rituals and bodily disposal methods traditionally embraced by a particular society. This chapter analyses the current methods of corpse disposal in England and Wales (and throughout the rest of the UK), examining the permissible variations on each method and the fate of any materials produced by a particular form of disposal. It also identifies alternative techniques that are available in other jurisdictions,[7] if the deceased's remains are transported elsewhere.[8]

Contemporary bodily disposal choices are increasingly shaped by three factors. As noted in the previous chapter,[9] the rituals associated with cremation and burial have changed dramatically, given the emergence of more secular and personalised funerals. However, this assumes sufficient funds to support the deceased or their family's funeral of choice. Bodily disposal methods are increasingly dictated by financial constraints, with significant cost variations between burial and cremation[10] (as well as regional fluctuations);[11] and while we might think of funerals as recession-proof industries, the reality is very different in an era of widespread austerity,

---

6   These two themes have always been central to the treatment of the dead – see Ch 3, pp 59–60.

7   Either because they are permissible in other countries, or the method contemplated is only available elsewhere.

8   Subject to the deceased's estate having sufficient funds to pay for this (see Ch 3, Pt II), and compliance with the legal requirements for moving bodies out of the jurisdiction (see Ch 1, Pt III).

9   See Ch 1, Pt V.

10   Burial costs are much higher than cremation because grave space is at a premium in many areas.

11   While individual estimates differ slightly, the average cost of a basic funeral (funeral director's fees, burial or cremation expenses and minister or celebrant's fees) is currently around £3,693 with London being well in excess of that (£5,068), and Northern Ireland recording the cheapest figure (£3,203) – Sun Life, *Cost of Dying: 2015* (13 October 2015) www.sunlifedirect.co.uk/press-office/cost-of-dying-2015/ (accessed 15 October 2015). The same report notes that the total cost of dying has fallen slightly to £8,126 (down £300 from the previous year, due largely to a drop in estate administration fees). However, the discretionary, add-on costs of funerals (e.g. flowers, catering and headstones, venue for the wake and a memorial) have risen by 15 per cent over the last 5 years, with an average figure of £2,000 – *ibid*. Anecdotal reports of funeral directors pressurising emotionally vulnerable consumers into giving their loved one a 'fitting send-off' can result in a much higher figure. See also R Jones, "Cost of Dying Outstrips Inflation", *The Guardian* (London, 5 October 2015) www.theguardian.com/money/2015/oct/05/cost-of-dying-outstrips-inflation-funeral-3500 (accessed 15 October 2015).

reduced consumer spending and an ageing population.[12] Recent estimates suggest that the cost of a basic funeral has increased by up to 80 per cent in recent years (well beyond the rate of inflation), and that further price rises are likely.[13] Finally, environmental concerns are playing an increasingly important role, with modern disposal techniques and their attendant rites being driven by the need to be ecologically sensitive.[14] New and innovative methods are also being developed,[15] with the core aims of reducing the environmental impact of corpse disposal and utilising the 'waste' products in a way that benefits both the living and the earth itself.[16]

I.   Burial

Burial of the dead has always been permissible at common law, and typically involves depositing a corpse[17] in the ground.[18] Like the rituals attached to it,

12   Families on low incomes, in particular, are struggling to afford funerals – a situation exacerbated by severe cuts to the social fund system (see Ch 3, pp 75–76) that helps with funeral costs – see K Woodthorpe, H Rumble and C Valentine, "Putting 'The Grave' into Social Policy: State Support for Funerals in Contemporary UK Society" (2013) 42 *Journal of Social Policy* 605. So-called 'funeral poverty' has also garnered political attention, with a Private Member's Bill to tackle rising funeral costs introduced by Labour MP Emma Lewell-Buckin in December 2014 – Funeral Services Bill 2014–15 located at http://services.parliament.uk/bills/2014-15/funeralservices.html (accessed 30 September 2015).

13   D Foster, "Too Poor to Die: How Funeral Poverty is Surging in the UK", *The Guardian* (London, 9 June 2015) www.theguardian.com/commentisfree/2015/jun/09/poor-die-funeral-poverty-costs-uk (accessed 30 September 2015).

14   See R Fegan, "Death to Life: Towards My Green Burial" (2007) 10 *Ethics, Place and Environment* 157, discussing the normative and practical reasons behind more environmentally sensitive disposal practices – including land-use issues around burial, increasing levels of chemical contaminants associated with bodily disposal methods (for example, preparatory acts of embalming), and the use of natural resources for shrouds, coffins and containers as well as for the upkeep of cemeteries. Similar issues are discussed in Canning and Szmigin (n 1), the authors noting that an increasingly populated global landscape has resulted in bodily disposal becoming a critical environmental issue.

15   In particular, resomation and promession – see Pt III. Although still at the 'ideas stage', US architect Katrina Spade is also working on the idea of 'human composting' as a low-cost and ecologically friendly means of bodily disposal – see K Herzog, "A Greener Afterlife: Is Human Composting the Future for Funerals?", *The Guardian* (London, 10 March 2015) www.theguardian.com/environment/2015/mar/10/a-greener-afterlife-is-human-composting-the-future-for-funerals (accessed 30 September 2015).

16   See H Rumble, J Troyer, T Walter and K Woodthorpe, "Disposal or Dispersal: Environmentalism and Final Treatment of the British Dead" (2014) 19 *Mortality* 243.

17   Or post-cremation ashes, though the emphasis here is on uncremated corpses. Occasionally, coffins may contain more than one body – usually in cases of simultaneous or temporally proximate deaths (for example, where a couple died in an accident, or mother and baby died in childbirth). Specific legal requirements apply here – see 24 *Halsbury's Laws of England* (n 3), para [1135].

18   Above-ground entombment in a mausoleum or vault is another possibility, though uncommon in this jurisdiction. Burial at sea is occasionally used, and is discussed below.

the reasons for choosing burial vary significantly, and include personal preference, adherence to social conventions, or the fact that certain religions or cultures insist on the committal of human remains.

In England and Wales, burial is now secondary to cremation in terms of disposal choices.[19] Ad hoc laws governing specific aspects of burial are located across a range of (sometimes archaic) statutory provisions,[20] and there are rules concerning the minimum depth of graves[21] and record-keeping requirements[22] (to list some examples). However, there is no legal requirement that the dead be interred in specially designated places.[23] As a result, burial is not confined to cemeteries or churchyards, and other permissible options are examined below. Other legal myths surrounding burial (and funerals more generally) can also be debunked. For example, there is no legal requirement to use a funeral director and no specific ceremony is necessary (unless burial is in consecrated ground).[24] Likewise, English law does not insist on

---

19 See Pt II.
20 The assortment of piecemeal burial (and exhumation) laws was identified over a decade ago in a Home Office consultation paper, which mooted the idea of a broad, cohesive framework within one single statute – Home Office Consultation Paper, *Burial Law and Policy in the 21st Century: The Need for a Sensitive and Sustainable Approach* (2004), Pt A. No action has been taken to date, though a debate on this and other issues is now taking place in Scotland – Scottish Government, *Consultation on a Proposed Bill Relating to Burial and Cremation and Other Related Matters in Scotland* (January 2015).
21 As well as the amount of soil between coffins where there are two or more separate interments in the same plot – 24 *Halsbury's Laws of England* (n 3), paras [1218]–[1220].
22 Namely, recording the location of individual graves within a particular burial ground – see for example, the Registration of Burials Act 1864, as well as art 11 of the Local Authorities' Cemeteries Order 1977. Recording the actual death itself was discussed in Ch 1.
23 See J Green and M Green, *Dealing with Death: A Handbook of Practices, Procedures and Law* (Jessica Kingsley Publishers, 2nd edn, 2008), p 105.
24 See S Wienrich and J Speyer, *The Natural Death Handbook* (Rider Books, 4th edn, 2003), pp 98–100.

embalming unless, for example, a corpse is being transported abroad or brought back into the jurisdiction,[25] and while a corpse must be "decently covered"[26] for burial, the use of a coffin is not mandatory.[27]

1. Burial in a churchyard or cemetery

Most interments still take place in either a churchyard or a cemetery.[28] Despite certain shared characteristics,[29] both locations are subject to different systems of law and regulatory frameworks because of their respective origins – and these are noted briefly below. From an environmental perspective however, churchyard and cemetery burial have come under increasing criticism, with concerns over soil contamination from embalming fluids, the use of natural resources in coffins and other containers which are left to disintegrate slowly in the ground, and the long-term sustainability of sole-occupied permanent graves from a land-use perspective.[30]

25 C Cowling, *The Good Funeral Guide* (Continuum, 2010), p 59. Embalming may also be a practical necessity where the deceased's family insist on having the body at home or the funeral is being delayed. Where embalming is carried out for viewing (either at the funeral director's premises or in a family home), preserving the deceased is secondary to restoring the body to a more life-like state to comfort the bereaved. In these circumstances, "[t]he processed body is a symbolic product, one that is ... both recognizable and reassuring" – G Sanders, "The Dismal Trade as Culture Industry" (2010) 38 *Poetics* 47, p 55.

26 *R v Stewart* (1840) 12 Ad & El 773, 778. According to *Gilbert v Buzzard* (1820) 3 Phillimore 335, 349–350, carrying a body in a state of "naked exposure" to the grave constitutes a "real offence to the living, as well as an apparent indignity to the dead".

27 Green and Green (n 23), p 25. A corpse can be interred in a shroud, coffin or other physically suitable container; restrictions are only imposed by the bodily disposal process itself (for example, specific requirements around sea burial or cremation – see below) or the rules of a particular cemetery. In *Seaton v Commonwealth* 149 Ky 498 (1912) a father buried his infant child in a paper box, placed inside a crude wooden box. According to the Supreme Court of Kentucky, there was "no rule of law defining how a corpse shall be dressed for burial, or the character of coffin or casket in which it shall be enclosed, or the material out of which the box, in which the coffin is placed, shall be made ..." – *ibid*, 501.

28 The term 'churchyard' denotes a smaller scale burial ground attached to a parish church of the Church of England or (prior to disestablishment) the Church of Wales. References to 'cemetery' denote public burial grounds operated by local authorities or commercial companies, as well as private graveyards affiliated with other religions.

29 Both are bounded sites with ordered internal layouts, which have strong personal and community meanings, are regarded as 'sacred' spaces by family and friends of the deceased, and protected from 'disrespectful' activities – J Rugg, "Defining the Place of Death: What Makes a Cemetery a Cemetery?" (2000) 5 *Mortality* 259.

30 See the points raised in Fegan (n 14).

(a)   Churchyard burial

Churchyards consecrated according to the rites of the Church of England are subject to ecclesiastical law[31] and protected under faculty jurisdiction.[32] At common law, every parishioner has a right to be buried in their parish churchyard as long as it remains open for interments[33] – an entitlement that extends to all inhabitants of that particular parish.[34] The right is confined to burial in an 'ordinary' manner,[35] and is little more than the right to be placed in a grave having been taken there "in a decent and inoffensive manner".[36] Although enforceable at common law, any questions surrounding the exercise of the right are within the exclusive cognisance of ecclesiastical courts.[37]

(b)   Cemetery burial

The nineteenth century saw the gradual erosion of the Church of England's monopoly on burial, as local authorities assumed responsibility for disposal of the dead and the provision of public burial grounds. An array of statutory provisions was enacted during this time,[38] and isolated parts remain in force despite the consolidating reforms introduced by the Local Government Act 1972, which still governs the provision and maintenance of cemeteries[39] by

---

31   See generally M Hill, *Ecclesiastical Law* (OUP, 3rd edn, 2007).

32   A faculty is a special licence or dispensation required for specific alterations or modifi-cations to churches and churchyards (including consecrated parts of municipal cemeteries as noted below) – see generally "Ecclesiastical Law", 34 *Halsbury's Laws of England* (5th edn, 2011), paras [1067]–[1084]. The distinction between consecrated and unconse-crated burial grounds "remains fundamental" to modern English burial law – Home Office Consultation Paper, *Burial Law and Policy in the 21st Century: The Need for a Sensitive and Sustainable Approach* (2004), p 3.

33   See also the authorities listed in 24 *Halsbury's Laws of England* (n 3), para [1238]. This appears to be a different species of entitlement from the common law 'right to a Christian burial' – see Ch 3, p 59.

34   Church of England (Miscellaneous Provisions) Measure 1976, s 6(1) with the same right extending to cremated remains under s 3 of the Church of England (Miscellaneous Provisions) Measure 1992, s 3.

35   *Winstanley v Manchester Overseers* [1910] AC 7, 16. No faculty is required here, either for burial of a body or the interment of cremated remains – 34 *Halsbury's Laws of England* (n 32), para [1084].

36   *Gilbert v Buzzard* (1820) 3 Phillimore 335, 352–353. However, there is no right to a particular grave plot (or to erect a headstone), and an 'ordinary' right of burial is very different from an 'exclusive' right of burial, which confers legal rights in a specific grave plot (alongside the entitlement to erect a memorial) and can only be obtained by grant of a faculty – see below.

37   24 *Halsbury's Laws of England* (n 3), para [1241].

38   Beginning with the Burial Act 1852 and culminating in the Burial Act 1906 (sometimes collectively referred to as the Burial Acts 1852–1906). For a full listing, see 24 *Halsbury's Laws of England* (n 3), para [1109].

39   And crematoria.

"burial authorities" in England and Wales today,[40] with more detailed burial provisions set out in ancillary statutory regulations.[41] Both sanitary concerns and, to a lesser extent, health and safety laws also regulate specific aspects of bodily disposal within cemeteries; and while there are no legal restrictions on who can be interred here, many cemeteries have physically demarcated areas that are reserved for particular religious denominations.[42] However, where part of a cemetery has been marked off and consecrated according to the rites of the Church of England, it is protected under faculty jurisdiction.[43]

Other types of cemetery include those founded as commercial enterprises by private operators,[44] and private burial grounds set aside by specific religions.[45] Although the 1972 Act does not apply here, other generic burial laws are still relevant.[46]

(c)   Exclusive rights of burial

Where someone is buried in a churchyard or cemetery, those responsible for making the funeral arrangements will acquire an 'exclusive right of burial' upon purchasing a grave plot[47] – in other words, a right to inter one or more sets of human remains in that specific plot, and to erect a tombstone or other

---

40   1972 Act, s 214. Certain local authorities now constitute "burial authorities" under the 1972 Act, following the legislation's restructuring of local government – for more detail see 24 *Halsbury's Laws of England* (n 3), para [1112]. However, overall supervisory powers are vested in the Secretary of State for Justice, though certain functions relating to burial (and cremation) in Wales have been transferred accordingly – see 24 *Halsbury's Laws of England* (5th edn, 2010), para [1113].

41   See the Local Authorities' Cemeteries Order 1977 and the Local Authorities' Cemeteries (Amendment) Order 1986. These regulations also apply to the interment of cremated remains in cemeteries – 1977 Order, art 2(2).

42   24 *Halsbury's Laws of England* (5th edn, 2010), para [1202]. Prior to doing so, the local authority must be satisfied that a sufficient area of the cemetery is still available for non-denominational and non-religious burials. Where an area has been set aside for specific groups, the Attorney-General can prevent others from being buried there – *Preston Corpn v Pyke* [1929] 2 Ch 338 (whether a non-Catholic could be buried within the Catholic portion of a cemetery).

43   24 *Halsbury's Laws of England* (5th edn, 2010), para [1201] – though the same source suggests that the faculty jurisdiction is "exercised sparingly" here. Again, before consecration, the local authority must be satisfied that a sufficient area of the cemetery remains unconsecrated.

44   See for example, P Collinson, "Are Cemeteries the New Safe Investment?", *The Guardian* (London, 16 October 2010) www.theguardian.com/money/2010/oct/16/cemeteries-burial-investment (accessed 30 September 2015).

45   For example, Catholic graveyards as well as dedicated Jewish and Muslim burial grounds.

46   Green and Green (n 23), p 107. These include public health and health and safety laws, as well as keeping a register of individual burials.

47   These rights are generated by faculty in the case of churchyards, and by statute in the case of cemeteries – see A Dowling, "Exclusive Rights of Burial and the Law of Real Property" (1998) 18 *Legal Studies* 438, p 439.

suitable memorial.[48] The result is a private, family grave that can accommodate several related interments, over a set period of time.[49]

Both the legal status of exclusive rights of burial and their consequent scope have been debated extensively, with opinions divided on whether they generate proprietary interests in or mere contractual rights over the grave in question.[50] In practical terms, this becomes an issue if the grave is damaged or interfered with in any way[51] (what the grantee can do about this), and if the burial ground is sold at a later date (can the exclusive right of burial be enforced against the purchaser). Problems can also arise within families where the grantee has promised to transfer the burial plot and all rights associated with it to another family member, but later changes his/her mind.[52]

The grantee will usually be the person who secured the exclusive right of burial, though contributions from different parties towards the cost of the grave and overall funeral expenses can create complex legal entitlements. For example, in *Re West Norwood Cemetery*,[53] the deceased's son and a maternal uncle had purchased the exclusive right to a burial plot in the consecrated part of a municipal cemetery following the death of the deceased's wife over a decade earlier; the rest of the family (the deceased and his other six children) had contributed to the cost of the funeral expenses. Following the deceased's death, the son objected to his father's ashes being placed in the same grave as those of his mother and wanted to exhume them, arguing that he (the son) was the registered owner of the grave plot and that the other siblings had forged his signature to enable their father's ashes to be interred there. While the court accepted that the son was the legal owner of exclusive rights of burial in the plot, the actual grant was not definitive as to ownership of these rights. Instead, the son held them on constructive trust for the other

---

48 However, cemetery or churchyard regulations invariably restrict the structure and design of any memorial – see Ch 8, Pt 1.

49 See immediately below.

50 For an in-depth discussion see Dowling (n 47), as well as P Sparkes, "Exclusive Burial Rights" (1991) 2 *Ecclesiastical Law Journal* 133 and RN Nwabueze, "Property Interest in a Burial Plot" [2007] *Conveyancer and Property Lawyer* 517. While the bulk of authority suggests that an exclusive right of burial does not generate a freehold estate, opinions are still divided over whether it constitutes an easement or a species of licence.

51 For example, if another corpse is accidentally interred in that particular grave as in *Reed v Maddon* [1989] 2 All ER 431.

52 As in *Vosnakis v Arfaras* [2015] NSWSC 625 where the defendant (the deceased's mother) allowed the plaintiff (the deceased's husband) to bury the deceased in a burial plot originally purchased by the defendant and having space for only two interments. The defendant subsequently reneged on her promise to transfer the plot to the plaintiff, so that he could be buried beside his wife. While there was no legally binding and enforceable contract to transfer what constituted a 'burial licence', the court held that the circumstances generated an estoppel; the defendant would be required to transfer the licence to the plaintiff.

53 [2005] 1 WLR 2176.

family members (including the deceased) as a result of their contributions towards the funeral expenses,[54] and could not object to his father's ashes being placed in the grave.[55]

While *Norwood* was essentially an exhumation case, the impact of ownership rights can also be felt in subsequent disputes around commemoration of the dead and who controls the type of memorial and any wording used.[56] Turning to duration, exclusive rights of burial are no longer granted in perpetuity but typically expire after a specified period (usually between 50–100 years).[57] This impacts on the potential re-use of existing graves.

(d)   Space constraints and issues of permanency?

Burial of the dead raises long-term land-use and management issues. Existing burial grounds are often viewed as 'sacred' places to be maintained and preserved in perpetuity – out of respect for those interred there, but also as repositories of important historical information recorded on individual gravestones. Some would argue that setting aside land in this way appropriates a finite and valuable resource that could be better utilised by the living.[58] The fact that "burial space is essentially mutable" and its "significance … alters as time accrues between the living and the dead"[59] also means that eternal preservation claims are less convincing after interred remains have disintegrated, and connections between the living and the dead span

---

54   The court's analysis of the issue is not entirely convincing, given the evidential uncertainties around who contributed to what aspects of the initial funeral expenses and on what understanding.

55   However, payment of funeral expenses may not be determinative if the grant is issued to someone else. In *Smith v Tamworth City County Council* (1997) 41 NSWLR 60 the deceased's adoptive parents arranged for burial of their adult son's remains, and engaged a funeral director who (on their behalf) obtained a burial plot from the local council. However, the undertaker's bill was paid by the deceased's natural parents who claimed that title to the grave should be transferred to them, despite the formal documentation to the plot having been issued to the adoptive parents (the essence of the dispute was the wording on the headstone erected by the adoptive parents – see Ch 8, pp 219–220). Young J held that payment of the funeral director's account was irrelevant, and that the exclusive right of burial belonged to the adoptive parents.

56   See Ch 8, Pt 1.

57   See the discussion in Home Office Consultation Paper, *Burial Law and Policy in the 21st Century: The Need for a Sensitive and Sustainable Approach* (2004), p 14. Local authority grants have a maximum period of 100 years but can be terminated prematurely if, for example, the exclusive right of burial has not been exercised within 75 years of being granted – Local Authorities' Cemeteries Order 1977, art 10 and Sch 2.

58   As Hansen hypothesises, "if every human being who ever lived on this earth were given his or her own burial plot, at some point the dead could crowd the living right off the earth" – K Hansen, "Choosing To Be Flushed Away: A National Background on Alkaline Hydrolysis and What Texas Should Know about Regulating Liquid Cremation" (2012) 5 *Estate Planning & Community Property Law Journal* 145, p 146.

59   Rugg (n 29), p 259.

several generations.[60] Meanwhile financial pressures also impact on older burial grounds, faced with reduced income from interments yet ongoing maintenance and repairing liabilities.

Established churchyards and cemeteries in England and Wales are facing a serious shortage of burial space, with many either full or approaching capacity – a situation which will exacerbate as populations increase,[61] and the lack of affordable grave sites drives funeral costs even higher. Adjacent land may not be available to meet increasing demand through enlargement,[62] while newly dedicated cemeteries on the periphery of towns and cities (where more land is available) may not appeal to those who want to maintain connections with their local community or with dead relatives interred else-where. To address these problems, a 2004 Home Office consultation paper advocated re-using older graves in cemeteries by means of a 'lift and deepen' technique whereby remains would be exhumed, and reinterred at a greater depth – thus freeing up more space. [63] Only graves of 100 years old or more would be eligible, to ensure (as far as possible) that "the remains had been reduced to skeletal material, and that there would be no immediate descendants of the deceased".[64] Stakeholder responses were favourable,[65] resulting in several London boroughs being granted the right to re-use graves.[66]

Intensifying the use of existing burial space seems like an obvious solution to an ongoing problem, in terms of land management and sustainability. Of course, disturbing human remains, combined with the spectre of multiple and unrelated burials in the same plot (albeit over a lengthy period of time)

---

60   See *AF Hutchinson Land Co v Whitehead Bros* 217 NYS 413, 423 (NY Sup Ct 1926): "When … the remains of those who lie buried in the soil have disintegrated and mingled with the dust beneath, when there is nothing left to identify the ashes that lie buried there, when the names of the dead are no longer heard in the ears of men, … no plausible reason suggests itself to the mind why such land should be withheld from serving the needs of a community solely for sentimental reasons".

61   Rising global populations have resulted in similar trends elsewhere, even in the United States with its vast areas of land – see PR Kehoe, "Cemetery Abandonment and Disinterment of Human Remains (1971) 35 *Albany Law Review* 320, p 322.

62   Some cemeteries have started using their car parks and pathways to free up burial space – see E Malnick, "Bodies 'Buried in Former Car Parks Due to Graves Shortage'", *The Telegraph* (London, 27 September 2013) www.telegraph.co.uk/news/religion/ 10337249/Bodies-buried-in-former-car-parks-due-to-graves-shortage.html (accessed 30 September 2015).

63   Home Office Consultation Paper, *Burial Law and Policy in the 21st Century: The Need for a Sensitive and Sustainable Approach* (2004), p 15.

64   *Ibid.* The paper also mooted the possibility of re-using or adding to existing tombstones or memorials on the grave in question – *ibid*, p 16.

65   See Ministry of Justice, *Burial Law and Policy in the 21st Century: The Need for a Sensitive and Sustainable Approach* (2004): Government Response to the Consolation Carried Out by the Home Office DCA (2007), p 16.

66   London Local Authorities Act 2007, s 74.

may be unacceptable to some – even if the practice is commonplace throughout the rest of Europe.[67]

2.   Natural burial

Sometimes referred to as 'woodland burial',[68] natural burial is a relatively modern trend in bodily disposal, which frames the decomposing body within a discourse of environmental sustainability. It involves placing an unembalmed corpse in a biodegradable coffin or shroud, and interring it in a shallow grave[69] in a field or woodland area; individual graves eventually become part of the natural landscape, and are 'marked' by a tree or shrub, as opposed to a traditional headstone or permanent memorial.[70] There are currently over 200 natural burial grounds in the UK with the vast majority spread across England,[71] and while there are no specific laws governing natural burial,[72] individual sites must have planning permission.

Although it still represents a small proportion of the country's overall disposal choices,[73] natural burial is an increasingly popular option for environmentally friendly individuals, which provides the deceased with a "route to ecological immortality" and the bereaved with "a pleasant and sympathetic environment" to remember their dead.[74] The latter point might also appeal to aesthetically minded individuals who want to play an active role in creating beautiful surroundings for future generations. It has been suggested that the rhetoric of natural burial as a form of nourishing and giving back to the earth challenges ingrained notions of toxicity around decaying remains, and blurs conventional boundaries between the living and

---

67   As noted, for example, in C Gittings and T Walter, "Rest in Peace? Burial on Private Land" in A Madrell and JD Sidaway (eds), *Deathscapes: Spaces for Death, Dying and Bereavement* (Ashgate, 2010).

68   Though this is a misnomer since natural burial is not restricted to woodland sites.

69   To promote decomposition.

70   See generally K West, *A Guide to Natural Burial* (Sweet & Maxwell, 2010). The absence of both a bounded gravesite and a marker can subsequently cause problems for bereaved relatives who wish to visit what effectively become anonymous graves.

71   The Natural Death Centre website contains a list of sites that belong to the Association of Natural Burial Grounds – see www.naturaldeath.org.uk/index.php?page=natural-burial-grounds (accessed 30 September 2015). At present there are only two such sites in Wales (and two in Scotland, with none in Northern Ireland).

72   24 *Halsbury's Laws of England* (n 3), para [1181].

73   See A Clayden, J Hockey and M Powell, "Natural Burial: The De-Materialising of Death" in J Hockey, C Komaromy and K Woodthorpe (eds), *The Matter of Death: Space, Place and Materiality* (Palgrave Macmillan, 2010).

74   Canning and Szmigin (n 1), p 1139. However, Cowling (n 25), p 27 offers a more cynical analysis: "[A] typical green funeral might be viewed by greener-than-thou fundamentalists as an act of atonement for a life which consumed far too much energy – and a gestural act at that. Only if the person who has died had made every effort to minimize their impact when they were alive does a green funeral make sequential sense".

the dead.[75] However, while the overall environmental impact may be lower, this method of disposal still requires extensive land use.

3. Burial on private land

While the vast majority of burials take place in designated burial sites, interment on private land is another option.[76] Usually favoured where there is some personal or family connection with the land in question,[77] there is nothing to prevent an individual from being buried on a family farm, or even in their own back garden, provided that the grave is of sufficient depth and the requisite distance from existing water supplies to prevent possible contamination.[78] While restrictive covenants affecting a piece of land might occasionally prevent it being used for private burial, there is no need to secure planning permission since a limited number of non-commercial interments on a given piece of land will not constitute a change in land use.[79]

Burial on private land is rare though isolated examples do occur.[80] Owners of neighbouring properties may not be comfortable with the idea, given society's preoccupation with death boundaries and the dead being "kept in their place, [in cemeteries] marked off from everyday life".[81] Those contemplating the burial of a loved one in their garden or on family land should also bear in mind the longer-term implications, especially if the property is sold at a later date. Beyond a likely depreciation in value due to the presence of a

75   See Rumble *et al* (n 16), pp 246–247 and the various sources cited therein.

76   In other words, burying corpses in their entirety as opposed to scattering ashes on private land (though the latter is much more common).

77   See generally Gittings and Walter (n 67).

78   For more detailed information, see Wienrich and Speyer (n 24), pp 95–98 and "Private Land Burial", *The Natural Death Centre*, located at www.naturaldeath.org.uk/index.php?page=home-burial (accessed 30 September 2015). The latter stipulates a minimum burial distance of 10 metres from any 'dry' ditch or field drain, 30 metres from any spring, running or standing water and 50 metres from any well, borehole or spring that supplies water, along with a minimum depth of one metre (not just the legally required two feet) between the settled soil level and the top of the coffin or shroud. A register of any burials should also be maintained (in accordance with the Registration of Burial Acts 1864), with both the date of burial and the exact location of the gravesite recorded on the title documents to the property.

79   Wienrich and Speyer (n 24), pp 97–98.

80   Perhaps the most famous private land burial was that of Diana, Princess of Wales in a grave on the Spencer family estate at Althorp in September 1997, though more recent examples do exist – see L Mangan, "One House For Sale, Two Bodies in the Garden – But No Discount", *The Guardian* (London, 13 January 2014) www.theguardian.com/lifeandstyle/shortcuts/2014/jan/12/house-for-sale-bodies-garden-no-discount (accessed 30 September 2015).

81   T Walter and C Gittings, "What Will the Neighbours Say? Reactions to Field and Garden Burial" in Hockey, Komaromy and Woodthorpe (n 73), p 167.

burial site,[82] would the family still be able to visit the grave if they no longer owned the land, and how would they secure access;[83] would the new owners be prepared to give some sort of guarantee that the buried corpse would remain in situ and not be disturbed in the future; failing that, could the remains be exhumed and moved to a new location?[84] These and other land management concerns might help to explain why the number of private land burials is so small.

4.   Burial at sea

While scattering post-cremation ashes at sea (or on any waterway) is relatively unproblematic,[85] a conscious decision to dispose of an uncremated corpse by boat or ship, off the coast of a particular country[86] is both costly and legally complex – even if there are environmental benefits.[87] Sea burial requires a special licence from the Department of the Environment, Food and Rural Affairs in England and Wales.[88] Any such permission is subject to strict specifications; aside from the necessary documentation,[89] the corpse must not be embalmed and there are restrictions on the composition and design of the coffin (solid softwood and weighted down, with holes bored

---

82   Green and Green (n 23), p 109 also suggest that the potential beneficiaries of any future sale of the land should indicate their consent to the burial in writing, to prevent legal action around losses incurred on sale.

83   Independent access would probably require some sort of easement, which the new owner might not be willing to grant since they are under no obligation to allow the deceased's family access to the grave.

84   This would be dependent on a disinterment licence from the Ministry of Justice. Likewise, there is nothing to prevent the new landowner from exhuming the deceased's remains and moving them elsewhere, though the views of the deceased's next-of-kin would also be taken into account here – see Ch 7.

85   No licence is required, and there are no specific legal restrictions on where ashes can be scattered – see p 47.

86   As opposed to burial at sea taking place by necessity where, for example, someone dies while on-board a ship and the distance from shore mandates immediate disposal of a decaying corpse. This would have been fairly common on long sea voyages centuries ago, but is much less likely today because larger-scale commercial cruise ships have morgues to store dead passengers or crew.

87   There is no impact on land use, and none of the pollutants associated with cremation – Canning and Szmigin (n 1), p 1130. Other environmental benefits include the non-use of embalming chemicals (a corpse must be unenbalmed for sea burial – see below) and this disposal method uses less energy than cremation.

88   Under the Food and Environment Protection Act 1985, Pt II with more detailed information available at www.gov.uk/burial-at-sea (accessed 30 September 2015). A marine licence for sea burial is only available for one of three coastal areas: off The Needles, to the west of the Isle of Wight; between Hastings and Newhaven; and off Tynemouth, North Tyneside.

89   Medical certification that the body is disease and infection-free, and (if necessary) coroner's confirmation where the body is being removed from the jurisdiction.

along the sides to allow rapid ingress of water).[90] Practical restraints will also play a much larger role in sea burial; the amount of space available on the boat means that fewer people can be involved, and (unlike land disposal) adverse weather conditions may result in the disposal being postponed or the off-shore element being cut short.

II. Cremation

Cremation involves the burning of human bodies at a high temperature to produce ashes, and now accounts for almost 75 per cent of bodily disposals in England and Wales.[91] However, this was not always the case.

Although widely practised by the ancient Greeks and Romans,[92] cremation fell into disfavour with the spread of Christianity, and its emphasis on resurrection of the body.[93] Religious and social taboos were reinforced by negative connotations around burning bodies. As White points out:

> [C]remation was regarded with repulsion by many, who knew it mainly either as a pagan method of disposing of corpses, or as a method of execution in former times for heretics, witches and female traitors, or as a means by which infanticides … attempted to conceal the evidence of their crimes.[94]

Yet, cremation began to be promoted as a socially acceptable and more sanitary method of corpse disposal towards the end of nineteenth century, and while it slowly began to assume popularity, cremation rates only started to rise significantly in the latter half of the twentieth century. Various socio-cultural factors have contributed to this, including the emergence of more secular societies, and the lifting of bans on cremation by certain faiths.[95] Practical considerations include growing urbanisation, land space constraints around the provision of graves and the lower costs of cremation relative to

---

90   See Wienrich and Speyer (n 24), pp 113–115.
91   Information provided by the Cremation Society of Great Britain and located at www.cremation.org.uk/ (accessed 30 September 2015).
92   The Cremation Society of Great Britain, *History of Modern Cremation in Great Britain From 1874: The First Hundred Years* (1974), located at www.cremation.org.uk/ (accessed 30 September 2015).
93   See P Vines and R Croucher, "Religion and Death in the Common Law" in P Radan, D Meyerson and R Croucher, *Law and Religion: God, the State and the Common Law* (Routledge, 2005), p 308.
94   S White, "A Burial Ahead of its Time? The Crookenden Burial Case and the Sanctioning of Cremation in England & Wales" (2002) 7 *Mortality* 171, p 173.
95   For example, the Catholic Church, which only removed its ban in the 1960s. Some religions still oppose cremation (for example, the practice is repugnant to both Orthodox Jews and Muslims), while other faiths and cultures (notably Hindus and Buddhists) insist on it as the only means of disposing of their dead.

burial. Cremation also offers more flexibility, producing ashes that can be divided among the bereaved or 'used' in various ways after the death of a loved one.[96]

1.   Modern cremation laws

Doubts existed over whether cremation was permissible under English law when this method of disposal emerged in the latter half of the nineteenth century.[97] The first authoritative ruling came in *R v Price*[98] in which a father was indicted for attempting to burn the remains of his five-month old son in a field nearby his house, two days after the child's death. Stephen J held that cremation was lawful so long as it did not constitute a public nuisance and no attempt was made to prevent a coroner's inquest where one was deemed necessary.[99] Any lingering doubts were removed by the Cremation Act 1902.

The 1902 Act still governs the burning of human remains in England and Wales, and allows cremation authorities[100] to establish crematoria under the terms of the legislation.[101] Supplementary directives are contained in the Cremation (England and Wales) Regulations 2008,[102] which prescribe the conditions in which cremation can take place.[103] On the whole, cremation is more heavily regulated than burial – hardly surprising given the finality of the process and the potential for concealing crime through destruction of the body and vital physical evidence. For example, there are stringent legal rules around who should apply for cremation and the documentation required.[104]

---

96   See pp 45–47.
97   The Cremation Society of Great Britain, *History of Modern Cremation in Great Britain From 1874: The First Hundred Years* (1974).
98   (1884) 12 QBD 247.
99   Both offences were discussed in Ch 1, Pt VI. After an extensive review of the authorities in *Price*, Stephen J concluded that "burning of the dead has never been formally forbidden ... in any part of our law" – (1884) 12 QBD 247, 249. For an analysis of the legal and historical significance of the decision, see White, (n 94), pp 179–185.
100   Defined by reg 2 of the Cremation (England and Wales) Regulations 2008, SI 2008/2841 (discussed immediately below) as a burial authority or any person who has opened a crematorium. Burial authorities are designated local authorities (see n 40), though crematoria can also be privately operated.
101   See also the Cremation Act 1952, as well as s 214 of the Local Government Act 1972 (the latter dealing specify with the provision and maintenance of crematoria by burial authorities in England and Wales). For an analysis of the rules governing the siting of crematoria, and distance from dwelling houses and public highways see S White, "Cremation Act 1902, s 5 (The 'Distance' or 'Radius' Clause): The Balloon and String Theory of Statutory Interpretation" (2013) 79 *Pharos International* 79.
102   These were introduced to modernise and consolidate the law. Although the present discussion focuses on the cremation of whole bodies, the Regulations also deal with the cremation of human body parts and other material (reg 19) and the cremation of stillborn infants (reg 20).
103   Though the deceased's consent is not a prerequisite.
104   Set out in Pt 4 of the 2008 Regulations.

An applicant must be either the deceased's executor or a "near relative" aged 16 or over,[105] and cremation cannot take place until the necessary paperwork has been completed[106] – including a designated application form, both medical and confirmatory medical certificates, a coroner's certificate (where necessary)[107] and overall authority to cremate from a medical referee.[108] As with burial, there is no legal requirement to use a funeral director[109] or to embalm a corpse prior to cremation.[110] And while the use of a coffin is not mandatory as long as the corpse is decently covered,[111] individual crematoria will probably insist on one to facilitate transport and handling.[112] Finally, cremation authorities are obliged to keep a permanent register of all cremations carried out by them, and to retain any preliminary certificates and documentation for at least 15 years.[113]

In the last two decades cremation has come under increasing environmental scrutiny,[114] with concerns about mercury emissions from vaporised amalgam fillings in teeth,[115] as well as being 'wasteful' in terms of both gas

---

105 Regulation 15(1), with "near relative" being defined in reg 15(3) – see Ch 3, p 71 for a more detailed analysis. Under reg 15(3), someone else may make the application if the medical referee accepts that they are a "proper person" to do so and the reason why the executor or near relative is not applying.

106 Not simply the registration of death requirements (see Ch 1, Pt III), but documentation which *must* be completed before cremation can take place.

107 As a general rule, however, the Cremation Act 1902 cannot interfere with the statutory jurisdiction of the coroner – 1902 Act, s 10. For example, the 2008 Regulations clearly stipulate that, where an inquest has been ordered, cremation cannot be authorised until the inquest has opened – reg 23(3).

108 2008 Regulations, reg 14 and regs 16–28, which also deal with authorisation for cremation and the duty to give reasons in the event of a refusal.

109 Though the more bureaucratic nature of the cremation application and rules laid down by individual crematoria might encourage the use of a funeral director who can advise and liaise accordingly.

110 Unless required where the deceased's remains are, for example, being kept at home prior to cremation.

111 See p 30.

112 Wienrich and Speyer (n 24), p 108. Restrictions are also imposed on the coffin (as well as its fittings and furnishings) to ensure that it is easily combustible and environmentally sound – *ibid*, pp 108–109. For example, plastic products would not be permitted, and while metal of a high ferrous content is acceptable the use of a zinc or lead-lined coffin would not be.

113 2008 Regulations, Pt 7.

114 Beginning with Pt 1 of the Environmental Protection Act 1990, which introduced controls on crematoria emissions. For an outline of the regulatory framework, see 24 *Halsbury's Laws of England* (n 3), para [1150].

115 See DEFRA, *Mercury Emissions from Crematoria* (2003) and "Crematoria Warned Over Mercury", *BBC News Online* (10 January 2005) http://news.bbc.co.uk/1/hi/health/4160895.stm (accessed 30 September 2015). Similar concerns have been raised elsewhere – see for example, PD Batchelder, "Dust in the Wind? The Bell Tolls for Crematory Mercury" (2008) 2 *Golden Gate University Environmental Law Journal* 118, discussing state government policies for dealing with crematoria emissions in the US.

energy consumption[116] and the destruction of natural resources through the burning of coffins (which also emits carbon dioxide).[117] Conventional arguments around the land-use benefits of cremation have also been challenged, given the widespread practice of interring ashes in designated burial sites.[118] Crematoria have been forced to remove harmful products from emissions, installing expensive abatement equipment to meet government targets for a 100 per cent reduction in mercury levels by 2020.[119] However, emissions control is only one aspect of a bigger ecological picture. For example, many crematoria participate in recycling programmes, which collect metal hip joints and similar objects that fail to vaporise during cremation; these are smelted down and re-used, with a portion of the annual profits being donated to death-related charities.[120] Recycling heat from the cremation process is another option; for example, in 2011 Redditch Borough Council announced plans to warm the water in its local swimming pool using heat from the municipal crematorium.[121] Thus cremation has moved beyond bodily disposal, to a "carefully managed process of reuse"[122] for the benefit of the living. At a more basic level, however, human remains are still converted into ashes – a dark powder, devoid of organic material and with no 'nutrient' value.[123]

2.   Licensed crematorium or 'DIY' cremation?

While there are a range of physical sites where burial of the dead can occur, cremation is much more restricted. In this country (as in most Western societies) human remains are invariably reduced to ashes in a crematorium,

---

116   The amount of energy required to cremate a single body is equivalent to that consumed by a living person in one month – Cowling (n 25), p 33.

117   See A Monaghan, "Conceptual Niche Management of Grassroots Innovation for Sustainability: The Case of Body Disposal Practices in the UK" (2009) *Technological Forecasting & Social Change* 1026, p 1037. Personal effects placed inside the coffin (e.g. teddy bears and personal mementos) also contribute to emissions levels.

118   See Canning and Szmigin (n 1), p 1136.

119   Rumble *et al* (n 16), p 248.

120   Noted in Rumble *et al* (n 16), pp 247–248. One of the leading companies specialising in this recycling process is Orthometals – see www.orthometals.com/ (accessed 30 September 2015). More detailed guidance can be found on the website of the Institute for Cemetery and Crematorium Management (ICCM), who promote the recycling scheme throughout the UK – see www.iccm-uk.com/iccm/index.php?pagename= recyclingmetal (accessed 30 September 2015).

121   The council envisaged annual savings of £14,500 on heating bills – "Crematorium Heats Swimming Pool", *The Telegraph* (London, 25 January 2011) www.telegraph. co.uk/news/newstopics/howaboutthat/8279648/Crematorium-heats-swimming-pool.html (accessed 30 September 2015).

122   Rumble *et al* (n 16), p 248.

123   Contrast this with the products of resomation, which can have a positive ecological impact – see Pt III.

and not on open-air funeral pyres or in private furnaces elsewhere.[124] However, the legality or otherwise of 'DIY' cremation has been called into question.

As already noted, *R v Price*[125] established that burning human remains outdoors was permissible at common law, as long as it did not constitute a public nuisance or obstruct the coroner in performing his/her statutory duty.[126] The same reasoning was applied as recently as 2005 in *R v Wriggle-worth*,[127] where a son continued to claim his dead mother's pension for two years despite having set fire to her body in the back garden of the family home some eight weeks after her death.[128] No public nuisance had been complained of, and the son was not guilty of preventing an inquest given that his mother had previously been diagnosed with terminal cancer.[129] While these rulings are relatively straightforward, the common law position is only part of the picture. Since cremation is heavily regulated by statute and ancillary regulations, an obvious question is whether these instruments prescribe the only conditions in which this method of corpse disposal can take place (rendering anything else unlawful), or whether open-air and other forms of private cremation fall outside this particular legal framework.

Section 8(1) of the Cremation Act 1902 makes it a criminal offence to "knowingly carry out or procure or take part in the burning of any human remains except in accordance with ... the provisions of this Act". At a cursory glance, this might seem to answer the question just posed. Yet while s 2 of the 1902 Act defines a 'crematorium' for the purposes of the Act as "any

---

124 Though DIY cremation does occur from time to time – see below.
125 (1884) 12 QBD 247 and discussed above.
126 See also *R v Stephenson* (1884) 13 QBD 331.
127 (Leeds Crown Court, 21 January 2005) and noted in Green and Green (n 23), p 113.
128 The son had kept his mother's body in the bedroom of the house.
129 However, the son had acted unlawfully in failing to register her death, and was guilty of 12 cases of deception for fraudulently claiming benefits – see "Body Burning Son Spared Jail Term," *BBC News Online*, (21 January 2005) http://news.bbc.co.uk/1/hi/england/west_yorkshire/4196447.stm (accessed 30 September 2015). It is interesting that the offence of outraging public decency does not appear to have featured strongly in any of these cases. The offence mandates visual and open elements, which may have been difficult to establish on the facts – even though human remains were burnt outdoors, and members of the public would (presumably) have been in a position to see this. In contrast, American courts have been willing to invoke broader concepts of public morality when confronted with the spectre of DIY cremation. For example, in *State v Bradbury* 136 Me 347 (1939) an elderly brother and sister lived together in the same house; neither sibling was married and they had no other family. When the sister died, the brother burned her body in a furnace in the cellar – a laborious process to start with, given that he had to wait until the head and shoulders were consumed by the flames, before inching the body slowly into the fire until the furnace door could be closed. Although the cremation took place in a basement away from public view, complaints by a next-door neighbour about the smoke and the smell prompted authorities to investigate. Bradbury was convicted of an indecent act which was *contra bonos mores* at common law.

building fitted with appliances for the burning of human remains", and subsequent versions of the accompanying regulations have provided that no cremation can take place except in a crematorium (the opening of which has been notified to the Secretary of State),[130] the long-term failure to define 'cremation' hinted at a legislative lacuna.[131] Shortly after the 1902 Act was passed, the issue was raised in *R v Byers*.[132] After children in her care had died, the defendant procured burial money from the children's mothers; instead of interring the remains, the defendant burnt the corpses on a fire in her kitchen grate and was charged under s 8 of the Cremation Act.[133] Interpreting 'cremation' as burning human remains in a crematorium *as defined by the Act* and directing an acquittal on the various criminal charges,[134] the decision in *Byers* seemed to imply that burning corpses outside a crematorium (i.e. otherwise than by 'cremation') was not necessarily unlawful. A century later, the same issues were explored in depth in *Ghai v Newcastle City Council*.[135]

In January 2006, Davender Ghai, founder President of the Newcastle-based Anglo-Asian Friendly Society, approached Newcastle City Council about dedicating land on the outskirts of the city and close to flowing water for traditional funeral pyres. These were commonplace in India, and Mr Ghai believed their absence in Britain prevented the transmigration of the deceased's soul while inflicting remorse on bereaved Hindu families living here. Several months later, the council declined the request on the basis that the current law prohibited funeral pyres; Mr Ghai challenged this decision.[136] In the High Court, Cranston J upheld the council's decision based on the wording of the legislation, which also included a new definition of 'cremation' in the 2008 Regulations:

> In my view..., the combined effect of the legislation and attendant regulations is plain: a cremation is the burning of human remains: regulation 2(1); all cremations must take place in a crematorium: regulation 13; a crematorium is a building: (1902 Act, section 2); and the burning of human remains other than in accordance with the provisions of the

130   See reg 3 of the original Cremation Regulations 1902 – now substantially re-enacted in reg 13 of the Cremation Regulations 2008.

131   This has now been remedied by the 2008 Regulations – see below.

132   (1907) 71 JP 205. For a detailed discussion, see S White, "An End to D-I-Y Cremation?" (1993) 33 *Medicine, Science and the Law* 151, pp 156–158.

133   Namely s 8(3) (now repealed), which deals with cremation with intent to conceal the commission or impede the prosecution of any offence.

134   The fact that burning the children's remains in the kitchen grate did not constitute a public law nuisance meant that no common law offences were committed either.

135   [2009] EWHC 978 (Admin).

136   Mr Ghai had presided over an open-air cremation in Northumberland in July 2006 – the first of its type in 70 years. See S Jones, "Police Say Sikh Funeral Pyre May Have Broken Cremation Laws", *The Guardian* (London, 13 July 2006) www.guardian.co.uk/uk/2006/jul/13/religion.world (accessed 30 September 2015).

2008 Regulations is a criminal offence: (1902 Act, section 8). Thus the burning of human remains, other than in a building, such as on an open air pyre, is an offence. [137]

Cranston J also rejected Mr Ghai's claim that the legal requirement to be cremated in a crematorium infringed his rights under Article 9 of the European Convention on Human Rights; while accepting that the claimant's right to practise his religion encompassed his entitlement as an orthodox Hindu to be cremated on an open-air pyre, any resulting interference was both proportionate and justified.[138] On appeal, however, the discussion shifted to what constituted a permissible 'building' for the purposes of s 2 of the 1902 Act, Mr Ghai having conceded that his religious beliefs could be satisfied by cremation within a proposed structure as long as the actual process was by 'traditional fire' and sunlight could shine directly on his body.[139]

In some ways, it is unfortunate that the ruling on whether the 1902 Act and accompanying regulations prohibit funeral pyres was overshadowed by the human rights aspect of the first instance decision in *Ghai*. Cranston J's observations on the subject would seem to provide a fairly definitive answer – though White has argued that, given the shift in emphasis at the Court of Appeal stage and consequent failure to address the issue in the higher court, "we still have no authoritative ruling about whether pyres outside crematoria are prohibited by [the legislation]".[140]

3.   Dealing with the ashes

Following cremation, there is the additional question of what happens to the ashes.[141] In England and Wales, these can be lawfully removed from the

---

137 [2009] EWHC 978 (Admin), [83]. In reaching this conclusion, Cranston J pointed to the legislative intent behind the 1902 Act (this would be undermined if the legislation "were to establish a series of detailed rules for the burning of human remains inside crematoria, but [left] ... the burning of human remains outside crematoria unregulated" – *ibid*, [84]).

138 The human rights aspect of the decision is discussed in Ch 5, pp 141–142.

139 *Ghai v Newcastle County Council* [2010] EWCA Civ 59, the Court of Appeal deciding that the relatively permanent and substantial structure proposed by Mr Ghai (similar to those used in Hindu cremations abroad) could be accommodated within the legislation.

140 S White, "Funeral Pyres in a Legal Limbo" (2010) *Pharos International* 30, pp 32–33.

141 See generally DJ Davies and MJ Guest, "Disposal of Cremated Remains" (1999) 65 *Pharos International* 26. Ashes do not have any distinct legal status – L Skene and B Masters, "What Legal Rights Do You Have Over Your Body After Your Death?" (2002) 81 *Australian Law Reform Commission Reform Journal* 38, p 40.

crematorium[142] and, under the 2008 Regulations, must be given to the person who applied for cremation.[143] However, there is no legal obligation to retrieve ashes or to deal with them in a particular way, resulting in unclaimed remains being interred or dispersed in the crematorium gardens.[144] Many families were content for this to happen in the past, though recent years have seen a growing percentage of ashes being removed from crematoria[145] as the twenty-first century trend towards individualised funeral rituals exerts its influence on contemporary ash disposal practices. Instead of simply entrusting ashes to communal and depersonalised sections of crematoria, they are increasingly dealt with in specialised ways which suggest a "more materially engaged connection with the remains of the dead".[146]

Ashes attract a broader range of disposal options than corpses, given the absence of public health concerns alongside their physical compactness, portability, fluidity and divisibility.[147] Conventional choices include interment in existing family burial plots[148] or in miniature graves in specially reserved sections of municipal cemeteries,[149] as well as placement in a columbarium or niche.[150] Ashes can also be kept at home in a commemorative urn – either as a transitory measure while a decision is made about a final resting place,[151] or

---

142   The UK (like Finland, France and Spain) exerts minimal legal controls over the final destination of ashes, whereas regulations in Belgium, Denmark, Germany, Italy and Slovenia confine ashes to the crematorium cemetery or a similar disposal site – D Prendergast, J Hockey and L Kellaher, "Blowing in the Wind? Identity, Materiality and the Destinations of Human Ashes" (2006) 12 *Journal of the Royal Anthropological Institute* 881, p 883.

143   Or someone nominated by the applicant – reg 30(1). However, the person with the legal right and duty of disposal of the deceased's remains has the strongest legal claim to the ashes (once released by the crematorium) if a dispute subsequently arises – see Ch 3, pp 72–73.

144   Individual crematoria are legally obliged to bury or scatter these ashes in a designated part of their grounds, having given the applicant 14 days' notice of their intention to do so – 2008 Regulations, regs 30(3)–(4). This avoids crematoria having to store accumulating quantities of ashes indefinitely.

145   Estimates suggest that, by 2005, around 60 per cent of ashes were being removed – see L Kellaher, J Hockey and D Prendergast, "Wandering Lines and Cul-de Sacs: Trajectories of Ashes in the United Kingdom" in Hockey, Komaromy and Woodthorpe (n 73), p 133.

146   Prendergast, Hockey and Kellaher (n 142), p 883 – the authors also noting that "incineration of the corpse is no longer a key point of physical separation between the dead and living".

147   The deceased's family can divide the ashes between them, allowing different individuals to deal with their portion in any of the ways outlined here.

148   Whether in a churchyard, cemetery or elsewhere.

149   See L Kellaher, D Prendergast and J Hockey, "In the Shadow of the Traditional Grave" (2005) 10 *Mortality* 237.

150   Though this tends to be a less common option.

151   Or as an indefinite holding arrangement where the intent is to combine the deceased's ashes with those of a close relative (for example, a spouse or partner) when he/she dies. In *Sopinka v Sopinka* (2001) 55 OR (3d) 529 the son's ashes were placed in his father's coffin, the latter having died some three months later.

as a long-term arrangement that symbolises an ongoing meta-physical connection to the dead. However, scattering ashes in places which the deceased had a strong personal and emotional attachment to in life[152] is a relatively recent phenomenon.[153] In England and Wales (as in the rest of the UK) there are few laws surrounding ash dispersal,[154] and no established social conventions. Scattering can be informal and unceremonious; it does not have to be a solemn affair,[155] and can extend beyond the earth's physical surface if, for example, a loved one's ashes are exploded inside a celebratory firework or launched into space.[156] The fact that cremated remains are "much more open to symbolic creativity"[157] has also lead to them being converted into more permanent (and increasingly innovative) memorials to the deceased. For example, the ashes of a loved one can be converted into a certified diamond as a "unique … and timeless" memento,[158] or used to create a memorial reef as a "meaningful contribution to future generations … [and] to the marine environment".[159] These and other creative practices are part of a growing movement towards what Davies describes as "consumerist death style".[160]

III. New and emerging methods: Resomation and promession

Although burial and cremation dominate the bodily disposal landscape, two new techniques are being developed which could alter this. Both have

---

152 For example, in a local park or forest that the deceased used to visit, or in an area that holds memories of family holidays with the deceased – see Kellaher, Hockey and Prendergast (n 145), p 133.

153 Cultural narratives may also come into play. For example, Prendergast, Hockey and Kellaher (n 142), p 885 have noted a movement towards establishing *ghats* (waterside sites for the disposal of ashes) for Hindu families in the UK who view releasing the ashes into flowing water as an essential death rite.

154 No specific permit is required, though ashes cannot be dispersed in a particular location if, for example, entering the land in question would constitute an actionable trespass or ash-scattering is prohibited there. For example, art 5(6) of the Local Authorities' Cemeteries Order 1977 allows local authorities to prohibit ash scattering in parts of cemeteries set aside for particular religions.

155 Whether a market develops in this country for specialist 'party boats' catering for celebratory ash disposals at sea, remains to be seen. These are popular in sunny California – see RE Haddleton, "What To Do With The Body? The Trouble With Postmortem Disposition" (2013) 20 *Property & Probate* 55, pp 58–59.

156 Again, these are popular in the US with specialist firms offering bespoke fireworks or space orbits. The remains of Star Trek creator, Gene Roddenberry, and James Doohan, who played the starship Enterprise's engineer Scotty in the original series, were both launched into space – NL Cantor, *After We Die: The Life and Times of the Human Cadaver* (Georgetown University Press, 2010), p 117.

157 Davies (n 2), p 121.

158 As offered by LifeGem Memorials (see www.lifegem-uk.com (accessed 30 September 2015)).

159 As offered by the American firm Eternal Reefs (see http://eternalreefs.com (accessed 30 September 2015)).

160 Davies (n 2), p 66.

environmental impact and sustainability at their core, and are being actively promoted within the deathcare industry on this basis.[161]

1.   Resomation

Resomation[162] is a liquefaction process which uses alkaline hydrolysis to dissolve the body's organic matter in a heated and pressurised steel container; the result is a sterile liquid which is either disposed of through the municipal water treatment system or put to some other use (for example, as a fertiliser), and bones which can be crushed and given to the deceased's family in the same way as ashes produced by conventional cremation.[163] Advocates of resomation emphasise that it is more environmentally friendly than cremation because the process uses less energy and does not release carbon dioxide and other trace chemicals (such as mercury from dental fillings) into the atmosphere.[164] Opponents argue that "flushing human remains down the drain" is "undignified, cold ... [and] unsanitary",[165] and that the actual physical process resonates with the brutal punishments meted out by drug lords, dictators and murderers who dissolved their victims in acid baths.[166] Less sensationalist, yet equally cogent, are concerns over post-resomation bodily liquids being recycled within the drinking water system or promoting food growth as fertilisers, and whether this would be "culturally acceptable".[167]

Resomation facilities have been introduced in a small number of US states,[168] and also in parts of Australia.[169] In Britain, the process is being

161   Rumble *et al* (n 16), pp 249–250.
162   Sometimes referred to as 'aquamation' or 'bio-cremation'.
163   See W Zukerman, "Dissolving Your Earthly Remains Will Protect the Earth", *New Scientist* (online) (19 August 2010) www.newscientist.com/article/dn19333-dissolving-your-earthly-remains-will-protect-the-earth.html (accessed 30 September 2015).
164   Zuckerman, *ibid*. See also M Kamanev, "Aquamation: A Greener Alternative to Cremation?", *Time Science* (online) (28 September 2010) http://content.time.com/time/health/article/0,8599,2022206,00.html (accessed 30 September 2015). Any metal implants located in the deceased will also be in a much better condition after the process than they would be with cremation, and can be recycled more effectively.
165   See Hansen (n 58), p 152 and the various sources cited therein.
166   Hansen, *ibid*. Of course, this overlooks the fact that resomation is a highly controlled, respectful and dignified chemical process – and not a method of killing or inflicting posthumous punishment on the dead.
167   Rumble *et al* (n 16), p 249.
168   See Hansen (n 58), as well as PR Olson, "Flush and Bones: Funeralizing Alkaline Hydrolysis in the United States" (2014) *Science, Technology and Human Values* 1.
169   For example, in both Queensland and New South Wales – "Queenslanders Can Now Have a Watery Grave with 'Aquamation' Centre Opening on the Gold Coast", *The Courier-Mail* (Brisbane, 12 August 2010) www.couriermail.com.au/news/queens-landers-can-now-have-a-watery-grave-with-aquamation-centre-opening-on-the-gold-coast/story-e6frep26-1225904435756 (accessed 30 September 2015).

driven by Glasgow-based company Resomation Ltd,[170] with plans to make the technology widely available over the next few years at a comparable cost to cremation.

2. Promession

Described by one American lawyer as a variant on freeze drying,[171] promession is a high-tech yet simplistic concept, which utilises the fact that humans are composed primarily of water. After being initially frozen to −18°C, the (coffined) body is super-cooled in liquid nitrogen at temperatures approaching −200 °C; the brittle remains are then shattered (using ultrasonic vibration) into an odourless, organic residue. After the water content has been evaporated and dispersed into the atmosphere, the resultant dry powder is placed in a small bio-degradable container and interred in a shallow grave where both 'coffin' and contents will turn into compost within 6–12 months.[172] Promoted as a form of "ecological burial",[173] exponents of promession highlight the environmental benefits of using liquid nitrogen instead of fossil fuels, alongside the fact that the end-product contributes to the soil and takes up much less space than conventional burial.[174] However, the prospect of freeze-drying corpses and smashing them into pieces on a vibrating table may be regarded as unpalatable by some.

As things currently stand, promession is still in the developmental stages, and is not commercially available as a bodily disposal option in the UK[175] or elsewhere.[176]

3. Moral status and legal status?

New methods of corpse disposal are always controversial, and it remains to be seen whether resomation and promession will be regarded as socially acceptable alternatives to burial and cremation – even if the technology becomes widely available. The environmental benefits are obvious, and neither option raises public health issues; instead, the wider debate is likely

170   Further information is available at www.resomation.com (accessed 30 September 2015). The fact that the company is majority controlled by the Co-operative Group, owner of the largest funeral home business in the UK, will assist in financing, marketing and 'rolling out' the procedure.

171   Haddleton (n 155), p 58.

172   See Monaghan (n 117), p 1038. Like resomation, metals imbedded in the deceased can also be removed for recycling, and will be in better condition than those retrieved after cremation.

173   See Rumble *et al* (n 16), p 251.

174   See Monaghan (n 117), p 1038.

175   Though two companies – Promessa UK www.promessa.org.uk (accessed 30 September 2015) and Cryomation www.irtl.co.uk/ (accessed 30 September 2015) – are trying to remedy this.

176   Rumble *et al* (n 16), p 251.

to focus on the dignity or 'sacredness' of human remains and whether the techniques employed in resomation and promession treat the dead with sufficient respect.[177] Winning over an initially apprehensive (and squeamish) public may take time, just as cremation gradually gained acceptance despite widespread opposition when first introduced.

If resomation and promession are to become mainstream forms of bodily disposal, specific regulatory frameworks will have to be introduced. Existing public health and environmental statutes may suffice for the developmental stages,[178] and it appears that both processes are currently legal in Britain as long as they do not infringe sanitation laws or offend public decency.[179] However, detailed provisions would have to be introduced in the longer term, confirming resomation and promession as lawful methods of corpse disposal and putting appropriate safeguards in place.[180]

IV.   Exposure and natural decomposition

Some cultures dispose of their dead by exposing them to the elements, allowing the corpse to decompose and/or be consumed by animals. For example, in a Tibetan 'sky burial' the body is incised and placed on a mountaintop to be devoured by birds of prey, and body exposure is still practised by a declining number of Aboriginal tribes in Australia.[181] Although customary within certain cultures and countries, this method of disposal is not regarded as socially acceptable in most Western societies. Leaving corpses to decay above ground raises serious public health and human dignity concerns,[182] and could also attract criminal law charges of preventing the lawful burial of a body or outraging public decency.[183]

---

177   There may also be religious or cultural objections around destruction of the body.

178   Currently the position in Queensland – see Queensland Law Reform Commission, *A Review of the Law in Relation to the Final Disposal of a Dead Body* (Report No 69, December 2011), pp 15–16.

179   Rumble *et al* (n 16), pp 251–252.

180   For example, regulating resomation and promession facilities, and controlling environmental and public health aspects of the process. There are two ways of achieving this. In New South Wales, for example, resomation is included in a new and expanded definition of cremation, and regulated accordingly (see Queensland Law Reform Commission (n 178), p 18) and a similar approach has been suggested for England and Wales (S White, "The Public Health (Aquification) (England and Wales) Regulations?" (2011) 77 *Pharos International* 10). In the US, different states adopt different practices with some incorporating resomation within existing cremation laws and others regulating it as a distinct method of corpse disposal – see Hansen (n 58), pp 153–158.

181   Green and Green (n 23), p 25.

182   See G McBain, "Modernising the Law on the Unlawful Treatment of Dead Bodies" (2014) 7 *Journal of Politics and Law* 89, p 92.

183   Common law offences, which were outlined in Ch 1, Pt VI. Interesting legal arguments could arise if, for example, exposure and subsequent decomposition of the deceased's body took place on private land hidden from public view; would the latter offence have been committed in these circumstances?

However, individuals hoping to solve crime and bring killers to justice can donate their remains to 'body farms'. Located in the United States, these are research facilities that study human decomposition in a variety of settings,[184] and have important applications within the realms of forensic anthropology and forensic science.[185] There are currently no designated body farms in the UK, though some forensic anthropologists support the idea.[186] If such facilities were introduced on this side of the Atlantic, this would constitute another method of corpse disposal – assuming that appropriate regulatory measures were put in place.[187]

V.   Preservation

Like conventional bodily disposal methods, preservation ensures a permanent fate for the dead while eliminating the public health risk surrounding a decaying corpse.[188] Yet, while the former reduce the cadaver to either waste products or corporeal matter bearing no physical resemblance to the deceased, preservation is a highly technical process, which prevents the body from decaying or disintegrating while retaining its ante-mortem form.[189] There are no funeral rites, and the preserved remains may be visible to others long after the person's demise. Describing it as a "vainglorious conceit", Cantor questions why the world would want or need a cadaver representing a "highly debilitated version of a previous self", which is "preserved for the indefinite future".[190] Yet, bodily preservation can be driven by a number of things – for example, reunification of body and soul in the religious

---

184   There are currently a handful of operational body farms in the United States. One of the best known is the University of Tennessee Anthropological Research Facility which dates back to 1981 and was the first to be established – see http://fac.utk.edu/ (accessed 30 September 2015).

185   See WM Bass and J Jefferson, *Beyond the Body Farm* (William Morrow, 2007).

186   J Clinton, "We Need Human Body Farms, Says Real-Life Dr Bones and Forensic Expert Anna Williams", *Sunday Express* (London, 22 September 2013) www.express.co.uk/ news/science-technology/431287/We-need-human-body-farms-says-real-life-Dr-Bones-and-forensic-expert-Anna-Williams (accessed 30 September 2015).

187   Namely legal requirements around donation, as well as practical measures to ensure that these open-air forensic labs were hidden from public view (and prying eyes) and secure from wild animals that might enter and remove decomposing remains. Whether the various 'controls' that have to be imposed on body forms would undermine the scientific viability of any data obtained is, of course, another matter.

188   The focus here is on deliberate acts of preservation (not remains that were accidentally preserved because the individual died or was buried in a river bank or peat bog, and is found centuries later).

189   Contrast this with standard embalming, which envisages short-term preservation of human remains to allow transportation and/or viewing of the deceased in the immediate post-mortem period.

190   Cantor (n 156), p 119.

afterlife;[191] medical training and teaching;[192] personal choice;[193] political symbolism;[194] educational displays (usually for financial gain) and a basic desire for immortality. The latter two have attracted attention in recent years, given the controversies surrounding the Body Worlds exhibit and the cryonics movement respectively.

One of the most technical methods of modern preservation is the plastination process pioneered by German anatomist Gunther von Hagens and made famous in his travelling Body Worlds exhibit.[195] Bodily fluids are replaced with a plastic solution, which creates a rigid structure; partially dissected corpses and body parts (including organs, muscles and soft tissues) can then be displayed in a variety of lifelike poses.[196] Despite its worldwide popularity, legal and ethical aspects of the exhibit have been called into question since it began in 1995. In the absence of a specific legislative framework for dealing with preserved cadavers, bodily donations to the Body Worlds exhibit must be legally akin to a form of anatomical donation and, as such, dependent on the written consent of the individuals who want their remains to be preserved and displayed in this manner. This is straightforward enough – though concerns were raised, in the past, about whether all of the displayed remains had been obtained legally.[197] Assuming the existence of a valid consent, what if an individual's request to be part of the exhibit is challenged by their family? Young notes

191   As in ancient Egypt where mummification was one of the earliest methods of preserving the dead for this purpose – see JH Taylor, *Death and the Afterlife in Ancient Egypt* (University of Chicago Press, 2001).

192   See generally DG Jones and MA Whitaker, *Speaking for the Dead: The Human Body in Biology and Medicine* (Ashgate, 2009).

193   As in the case of Jeremy Bentham, whose auto-icon is still on display in University College London – see N Naffine, "When Does the Legal Person Die: Jeremy Bentham and the Auto-Icon" (2000) 25 *Australian Journal of Legal Philosophy* 79. A more recent example is the preserved body of Edward MacKenzie (a homeless tramp known as 'Diogenes'), found in a cupboard drawer in the Plymouth studio of controversial artist Robert Lenkiewicz after the latter's death in 2002. Mackenzie was a friend and model for Lenkiewicz, and the two men had made a pact before Mackenzie's death in 1984 that the artist would embalm Mackenzie's body rather than give it up for burial – see S de Bruxelles, "Artist Kept Tramp's Corpse in a Cupboard", *The Times* (London, 12 October 2002).

194   For example, the embalmed body of former Russian leader Vladimir Lenin has been on public display since his death in 1924 – see I Zbarsky and S Hutchinson, *Lenin's Embalmers* (Harvill Press, 1999).

195   For up-to-date details, see www.bodyworlds.com/en.html (accessed 30 September 2015).

196   From an anatomical perspective, Body Worlds has broken basic conventions since "[w]hat has traditionally been private is made public, and the deadness of cadavers appears to have been replaced by a disconcerting life-likeness" – DG Jones and MI Whitaker, "Engaging with Plastination and the Body Worlds Phenomenon: A Cultural and Intellectual Challenge for Anatomists" (2009) 22 *Clinical Anatomy* 770, p 771.

197   See T Patterson, "Body Worlds Impresario 'Used Corpses of Executed Prisoners for Exhibition'", *The Telegraph* (London, 25 January 2004).

that there appear to be no examples of such litigation, though suggests that Body Worlds would "relinquish any legal claim to a cadaver rather than risk the public relations nightmare of fighting with grieving survivors over the right to inject the deceased with polymers and put her on public display".[198]

While the actual plastination process has not provoked widespread public disapproval, the display aspect has.[199] Exhibiting whole-body, plastinated figures in what some regard as degrading poses has been criticised as "intrinsically undignified and ... a moral abuse of the human cadaver",[200] regardless of the educational potential and the informed consent of individual donors.[201] Commodifying the dead and profiting from the display of human remains also raises issues,[202] and feeds into broader notions of societal harm.[203] At a more basic human level, those on display (despite being anonymised) will have died relatively recently and probably still have living relatives, some of whom are uncomfortable with their loved one's remains being part of a cadaver exhibit.

In contrast to plastination where the legal and ethical issues are centred on treatment of the dead, those surrounding cryonic preservation focus on the dubious assurances given to individuals who decide to undergo the process. Cryonics is a speculative technology that uses extremely cold temperatures[204] to place dead bodies in a state of long-term preservation until medical advancements enable reanimation.[205] The science behind it is suspect,[206] and a convincing case has yet to be made that someone who is frozen or who has had their head frozen (the intent being to reconstruct

---

198 H Young, "The Right to Posthumous Bodily Integrity and Implications of Whose Right It Is" (2013) 14 *Marquette Elder's Advisor* 197, pp 249–250.

199 See T Walter, "Plastination for Display: A New Way to Dispose of the Dead" (2004) 10 *Journal of the Royal Anthropological Institute* 603.

200 Cantor (n 156), p 279.

201 In response to these and other concerns, Hawaii has banned Body-Worlds-type cadaver displays, regardless of the posed individuals having consented to this – *Haw Rev Stat* § 327–338 (2013).

202 LA Giunta, "The Dead on Display: A Call for the International Regulation of Plastination Exhibits" (2010) 49 *Columbia Journal of Transnational Law* 164 and DM Baker, "Cryonic Preservation of Human Bodies: A Call for Legislative Action" (1994) 98 *Dickinson Law Review* 677.

203 Young (n 198), p 198.

204 Around –196°C.

205 See generally C Knight, "A Science Without A Deadline" (2008) 19 *Engineering and Technology* 28 and D Shaw, "Cryoethics: Seeking Life After Death (2009) 23 *Bioethics* 515. Cryonic preservation is not (for those who opt for it) a method of final disposal.

206 According to cryobiologist Dr Arther Rowe, "[b]elieving cryonics could reanimate somebody who has been frozen is like believing you can turn hamburger back into a cow" – quoted in "Frozen Future", *National Review*, 9 July 2002.

the individual from their brain) can be regenerated and/or cured of some terminal illness when this becomes medically possible.[207] While cryonic preservation is an (albeit expensive) option in the United States[208] and its courts may be protective of an individual's choice to be placed in a state of suspended animation,[209] many European countries[210] do not recognise it as a lawful means of dealing with the dead and the technology is not widely available. Cryonics UK is a non-profit organisation that assists people living in the UK to have their remains cryopreserved on 'death' by shipping the body (in a state of preservation) to a chosen cryonics storage provider in another country.[211] This in itself is permissible, though it remains to be seen whether similar licensed facilities will be ever opened in the UK (and if so, how they will be regulated). In the meantime, however, long-term cold storage of a corpse in purpose-built refrigeration facilities is permissible under English law,[212] though

207   "Some people … view cryonics as offering the best opportunity for people finally to be able to transcend death. Others see cryonics as a pipe dream and the cryonics industry as charlatanism at its worst" – RD Madoff, *Immortality and the Law: The Rising Power of the American Dead* (Yale University Press, 2010), p 2.

208   Though it appears that there are only two such facilities (Cantor (n 156), p 131), the most well-known of which is the Alcor Life Institute in Scottsdale, Arizona (see www.alcor.org/ (accessed 30 September 2015)). This is where the head of Boston Red Sox baseball legend Ted Williams is currently stored, following a bitter dispute between his children over what should happen to their father's remains – see AA Bove and M Langa, "Ted Williams: Is He Headed for the Dugout or the Deep Freeze? Property Rights in a Dead Body Resurrected", *Massachusetts Lawyers Weekly*, 19 August 2002.

209   *Alcor Life Extension Foundation Inc v Mitchell* 7 Cal Rptr 2d 572 (1992). However, this has not prevented calls for a specific regulatory framework for cryonics – see D Friedman, "Does Technology Require New Law" (2001) 25 *Harvard Journal of Law & Public Policy* 71 and AA Perlin, "To Die in Order to Live: The Need for Legislation Governing Post-Mortem Cryonic Suspension" (2007) 36 *Southwestern University Law Review* 33.

210   For example, France.

211   See the Cryonics UK website located at www.cryonics-uk.org/ (accessed 30 September 2015).

212   As highlighted by newspaper reports of two sisters who kept their mother's remains refrigerated in a London funeral director's premises for 10 years and paid weekly visits to the body – F Barton, "Sisters Keep Mother's Body in the Fridge for Ten Years: And Visit Her Every Weekend", *Daily Mail* (London, 6 September 2007) www.dailymail.co.uk/news/article-480276/Sisters-mothers-body-fridge-years—visit-weekend.html (accessed 30 September 2015).

DIY preservation in a freezer at home would probably raise public health concerns.[213]

Technical issues and availability aside, all sorts of legal issues would arise if successful reanimation of a corpse were to occur – beyond the fact that the individual had already been declared legally dead.[214] Cantor, for example, poses the following questions:

> How much regained brain content would be necessary to make the regenerated being the same person as the frozen decedent…? Assuming that the reconstituted person will be the same as his or her progenitor, what is the legal status of the cryonically preserved corpse in the interim period? What happens to the former (and perhaps continuing) spouse, children and other survivors? Are the heirs free to dispose of the decedent's property while the corpse is in frozen limbo? Could they decide to thaw the corpse without revitalization?[215]

At a basic human level there is also the issue of how an individual would cope with being revived many years after their 'death', when confronted with a new social landscape devoid of family and friends.[216]

213   A situation that arose in France several years ago, when Remy Martinot lost a legal battle to keep his parents' frozen bodies after the freezer system in a cellar of the Loire valley chateau in which the bodies were stored malfunctioned causing a temperature rise. Mr Martinot was following the wishes of his father (a cryonics enthusiast) who had hoped that science might enable his own body, and that of his wife's, to be revived at a later date; however, France's highest court ordered that the bodies be either cremated or buried because French law only permits these two alternatives (the son eventually chose cremation) – see "Frenchman Cremates Frozen Parents", *BBC News Online* (16 March 2006)   http://news.bbc.co.uk/1/hi/world/europe/4814540.stm   (accessed   30 September 2015) and A Christafis, "Freezer Failure Ends Couple's Hopes of Life After Death", *The Guardian* (London, 17 March 2006). French authorities had made a similar ruling several years earlier, rejecting an attempt by siblings in the French territory of Réunion to keep their dead mother's remains in a glass-topped freezer in the cellar at home – "Keeping Body Frozen Ruled Illegal", *BBC News Online* (31 May 2000) http://news.bbc.co.uk/1/hi/world/europe/771811.stm (accessed 30 September 2015).

214   Even if the very concept of cryogenics is based on suspending death and reanimating the individual.

215   Cantor (n 156), p 134. Mims poses similar questions; for example, does a frozen corpse have legal rights, what happens to the soul during freezing, and whether we really want a world "filled to capacity with the thawed remnants of previous generations" – C Mims, *When We Die: The Science, Culture and Rituals of Death* (Robinson Publishing, 2000), p 215.

216   While Davies (n 2), p 67 suggests this is not something that people would value, Shaw (n 205), p 516 counters the "loneliness argument" on the basis that the thawed-out individual can make new friends and trace family descendants.

VI.   Specific body disposal offences

The criminal law also prescribes a number of specific offences around disposal of the dead,[217] in addition to those which fall under the rubric of interfering with a corpse.[218] Cremation otherwise than in accordance with the legislation is probably a criminal offence,[219] as is the making of a false representation or providing a false certificate with the aim of procuring the burning of human remains.[220] Meanwhile, it is a common law offence to prevent the lawful 'burial' of a body – a misdemeanour with a wide remit, which has enjoyed something of a legal renaissance since its existence was confirmed by the Court of Appeal in *R v Hunter*.[221] The act of concealing a body has attracted this offence where the defendant was clearly with the deceased when that person died but there are question marks over whether the defendant actually caused the death;[222] it has also been committed where a funeral director, who forgot to place the remains of an infant in its coffin before burial, attempted to hide the error by placing the body in someone else's coffin,[223] and where a husband concealed his wife's body in a room in

---

217   These are a combination of common law and, to a greater extent, statutory offences.

218   Discussed in Ch 1, Pt VI.

219   Cremation Act 1902, s 8(1) – though see the discussion in Pt II. As already noted, burning human remains outside a crematorium may also attract the common law offences of public nuisance and preventing an inquest.

220   Cremation Act 1902, s 8(2).

221   [1973] 3 WLR 374, the court confirming that it is a criminal offence to prevent the decent and lawful burial (or cremation) of a corpse without lawful excuse, and attributing the offence to *R v Lynn* (1788) 2 Term Rep 733. According to the Court of Appeal in *Hunter*, the offence is not confined to those who have the legal duty of disposal in respect of the deceased (see generally Ch 3); it extends to anyone who prevents lawful burial of the dead. See generally M Hirst, "Preventing the Lawful Burial of a Body" [1996] *Criminal Law Review* 96.

222   As in *R v Hunter*, *ibid* where the deceased girl's body was hidden under a pile of paving stones in a playing field for four months, after being accidentally strangled by her own scarf in a bout of 'horseplay' with the defendants. See also *R v Swindell* (1981) 3 Cr App (S) 225.

223   *R v Skidmore* [2008] EWCA Crim 15393. The infant's body had been accidentally left in the temporary coffin that had brought him from the hospital. Instead of alerting the funeral cortege (the error was discovered shortly after the empty casket had left the premises) the funeral director tried to conceal the mistake by placing the infant's body inside the coffin of an elderly lady which was also at the funeral home, and both bodies were cremated. According to the Court of Appeal, the fact that the defendant undertaker had acted out of panic (rather than from some dishonest or improper motive) was irrelevant.

their home when under the effects of a drug-induced haze and unable to come to terms with her death.[224]

Turning to burial grounds, any damage to or destruction of property (such as headstones and memorials) would probably constitute criminal damage.[225] However, there are also specific offences around anti-social or otherwise offensive conduct in churchyards and cemeteries, as well as the desecration of burial grounds more generally. For example, riotous, violent or indecent behaviour in a churchyard is a statutory offence, as is interfering with a clergyman in the execution of his/her office.[226] In municipal cemeteries, statutory offences include wilfully causing a disturbance or interfering with any burial taking place there, as well as entering or remaining in the cemetery after closing hours.[227] Although these offences have a clear public order element, they also recognise the special significance of burial grounds as repositories of the dead.[228]

Conclusion

Disposal of the dead is a basic societal need.[229] Driven by a combination of public health and pragmatism, and underpinned by fundamental beliefs around appropriate treatment of the dead, it can be carried out in a number of different ways. New methods of corpse disposal are being increasingly shaped by environmental concerns, and may become popular in years to

---

224 As in the high profile case of billionaire Hans Rausing who hid his wife's body under a pile of clothing and bin bags in a room at the couple's £70 million Chelsea mansion for some two months after her death from a drugs overdose (Mr Rausing had smoked a crack pipe while he watched his wife die). See M Evans and S Marsden, "Billionaire Heir Hans Rausing Escapes Jail for Preventing His Wife Eva's Burial", *The Telegraph* (London, 1 August 2012) www.telegraph.co.uk/news/uknews/crime/9443573/Billionaire-heir-Hans-Rausing-escapes-jail-for-preventing-his-wife-Evas-burial.html (accessed 30 September 2015).

225 As defined by s 1 of the Criminal Damage Act 1971.

226 Ecclesiastical Courts Jurisdiction Act 1860, s 2 and see *R v Cheere* (violent interruption of a burial service) for an illustration of the latter. These and other offences in churchyards are discussed in 24 *Halsbury's Laws of England* (n 3), para [1354].

227 Prescribed by s 18 of the Local Authorities' Cemeteries Order 1977 and outlined in 24 *Halsbury's Laws of England* (n 3), para [1354].

227 While these are essentially public order offences, the 1977 Order also lists a number of offences that relate to the orderly management of cemeteries. These include burial without the requisite permission, and non-compliance with minimum depth of graves – 1977 Order, Sch 2, Pt 1.

228 The basic principle of respect for the dead is also reflected in the fact that, once human remains have been interred, any unauthorised disinterment is an offence – see Ch 7, Pt VI.

229 "As long as a society exists, the need to dispose of the dead will be in as much demand as … food, shelter and transport" – Monaghan (n 117), p 1032.

come. In the meantime, burial and cremation will remain the most widely practised methods, and are discussed accordingly throughout this book.[230]

The law places inevitable restrictions on the disposal of the dead, the places where this can occur and the underlying conditions. However, this is only part of the legal picture. The following chapter addresses the all-important question of who controls disposal, and who has the right to make decisions about what happens to the deceased's remains.

---

230   Likewise, subsequent references to 'human remains' denotes both corpses and post-cremation ashes – but would include the by-products of resomation and promession when these technologies become available.

3 Disposal of the dead

Legal rights and responsibilities

Disposal of the dead is a necessary, however unpleasant, task imposed upon the living.[1]

Introduction

Historically, English law granted everyone the right to a "Christian burial"[2] with a full Church of England burial service and attendant funeral rites.[3] A modern variant is still recognised in most common law jurisdictions today – though reinterpreted as including both interment and cremation,[4] and denoting disposal of the body in a dignified manner as opposed to imparting any specific religious dimension.[5] The underlying reasons for this basic entitlement were twofold and remain central to the law's treatment of dead

---

1   T Stueve, *Mortuary Law* (Cincinnati Foundation for Mortuary Education, 6th edn, 1984), p 18.

2   *R v Stewart* (1840) 113 ER 1007. See also *Chapple v Cooper* (1844) 13 M & W 252 and *R v Vann* (1851) 2 Den 325.

3   See DJ Davies, *A Brief History of Death* (Blackwell, 2005), pp 153–155. Although the relevant authorities are not entirely clear, this appears to be something different from the common law right of burial in individual churchyards noted in Ch 2, p 31. While the two are probably interlinked, the original entitlement to a Christian burial related to funeral rites and forms of burial service, as opposed to interring the deceased's remains in his/her parish churchyard.

4   See for example, the comments of Wilson J in *Reid v Crimp* [2004] QSC 304, [21] ("[c]remation is nowadays equivalent to burial"). Similar reasoning would, presumably, apply to other permissible disposal methods discussed in Ch 2.

5   The latter point was initially raised in *R v Price* (1884) 12 QBD 247, the court noting that the right to a 'Christian' burial was inapplicable to non-Christians and see also HY Bernard, *The Law of Death and Disposal of the Dead* (Oceana Publications, 1966), p 15 (the right to a Christian burial does not denote religious rites, but "burial comporting with the prevailing sense of decency in the community").

bodies: ingrained notions of human dignity and respect for the dead,[6] and public health concerns around the prompt disposal of a decaying corpse.[7]

To ensure these objectives were met, certain individuals were under a common law duty to dispose of the dead, with a correlative right to possession of the body for this purpose and full decision-making powers over what form the deceased's funeral should take. This possessory entitlement is an exception to the general rule that there is no property in a corpse.[8] It also generates important legal entitlements, which form the basis of this chapter and map onto other issues surrounding disposal of the dead.[9]

6   "At death, the right to a decent burial can be seen as an instance of the right of all persons to be treated with dignity" – New Zealand Law Commission, *The Legal Framework for Burial and Cremation in New Zealand: A First Principles Review* (Issues Paper 34, October 2013), p 77. This is a basic entitlement; for example, international humanitarian laws forbid the desecration of the bodies of those killed in armed conflict and require the dead to be respectfully interred where possible (see Article 16 of the Fourth Geneva Convention for an illustration).

7   *R v Newcomb* (1898) 2 CCC 255, and see also the comments of Scott LJ in *Rees v Hughes* [1946] KB 517, 523–524. Prompt disposal is always desirable, but becomes essential in outbreaks of infectious disease, and other emergency situations (e.g. major earthquakes and other natural disasters) where high numbers of deaths overwhelm local systems and require large-scale, coordinated methods of storing, identifying and ultimately disposing of the dead. For example, ss 43–45 of the Public Health (Control of Disease) Act 1984 deal with notifiable diseases (including isolation of bodies and prohibitions on wakes), while the same legislation also authorises the Secretary of State to make regulations restricting or imposing conditions on the disposal of bodies in the interests of public health or safety (s 47), and sanctions the immediate burial or cremation of bodies where retention would cause a threat to public health (s 48). All councils in England and Wales are also required to draft civil contingencies plans (Civil Contingencies Act 2004, ss 1–2 and Sch 1); these typically include provisions for disposal of the dead in the event of an "emergency" (defined by s 1 as an event or situation that threatens serious damage to human welfare, the environment or security).

8   This is one of two exceptions to the 'no property' rule (see Introduction, pp 2–3) devised over the centuries. The second originated in Griffith CJ's judgment in the Australian case of *Doodeward v Spence* (1908) 6 CLR 406, and confers property status on corpses that have been subject to the "lawful exercise of work or skill" where the end product has "acquired some attributes differentiating it from a mere corpse awaiting burial" – *ibid*, 414 (two-headed baby, preserved in a glass jar and put on public display constituted property). Preparatory acts (such as embalming) are not enough to attract property status, since the end result is still a corpse awaiting disposal – M Pawlowski, "Property in Human Body Parts and Products of the Human Body" (2009) 30 *Liverpool Law Review* 35, p 45 (rejecting what appears to be a contrary statement in the judgment of Peter Gibson LJ in *Dobson v North Tyneside Health Authority* [1996] 4 All ER 474, 479). However, museum exhibits or anatomical specimens in medical collections have an independent use value and would be classed property under this particular exception.

9   In particular, the resolution of funeral disputes – see Ch 4.

I. Duty of disposal and right to possession of the corpse

In England and Wales the duty of disposal and right to possession of the corpse (or its ashes following cremation[10]) is still governed by old common law rules.[11] These derive from long-standing principles around estate administration and liability for funeral expenses;[12] and while the underlying rationale has occasionally been questioned, the rules contemplate a clear order of entitlement.

1. The legal framework

Primary responsibility falls on the deceased's personal representatives as dictated by succession law rankings.[13] Where the deceased made a will, this is the executor[14] who is entitled to possession of the body before the grant of probate,[15] unless there are doubts over the will's validity[16] and a grant of probate is doubtful due to non-compliance with testamentary formalities or lack of capacity. A good illustration is *Privet v Vovk*,[17] in which an elderly stroke patient in a nursing home married a young male carer shortly before her death, and made a new will appointing him as executor and principal beneficiary of her estate. The court ruled that the deceased's son should make the funeral arrangements; given the deceased's mental state, there were serious doubts over the validity of the will and the marriage.[18] Technicalities aside, the executor rule is also subject to practical limitations – for example, the executor knowing that they are entitled to make the funeral arrangements, or not being formally identified until days or weeks after the deceased's death by which time the funeral has already taken place.

---

10 Since broadly the same principles apply – see pp 71–75.

11 This is also the case in many derivative common law legal systems. However, the Canadian provinces of British Columbia, Alberta and Saskatchewan (and numerous American states) have introduced specific statutory frameworks detailing who has the right to control the disposition of the deceased's remains. These are noted in Ch 4, Pt VI.

12 See the discussion in SG Hume, "Dead Bodies" (1956) 2 *Sydney Law Review* 109, pp 110–114.

13 This is a legal paradox: corpses are not property, yet responsibility for disposing of the dead is based on rules that govern the disposal of property after death.

14 See *Williams v Williams* (1882) 20 Ch D 659, *Hunter v Hunter* (1930) 65 OLR 586, *Murdoch v Rhind* [1945] NZLR 425, *Schara Tzedeck v Royal Trust Co* [1951] 2 DLR 228, *Re Clarke (Deceased)* [1965] NZLR 182, *Grandison v Nembhard* [1989] 4 BMLR 140 and *Re Waldman and City of Melville* (1990) 65 DLR (4th) 154. However, the position is different in America where an executor is not automatically entitled to possession of the deceased's remains (see *Wales v Wales* 190 A 109, 110 (1936)).

15 Since an executor derives title from the will itself (*Buchanan v Milton* [1999] 2 FLR 844), his/her authority dates from the deceased's death.

16 *University Hospital Lewisham Trust v Hamuth* [2006] EWHC 1609 (Ch).

17 [2003] NSWSC 1038.

18 Contrast this with *Abeziz v Harris Estate* [1992] OJ No 1271 (lack of capacity and undue influence not established on the facts).

If the deceased died intestate, the law looks to the highest ranked next-of-kin as the person entitled to a grant of administration over the deceased's estate if one were sought[19] – what we might term the 'presumptive' administrator.[20] Under s 46 of the Administration of Estates Act 1925 in England and Wales, this is the deceased's spouse[21] or civil partner,[22] followed by

19   See *Dobson v North Tyneside Area Health Authority* [1996] 4 All ER 474, as well as *Brown v Tullock* (1992) 7 BPR 15,101, *Saleh v Reichert* (1993) 104 DLR (4th) 384 and *Smith v Tamworth City Council* (1997) 41 NSWLR 680.

20   Unlike executors, administrators do not take title from the date of death, but from the grant of letters of administration (at least 7 days, or 28 days in the case of a surviving spouse or civil partner due to the survivorship stipulation in s 46(2A) of the Administration of Estates Act 1925). Despite suggestions to the contrary in *Dobson v North Tyneside Area Health Authority* [1996] 4 All ER 474, it is irrelevant that no grant has actually been made or applied for; the person due to be granted letters of administration or who would be entitled on applying has the right to possession of the deceased's remains (see *Holtham v Arnold* [1986] 2 BMLR 123, *Brown v Tullock* (1992) 7 BPR 15,101 and *Meier v Bell* (Supreme Court of Victoria, 3 March 1997). Where applying for a grant is unlikely because the deceased's estate is very small (or non-existent), the position is still the same according to Cummins J in *Dow v Hoskins* [2003] VSC 206 (rejecting comments made by Perry J in *Jones v Dodd* (1999) SASC 125, [50] that, in these circumstances, a notional entitlement to apply for a grant "takes on an air of unreality" – though support for Perry J's approach can be found in *State of South Australia v Smith* [2014] SASC 64, [53]). The general reasoning around presumptive administrators appears to have been overlooked in R Hardcastle, *Law and the Human Body: Property Rights, Ownership and Control* (Hart, 2007), p 48.

21   Being estranged or having informally separated does not alter this – see *Holtham v Arnold* [1986] 2 BMLR 123. Of course, any spousal entitlement depends on a valid marriage having taken place. For example, in *R (on the application of Haqq) v HM Coroner for Inner West London* [2003] EWHC 3366 (Admin) the deceased was domiciled in England and already married when he entered into a second marriage in Bangladesh; since the second marriage was void under English law, the first wife was entitled to her husband's remains. However, the position may be different for a voidable marriage, which cannot be challenged after one of the parties is dead – see *Saleh v Reichert* (1993) 104 DLR (4th) 384.

22   Same-sex civil partners in England and Wales (and the rest of the UK) have exactly the same legal rights as spouses under the Civil Partnership Act 2004. Where same-sex partners cannot formalise their relationship (or have failed to do so), this has far-reaching consequences when one dies and the survivor may be denied any say in their partner's funeral – see JE Horan, "'When Sleep at Last Has Come': Controlling the Disposition of Dead Bodies for Same-Sex Couples" (1999) 2 *Journal of Gender, Race and Justice* 423 and E Wojcik, "Discrimination After Death" (2000) 53 *Oklahoma Law Review* 389.

children,[23] parents and siblings, then other specified relations in descending order of consanguinity. The highest ranking relative has the final say on disposal of the deceased's remains.[24] However, an intestacy-linked classification has two main drawbacks, the first of which is its potential to exclude certain individuals where legal constructs of family based on 'blood' or normative views of kinship do not match social or cultural realities, or reflect "the closeness of the deceased's relationships in life".[25] For example, step-families and cohabiting partners do not qualify as next-of-kin under English law, and would not be entitled to possession of the deceased's remains if someone within the 'traditional' nuclear family model insisted on different funeral arrangements;[26] in multi-ethnic societies, cultural notions of kinship and decision-making powers within families may also be at variance with this legal classification.[27] Second, the intestacy framework is ineffective where two or more persons fall within the same kinship tier and have equal rights but different views on the deceased's funeral; where parents, siblings or children of the deceased cannot agree, courts must find another way of ranking the competing claims.[28]

23   A 'child' of the deceased is not restricted to natural or biological children, but includes adopted children (Adoption Act 1976, s 39), unborn children who were conceived before the death of the deceased (Administration of Estates Act 1925, s 55(2)) and children born by assisted reproduction techniques where parentage is assigned according to the Human Fertilisation and Embryology Act 2008, Part II (including posthumous reproduction). Although infants and minors would be included under this statutory classification, the common law position appears to be that they are not under any specific duty to bury their parent – *Chapple v Cooper* (1884) 13 M & W 252 (again tied to the rules around funeral expenses and a child's inability to pay these, if their parent had limited means). However, infants and minors can be represented by their surviving parent or legal guardian for estate administration purposes, and granted the notional right to decide the fate of a parent's remains – see *Meier v Bell* (Supreme Court of Victoria, 3 March 1997). Much depends on the wording of the relevant intestacy framework and accompanying statutory rules.

24   Where the presumptive administrator is unable or unwilling to arrange the funeral, responsibility passes to someone further down the pecking order (or another person within the same kinship tier) – see *Mourish v Wynne* [2009] WASC 85 (mother of a 16-year-old child was in prison and unable to make the funeral arrangements; court had to decide whether possession of the child's remains should be given to the maternal or paternal grandmother as next-of-kin after the mother). The common law forfeiture rule (as modified by the Forfeiture Act 1982 in England and Wales) also comes into play where the presumptive administrator is criminally responsible for the deceased's death – see *Scotching v Birch* [2008] EWHC 844 (Ch).

25   New Zealand Law Commission (n 6), p 201.

26   While step-families are not generally regarded as next-of-kin for succession law purposes, cohabitants are in some jurisdictions (see for example, the respective positions in Australia and Canada noted in Ch 4, n 52) though not under English law as it currently stands.

27   See for example, P Vines, "Consequences of Intestacy for Indigenous People: The Passing of Property and Burial Rights" (2004) 8 *Australian Indigenous Legal Reporter* 1, discussing the issues facing Aboriginal peoples in Australia.

28   See Ch 4, Pt III. Similar issues arise where two or more joint executors under a will disagree over funeral arrangements.

At common law, a husband was under a duty to bury his wife's remains,[29] though this is probably a residual category today given the spousal doctrine of separate property, the economic independence of women and the superseding role of the executor or administrator when there is a (potential) estate to administer.[30] The result is that personal representatives can determine the funeral arrangements in the majority of adult deaths. As regards infants or minors, the duty of disposal falls jointly on the parents[31] though courts will differentiate between different *sets* of parents. The rights of adoptive parents prevail over those of natural or biological parents[32] – hardly surprising given the permanency of the process and its legal consequences.[33] And while we might assume that natural parents outrank foster parents (given that fostering tends to be a temporary care arrangement), case law suggests a fact-specific approach. In *R v Gwynedd County Council, ex parte B*,[34] the court ruled in favour of the natural mother who wanted to bury her daughter beside the child's father in the family burial plot, despite objections from the girl's foster

29   *Jenkins v Tucker* (1788) 1 H Bl 90, *Ambrose v Kerrison* (1851) 10 CB 777 and *Bradshaw v Beard* (1862) 12 CBNS 344.

30   Though the practical outcome is still the same in intestate deaths because, as the highest ranking next-of-kin, the husband would qualify as administrator of his wife's estate. The husband's common law duty to bury his wife was important prior to the enactment of the Married Women's Property Act 1882 when a wife's property automatically passed to her husband on marriage; since the wife left no estate, this duty was placed on the husband as the person liable for his wife's funeral expenses. Hume (n 12), p 112 has questioned the existence of a reciprocal obligation on wives, a proposition that finds support in more recent cases – see the comments of William Young J in *Takamore v Clarke* [2012] NZSC 116, [199] (widow's common law duty to bury her husband where he leaves no estate is "distinctly arguable").

31   *R v Vann* (1851) 2 Den 325 (affirmed in *Clark v London General Omnibus Co Ltd* (1906) 2 KB 648), and *Grovey v Moore* [1935] NZLR 739. Alternatively, parents can effectively be classed as the child's next-of-kin under intestacy rankings; although there is no estate to administer here, the parents would be jointly entitled to a grant of administration if one were necessary – see *Scotching v Birch* [2008] EWHC 844 (Ch). If one parent is dead, the right devolves on the sole surviving parent (*R v Gwynedd County Council, ex parte B* [1992] 3 All ER 317); if both are dead, the persons with legal responsibility for the child have the duty to bury (*Watene v Vercoe* [1996] NZFLR 193).

32   See *Buchanan v Milton* [1999] 2 FLR 844 (discussed at Ch 4, p 103), as well as *Waskewitch v Hastings* (1999) 184 Sask R 79.

33   Adoption Act 1976, s 39 (legal status conferred by adoption). Where the adoptive parents are dead, other members of the deceased's adoptive family can claim possession of the deceased's remains in order of entitlement. In *Re Schubert* (Supreme Court of Queensland, 5 November 2010) Byrne SJA ruled in favour of the deceased's adoptive brother. A significant feature of this case is that the deceased was an Aboriginal man who had been 'informally' adopted out by his biological parents before he was two years old; the adoption had never been legally formalised, although the deceased (who was 29 when he died) regarded himself as a member of his adoptive family and had had very little contact with his biological family. While other factors were at play, the adoptive brother would have been the highest ranking next-of-kin had the adoption formalities been completed (decision upheld on appeal as *Frith v Schubert* [2010] QSC 444).

34   [1992] 3 All ER 317.

parents who had cared for her since she was a few weeks old.[35] The role of the local authority (which had placed the child with foster parents) ceased on death, at which point the duty to bury reverted to the natural parents.[36] A slightly different factual scenario arose in *Re LL (Application for Judicial Review)*,[37] which involved a dispute over a terminally ill 11-year-old boy.[38] The child's mother wanted to bury him in the same grave as her grandfather; the foster parents wanted to bury him in a cemetery close to where the boy had lived with them and their three daughters for seven years. Unlike *Gwynedd*, the court ruled in favour of the foster parents here; the child had been freed for adoption without parental agreement two years earlier, with all parental rights and duties being extinguished from the date of that order. As a result, the health trust with responsibility for the child could allow the foster parents to arrange his funeral.[39]

Moving beyond minor children, where all other options have been exhausted and there is no-one with a higher ranking claim, responsibility for disposing of the deceased's remains falls on the householder in whose premises death occurred.[40] Though probably not all that relevant today, given the ease with which family members can be identified and contacted,[41] it has been suggested that a "logical extension"[42] of the householder rule is a similar duty of disposal on a hospital or equivalent institution if the individual died there and no family members claim the body.[43] Finally, the obligation to bury or cremate unclaimed remains falls on the local authority

---

35   The deceased was disabled and was voluntarily placed into care because her parents were unable to cope; her father died when she was four.

36   Though the position might be different if the natural parents could not be found or were unwilling to act. See also *Warner v Levitt* (Supreme Court of New South Wales, 23 August 1994) (natural parents do not lose the right to bury their child because of separation or maltreatment, or both).

37   [2005] NIQB 83.

38   The possibility of making a pre-emptive ruling here is discussed in Ch 4, Pt VIII.

39   See also *L.A.W. v Children's Aid Society of the District of Rainy River* [2005] OJ No 1446 where the deceased (a 16-year-old boy) was a Crown ward and under the care of the defendant; the boy's estranged mother sought possession of his remains for burial in accordance with First Nation customs. However, the court ruled for the defendant, based on the "very unique circumstances" before it (*ibid*, [42]). Maternal estrangement aside, the deceased had wanted to be buried in a particular cemetery and not taken away for burial on the reserve, and the defendant was intent on carrying out these wishes.

40   *R v Stewart* (1840) 12 Ad & El 773. Expenses are recoverable from the deceased's estate – see Part II.

41   Not to mention the possibility of 'refrigerating' or embalming human remains, and moving them elsewhere if necessary.

42   William Young J in *Takamore v Clarke* [2012] NZSC 116, [187], citing *University Hospital Lewisham Trust v Hamuth* [2006] EWHC 1609 (Ch).

43   As affirmed by Gage J in *AB v Leeds Teaching Hospitals NHS* [2005] 2 WLR 358, 391. (in hospital deaths, "the hospital has the legal right to possess the body at least initially"). Unlike local authorities (see below), the hospital's duty is a non-statutory one.

with control over the city, town or district in which the body is found[44] – a last resort option, which ensures that the dead are disposed of where there are no relatives to make the arrangements.[45]

The duty of disposal generates a legal right to possession of the corpse for this purpose, which courts will protect and uphold if necessary.[46] And in family disputes concerning the fate of the dead, the legal framework outlined above provides an essential reference point; when warring factions cannot agree, the highest ranked individual usually has the strongest entitlement.[47] In most cases the executor or presumptive administrator will have the final say, though the origins of this particular rule were heavily criticised by two members of the Supreme Court of New Zealand in its recent decision in *Takamore v Clarke*.[48] Both Elias CJ and William Young J suggested that supporting authority[49] for the executor rule was limited and more concerned with payment of funeral debts than the fate of the dead; as such, there was "no clear basis"[50] for an executor being entitled to possession of the deceased's remains above everyone else (including the deceased's family, where the executor was a non-relative). The presumptive administrator's role was also open to question, given its origins in estate distribution[51] and the fact that the executor rule was simply extrapolated onto intestate deaths.[52]

44   Public Health (Control of Disease) Act 1984, s 46(1). However, cremation is prohibited where the authority has reason to believe it would contravene the deceased's wishes – s 46(3).

45   The local authority *must* take charge of the funeral here – *Secretary of State for Scotland v Fife County Council* 1953 SLT 214 (local authority tried to force hospital authorities to arrange the funeral of a patient in a mental institution).

46   For example, by allowing the person with the duty of disposal to recover possession of a corpse that has been removed from a funeral home, hospital or family residence by someone else. In *Awa v Independent News Auckland Ltd* [1995] 3 NZLR 701, the deceased's body was forcibly taken from his home by relatives who felt that the funeral arrangements made by the deceased's widow and immediate family were not honouring Maori custom. Although dealing with other substantive issues, the court confirmed that the deceased's wife (as executor) was entitled to the body for burial. Turning to the criminal law, it is a common law misdemeanour to prevent burial (or other lawful disposal) of a corpse by refusing to deliver it to this particular individual – *R v Fox* [1841] 2 QB 246 and *Williams v Williams* (1882) 20 Ch D 659. Likewise, the person with the duty of disposal may be guilty of an offence if he/she fails to discharge this duty, despite having the means to do so – *R v Vann* (1851) 2 Den 325.

47   See Ch 4.

48   [2012] NZSC 116. Closer to home, the executor rule may not apply automatically in Scotland – see *C v M* [2014] SCRFOR 22 (the court reviewing a number of authorities and suggesting that an executor does not have priority over close family members).

49   Such as *R v Fox* [1841] 2 QB 246 and *Williams v Williams* (1882) 20 Ch D 659.

50   [2012] NZSC 116, [53] (Elias CJ). The Chief Justice even went so far as to suggest that the executor rule may not apply in New Zealand – *ibid*, [90].

51   *Ibid*, [206] (William Young J).

52   *Ibid*, [55] (Elias CJ). However, the other members of the Supreme Court (Tipping, McGrath and Blanchard JJ) did not share these concerns, accepting that the duty of disposal falls on the deceased's personal representative – *ibid*, [152].

Other decision-making powers around the fate of the dead are firmly entrenched within a family or kinship paradigm.[53] However, the common law rules on bodily disposal are well-established, and unlikely to change without a comprehensive re-evaluation of the law in this area.[54] In the meantime, what is becoming increasingly clear is that courts are unwilling to regard the executor or presumptive administrator's right to possession of the deceased's remains as an absolute entitlement that overrides all other competing claims to take charge of the funeral arrangements. The entitlement now appears to be a *prima facie* one, which can occasionally be displaced.[55]

2.   Content and scope of the right to possession

The right to possession is purposive and transient; it is a custodial right that exists solely for disposal,[56] arises at the moment of death and lasts for a short period afterwards.[57] Possession may be actual if the deceased's body is in the home of the person with the duty of disposal, or constructive where the body is elsewhere (for example, in another relative's home, a hospital mortuary or funeral home).[58] More importantly, the right to possession also confers extensive decision-making powers about the method of disposal and what form the deceased's funeral should take. However, Young J in *Smith v Tamworth City County Council*[59] suggested that a person with the duty of

---

53   See for example, the relevant provisions of the Human Tissue Act 2004 dealing with organ donation – see Ch 6, Pt I.

54   Change has been mooted. In response to a previous Home Office consultation paper, the then government identified potential benefits in creating a statutory duty to dispose of the dead, which would clarify the responsibilities placed on the deceased's executor and family – see Ministry of Justice, *Burial Law and Policy in the 21st Century: The Need for a Sensitive and Sustainable Approach* (2004): *Government Response to the Consultation Carried Out by the Home Office DCA* (2007), p 10.

55   See Ch 4, pp 101–104.

56   It does not include a right to exhume the dead at a later stage, which raises separate legal issues – see Ch 7.

57   See D Mortimer, "Proprietary Rights in Body Parts" (1993) 19 *Monash University Law Review* 216, p 238. Possessory rights terminate when lawful disposal has taken place, as discussed below.

58   See Stueve (n 1), p 38. However, where a coronial inquiry is ordered, the coroner's right to possession of the body is paramount and all other entitlements are temporarily suspended – see Ch 1, p 15. Possessory rights can also be overridden where the deceased died from an infectious disease or in a public emergency, and public health regulations take effect (see n 7) or where the deceased's body is donated for transplant or research purposes under the Human Tissue Act 2004 (see Ch 6, Pt I).

59   (1997) NSWLR 680, 694.

disposal and concomitant right to possession is "expected to consult with other stakeholders, but is not legally bound to do so".[60]

What about excluding others from the funeral ceremony? Several US cases suggest that this is possible, though much depends on the type of funeral and where it is being held. For example, in *Rader v Davis*[61] spouses had divorced due to the husband's aggressive behaviour and the wife (who had custody of the couple's young son) had returned home to live with her father. When the boy died, the husband was prevented from attending the funeral by former father-in-law who owned the house where the private funeral was being held; there was nothing that the husband could do, even if the underlying motive was spite.[62] Yet while a private funeral suggests personal invitations and potential exclusions, there is a world of difference between a private ceremony in a family home[63] and one held in a public place such as a church or cemetery, which – by its very nature – is open to everyone who wants to attend. Excluding 'unwanted' attendees is not an option; the funeral is not taking place in a private space where the tort of trespass to land would trump any perceived entitlement to attend. Compared to *Rader*, the language in the recent Australian case of *Manktelow v The Public Trustee*[64] is much more conciliatory and inclusive, the court suggesting that the person with the right to possession of the deceased's remains "cannot use [it] ... in such a way as to exclude friends and relatives from expressing their affection for the deceased in a reasonable and appropriate manner".[65] What this means is unclear, though one might assume that it envisages participation in the funeral, or at least the ability to be there. However, attending a funeral has

---

60   See also *Keller v Keller* [2007] VSC 118 and *Milenkovic v McConnell* [2013] WASC 421. In the latter case, the deceased's de facto partner had consulted with the deceased's mother about where to place the ashes, and an appropriate memorial. This sufficed, the court noting that the principle set out in *Smith* generates nothing more "than an expectation of consultation" – [2013] WASC 421, [40].

61   134 NW 849 (Iowa 1912).

62   "[The father-in-law] was not required to invite anyone onto his premises simply to see the dead body or to have any sort of funeral services for the public...There is no implied invitation to anyone to attend a funeral conducted from a private dwelling unless it be announced that such funeral is public, and even if so announced the invitation may be revoked and anyone denied the right to attend whose presence may be objectionable" – *ibid*, 851 (Iowa 1912). See also *Seaton v Commonwealth* 149 SW 871 (Ky Ct App 1912) noted below, and *Tully v Pate* 372 F Supp 1064 (DSC 1973).

63   Or a private viewing of the deceased's remains in a funeral home, where certain individuals can be excluded by the person with duty of disposal and/or the person who had contracted with the funeral provider. See for example, *Sopinka v Sopinka* (2001) 55 OR (3d) 529 (deceased's former wife accused of violent conduct towards him; prevented by the deceased's father (also acting as his executor) from viewing the body at the funeral home).

64   [2011] WASC 290.

65   *Ibid*, [23].

been described as a "social rather than a legal right",[66] and while an executor or presumptive administrator cannot be liable for failing to notify other family members about the deceased's funeral,[67] the position may be different where that person is deliberately withholding information.[68] The court may prevent the funeral from going ahead in these circumstances.[69]

Modern funeral practices may raise further legal issues around funeral participation. In today's technology-obsessed world, being physically present at a funeral is not the only option, with an increasing number of funeral homes and crematoria offering live webcasts of the ceremony, so that absent relatives and friends can pay their last respects. While an 'internet funeral' facilitates virtual attendance by family and friends of the deceased living in different countries,[70] similar questions could arise as to whether the person entitled to possession of the deceased's remains could effectively exclude certain individuals from the webcast – given that the ceremony is broadcast via a secure web link which is password protected, and would have to be disclosed to all virtual attendees.

Possessory rights to the remains terminate when lawful disposal of the corpse has taken place.[71] Where the deceased is buried,[72] it seems logical that the executor or presumptive administrator's entitlement ceases on interment. However, this is subject to two caveats, the first of which is that the deceased must have been "properly buried"[73] so that the right to possession persists as a cause of action where others have taken the body unlawfully and buried it

---

66   Stueve (n 1), p 41.
67   As in *Seaton v Commonwealth* 149 SW 871 (Ky Ct App 1912) where the father of a dead infant failed to notify the extended family about the child's death or subsequent funeral arrangements. Even though their feelings were hurt, the relatives had no legal right to attend the funeral.
68   The executor or presumptive administrator must provide details of the funeral arrangements if the deceased's next-of-kin make reasonable requests for such information – *Sopinka v Sopinka* (2001) 55 OR (3d) 529.
69   See *Re Lochowiak (Deceased)* [1997] SASC 6301 (deceased's son granted an interlocutory injunction against the deceased's de facto partner where the latter was refusing to disclose details of the funeral and there was a strong chance of it proceeding without the deceased's family and friends being there).
70   Or those living nearby who are unable to attend – for example, elderly relatives and friends, or those who are too frail or ill.
71   "The personal representatives have the right to custody and possession until the body is buried [or otherwise disposed of], whereupon it plainly ceases" – *Re Campbell (Judicial Review)* [2013] NIQB 32, [17].
72   The position as regards cremation is discussed below.
73   *Williams v Williams* (1882) 20 Ch D 659, 665.

elsewhere.[74] The second is that the right to possession may extend beyond burial, thus preventing an aggrieved relative from disinterring the remains immediately afterwards and reburying them elsewhere – something that has been suggested in a number of Canadian cases.[75] Under English law, however, disinterment in either of these scenarios (both of which are unlikely) would be dependent on legal authorisation.[76]

Several cases have suggested that an interred corpse becomes part of the land, so that any unlawful interference is actionable in trespass, but only at the instance of the landowner (or holder of the right of burial).[77] How and when this assimilation occurs is open to question. If, as these cases seem to suggest, a dead body becomes part of the realty as soon as it is buried, the landowner acquires immediate rights over the body according to the general principle that chattels affixed or annexed to the land become the property of the owner of the soil.[78] The difficulty, however, with this approach is explaining how the annexation principle can apply to a corpse when, in legal terms, it does not constitute property before burial.[79] If the body is placed in a coffin or casket, the annexation argument becomes stronger because the corpse is inside a container which is an item of property and capable of annexation – thereby circumventing the non-property status of its contents. Alternatively, it could be argued that a dead body does not merge with the land until it ceases to exist in any recognisable form – in other words, when it is fully decomposed and becomes "indistinguishable from the soil".[80] This avoids the annexation problem but raises another key issue. Where the corpse is

74   See the comments of Fogarty J at first instance in *Clarke v Takamore* [2009] NZHC 901, [47]–[53], citing this as one reason why the deceased's wife (as his executor) was entitled to her husband's remains for burial in the cemetery of her choosing, the body having been taken unlawfully and buried elsewhere by other members of the deceased's family. Both the Court of Appeal and the New Zealand Supreme Court reached the same overall conclusion, though with different reasoning (see [2011] NZCA 587 and [2012] NZSC 116).

75   *Waldman v Melville (City)* (1990) 65 DLR (4th) 64 and cited with approval in *Sopinka v Sopinka* (2001) 55 OR (3d) 529. See also *Heafy v McRae* (1999) 5 ETR (3d) 121. However, this proposition has been criticised – see the comments of Elias CJ in *Takamore v Clarke* [2012] NZSC 116, [87].

76   See Ch 7.

77   See the respective comments of Foster J in *Meagher v O'Driscoll* 99 Mass 281, 284 (1868) ("after burial [a dead body] ... becomes a part of the ground to which it has been committed, 'earth to earth, ashes to ashes, dust to dust'") and Griffith CJ in *Doodeward v Spence* (1908) 6 CLR 406, 412 ("after burial a corpse forms part of the land in which it is buried, and the right of possession goes with the land"), as well as the decision in *O'Connor v City of Victoria* (1913) 4 WWR 4. The personal representative's right to possession obviously no longer exists at this point – see the comments of Waddell CJ in *Robinson v Pinegrove Memorial Park* (1986) 7 BPR 15,097.

78   As laid down by Chitty J in *Elwes v Brigg Gas Company* (1886) 33 Ch D 562, 567.

79   An issue raised in both P Matthews, "Whose Body? People as Property" (1983) 36 *Current Legal Problems* 193, p 203 and Pawlowski (n 8), p 39.

80   Pawlowski (n 8), p 39. See also *R v Jacobson* (1880) 14 Cox CC 522.

interred in a coffin or casket (as is usually the case), the body would not form part of the realty until the container also disintegrated into the earth, effectively creating a legal vacuum (in private law terms) between the end of the personal representative's right to possession and the vesting of legal rights in the landowner following a suitable period of decomposition. An action for trespass would not be available if someone disturbed the land to remove the corpse in the interim, though criminal law sanctions for unauthorised exhumation would still apply.[81]

3. Pre and post-cremation possessory rights

As with burial, the executor or presumptive administrator is entitled to possession of the deceased's body for cremation purposes and has ancillary decision-making powers over any funeral ceremony.[82] When applying for cremation in England and Wales, this common law proposition is statutorily buttressed by regulation 15 of the Cremation (England and Wales) Regulations 2008,[83] which requires the application to be made by the deceased's executor or a "near relative" aged 16 or over[84] unless a satisfactory explanation is given for it being made by some other person.[85] A "near relative" is defined as the surviving spouse or civil partner of the deceased, a parent or child of the deceased, "or any other relative usually residing with the deceased person".[86] Although broadly similar to the presumptive administrator rankings, this final caveat suggests a wider categorisation of familial relationships and could include unmarried cohabitants who do not currently qualify under English intestacy laws.[87] Where the cremation application is made by another person, the standard forms inquire whether the executor or near relative has been informed and expressed any objections.[88] This (aligned with the common law right to possession of the deceased's remains) suggests that the executor or presumptive administrator could prevent cremation from taking place where the application was made by someone else.

---

81   See Ch 7, Pt VI. Matthews notes that a conviction is unlikely the longer a corpse decomposes in the ground, with trespass to land providing a more likely remedy – Matthews (n 79), p 205.

82   See *Re Korda* (*The Times*, 23 April 1958) and *Fessi v Whitmore* [1999] 1 FLR 767 as well as *Robinson v Pinegrove Memorial Park* (1986) 7 BPR 15,097.

83   SI 2008/2841.

84   Regulation 15(1).

85   Regulation 15(2).

86   Regulation 15(3), which also refers to the parents of a stillborn child.

87   The 2008 Regulations do not define "relative". However, the term clearly suggests some sort of familial tie as opposed to, for example, someone living with the deceased as a close friend or carer.

88   See www.justice.gov.uk/downloads/burials-and-coroners/cremations/cr1.pdf (accessed 30 September 2015).

Unlike burial, where disposal is complete when the deceased's body is interred and the right to possession terminates, the position is less straightforward with cremation. Ashes have a lasting and tangible presence; what happens to them raises issues around "secondary decision making",[89] and may be a further source of contention within families. Initial rights to possession of the ashes are determined by prevailing cremation laws – for example, in England and Wales, the crematorium must return the ashes to the applicant or person nominated by them to collect the ashes.[90] However, it appears that this entitlement is subordinate to the common law right of the deceased's executor or presumptive administrator, who can insist on having the ashes returned to them and despite not having applied for cremation. This is exactly what happened in *Robinson v Pinegrove Memorial Park*[91] where a son had arranged for his father's remains to be cremated (the widow and the other children all supported this), but subsequently contracted with the crematorium to place half the ashes in a rose garden at the cemetery; the other half was to be given to the widow for scattering in a park in Birmingham close to where the family had lived before moving to Australia. However, the widow wanted to scatter all of the ashes in England, according to her dead husband's wishes. The deceased's executor intervened on the widow's behalf, claiming that the ashes should be released to him (at which point the executor would pass them to the widow). Waddell CJ held that the executor's right to possession extended to the ultimate disposal of the remains, and where the deceased had been cremated this included the fate of the ashes – especially where the aim was to carry out the deceased's wishes.[92] The son's contractual arrangement with the crematorium was subject to the executor's right to decide how the deceased's ashes should be disposed of.[93]

The fact that the executor is entitled to possession after the ashes have been released by the crematorium is hardly surprising, and was confirmed in *Leeburn v Derndorfer*[94] in which Byrne J suggested that the executor holds the ashes on trust "for the purpose of disposing or dealing with them in [an appropriate] way".[95] Where there is no executor, the decision in *Doherty v*

---

89   New Zealand Law Commission (n 6), p 228.

90   2008 Regulations, reg 30.

91   (1986) 7 BPR 15,097 and see the comment in S White, "Rights to (Buried?) Cremated Remains" (1998) 4 *Pharos International* 37.

92   There were no legal restrictions on exhuming the ashes. This is not a universal stance; for example, in England and Wales, disinterring ashes requires legal permission – see Ch 7.

93   See also *Doherty v Doherty* [2006] QSC 257, [18] (such contractual arrangements are "subject to the right of the executor to decide how, ultimately, a deceased person's remains shall be disposed of").

94   [2004] VSC 172.

95   *Ibid*, [28]. The trust analysis is not strictly necessary, since the executor's entitlement is an extension of the duty of disposal and corresponding right to possession the deceased's remains (in whatever form).

Doherty[96] establishes that the presumptive administrator is at liberty to deal with the ashes after cremation – once again, holding them on trust to dispose of appropriately.[97] Both cases define 'appropriate' as "hav[ing] regard to the claims of relatives or others with an interest",[98] which denotes some sort of collective input into the fate of the deceased's ashes. The decision in *Doherty* is particularly strong on this point, suggesting that where the ashes are to be disposed of, the presumptive administrator (or executor) must stipulate when this is going to take place – especially where other relatives of the deceased are keen for the ashes to have a permanent resting place, thus bringing about some sense of closure.[99]

In most situations, final disposal takes place when ashes are either interred or scattered.[100] For obvious reasons, possessory rights terminate immediately where ashes have been dispersed, but what rights (if any) persist where ashes have been placed in the ground and remain physically intact inside an urn or container? Again, by analogy to corpse burial, we might assume that the common law right to possession terminates at this point. However, the same two caveats probably apply here:[101] the ashes must be 'properly' interred, and the executor or presumptive administrator's right to possession continues for a short period of time afterwards to prevent any unauthorised removal.[102] Finally, if buried ashes are to be treated in the same way as buried corpses, the ashes might become part of the land from the moment of interment,

---

96   [2006] QSC 257 (deceased's wife preferred to other family members, as highest ranking next-of-kin).

97   See also *Milenkovic v McConnell* [2013] WASC 421 (deceased's de facto partner entitled to his ashes as presumptive administrator).

98   *Doherty v Doherty* [2006] QSC 257, [30], echoing almost identical comments in *Leeburn v Derndorfer* [2004] VSC 172, [28].

99   Again, this ties in with the duty to consult other stakeholders and take their interests into account – see pp 67–68. In *Doherty*, the deceased's wife had not fully considered the views of the deceased's mother and siblings, by failing to set a time when the ashes would be returned to New Zealand for burial; the court suggested that a final decision be made within 12 months. See also *Re Popp Estate* 2001 BCSC 183 (court critical of a husband's failure to disclose what he intended to do with his wife's ashes some five and a half years after her death, despite repeated requests from the wife's family).

100   Ashes can also be divided between the deceased's relatives, to keep or dispose of them as they please. There is no legal impediment to this, and public health issues do not come into play; as Byrne J pointed out in *Leeburn v Derndorfer* [2004] VSC 172, [28] "it is within the powers of the executors [or presumptive administrators] in possession of the ashes to deal with them in this way".

101   See pp 69–70.

102   Like corpses, exhuming ashes requires appropriate authorisation under English law – see Ch 7.

giving the landowner (or holder of the right of burial) an action in trespass if any attempt is made to disturb them.[103]

Unlike corpses however, ashes can also be retained *in specie* and kept, unburied, in an urn or container.[104] This generates further legal issues, with the court in *Doherty v Doherty*[105] suggesting that the personal representative has a "continuing" obligation to deal with the ashes appropriately for as long as they are retained and not disposed of "in some final way". Of course, it is not uncommon for an executor or presumptive administrator to relinquish the ashes to other members of the deceased's family after cremation, at which stage the former's right to possession presumably comes to an end – especially where the ashes are to be retained permanently by someone else. More importantly perhaps, it seems that retained ashes lend themselves to some sort of property analysis – despite their "notional 'identity' as 'the deceased'".[106] As Byrne J explained in *Leeburn v Derndorfer*[107]:

> [A]shes may be dealt with in a way that would not be possible with respect to a dead body. ... Moreover, so long as they are not dispersed or otherwise lose their physical character as ashes, they may be owned and possessed. ... Ashes which have in this way been preserved in specie are the subject of ordinary rights of property, subject to one possible qualification. In this way, ownership in the ashes may pass by sale or gift or otherwise. The only qualification ... arises from the fact that ashes are, after all, the remains of a human being and for that reason they should be treated with appropriate respect and reverence.[108]

Again, the obvious question is how a corpse as an item of non-property becomes the opposite when converted into ashes. Byrne J suggested that ashes should have the same legal status as the preserved body in *Doodeward v Spence*,[109] taking the view that the "application of fire to the cremated [*sic*]

---

103   Again, by analogy to corpse burial, interred ashes would probably become part of the land immediately with the urn or container circumventing the rigours of the 'no property' rule (assuming ashes, *per se*, do not constitute property – though note the argument presented below about *retained* ashes constituting property). If ashes must be indistinguishable from the soil to become part of the realty, then decomposition of the urn or container would be essential and the ashes would not become the property of the realty until they were released into the earth – applying the arguments at pp 70–71.

104   Retention can be temporary or permanent, and may be all or only part of the ashes.

105   [2006] QSC 257, [26].

106   R Croucher, "Disposing of the Dead: Objectivity, Subjectivity and Identity" in I Freckleton and K Peterson (eds), *Disputes and Dilemmas in Health Law* (Federation Press, 2006), p 336.

107   [2004] VSC 172, [27].

108   The same general view (i.e. that retained ashes constitute property) is put forward in Matthews (n 79), pp 206–207 and *Derwin v Ling* [2008] Fam CA 644.

109   (1908) 6 CLR 406. In other words, the second exception to the 'no property' rule is triggered because work and skill has been applied – see n 8.

body is to be seen as the application of work or skill which has transformed it from flesh and blood to ashes, from corruptible material to material which is less so".[110] While attractive as a means of circumventing the 'no property' rule, this line of reasoning is unconvincing and has not been adopted elsewhere;[111] even if we accept that cremation constitutes a skilled process within the *Doodeward* remit, the end product is a substance that does not have a function beyond stored ashes.[112] The position would, of course, be different where the person in lawful possession of the ashes has them converted into a permanent memorial to the deceased with some sort of independent use value (for example, an item of jewellery or other keepsake[113]). Any such item would have property status.

II.   Liability for funeral expenses

This is an important issue, given the sums of money involved.[114] While the deceased's personal representative has legal responsibility for the funeral arrangements, these must made within a short period of time – and that person's identity may not be clear from the outset. This can result in the deceased's family organising the funeral without questioning who has decision-making authority,[115] or legal responsibility for disbursing costs. Where the deceased contracted for particular services under a pre-paid funeral plan, certain charges are automatically covered.[116] Alternatively, close relatives may agree to pay the costs, with social fund payments providing

---

110   [2004] VSC 172, [27].
111   Though Groves sees some merit in applying the *Doodeward* reasoning, given the portability of ashes and the fact that the "physical transformation caused by cremation lessens their corporeal quality" – M Groves, "The Disposal of Human Ashes" (2005) 12 *Journal of Law and Medicine* 267, p 272.
112   See n 8.
113   See Ch 2, p 47.
114   See Ch 2, pp 27–28.
115   Unless a dispute arises over this – see Ch 4.
116   Pre-paid funeral plans allow individuals to finance their funeral in advance, by paying a lump sum or regular instalments into the designated scheme. Often marketed as means of protecting against rising funeral costs, pre-paid plans can also relieve some of the selling pressures placed on the bereaved by funeral directors – see BD Sher, "Funeral Prearrangement: Mitigating the Undertaker's Bargaining Advantage" (1963) 15 *Stanford Law Review* 414, p 426 ("[p]rearrangement on an individual basis ... [has] potential to minimize the inequalities in the undertaker–customer relationship"). While a few articles have analysed such plans from a consumer perspective (see for example, SW Kopp and E Kemp, "The Death Care Industry: A Review of Regulatory and Consumer Issues" (2007) 41 *Journal of Consumer Affairs* 150), little attention has been paid to their legal status – something that the current author hopes to rectify soon!

financial assistance if the eligibility criteria are satisfied.[117] However, this is usually an interim measure, and individuals who incur funeral expenses[118] often expect to be reimbursed for the full amount – something that can be a source of contention, long after the deceased's remains have been disposed of. A number of rules have evolved.[119]

1.   Expenses recoverable from the deceased's estate

A personal representative or other individual who orders the funeral (and any associated services and goods[120]) is liable in contract to the undertaker or relevant service provider.[121] However, contractual liability is tempered by the fact that reasonable funeral costs can be recovered from the deceased's estate.[122] The personal representative has an indemnity for expenses as the

---

117   These are set out in the Social Fund (Maternity and Funeral) General Regulations 2005 (SI 2005/3061), Pt III. Where the deceased was ordinarily resident in the UK at the time of death, and the person who accepts responsibility for the funeral payments is in receipt of an income-related benefit or specific tax credit, the latter can claim financial assistance. The definition of "responsible person" is widely drafted under reg 7(8), and includes (again, in descending order) the deceased's partner, an immediate family member, or a close relative or friend of the deceased. In this respect, there is an obvious disconnect between who can apply for funding, and who has the legal duty of disposal. Regulation 9 outlines the costs available, while specified deductions are outlined in reg 10. See generally, K Woodthorpe, *Affording a Funeral: Social Fund Funeral Payments* (Axa Sun Life Direct Report, 2012) and C Valentine and K Woodthorpe, "From the Cradle to the Grave: Funeral Welfare from an International Perspective" (2013) 48 *Social Policy & Administration* 515.

118   Whether the entire amount, or those not covered under a pre-paid plan or through social fund payments.

119   For an overview, see R Kerridge and AHR Brierly, *Parry and Kerridge: The Law of Succession* (Sweet & Maxwell, 12th edn, 2009), pp 509–510 and DA Smale, *Davies' Law of Burial, Cremation and Exhumation* (Shaw & Sons, 7th edn, 2002), ch 4.

120   For example, purchasing a right of burial in a churchyard or cemetery, providing funeral flowers and funeral food, and erecting a suitable memorial.

121   Kerridge and Brierly (n 119), p 509 citing *Corner v Shaw* (1838) 3 M & W 350, 356 and *Brice v Wilson* (1834) 8 Ad & E 349. The authors also note situations in which the personal representative can be liable in quasi-contract, where someone else orders the funeral without accepting liability for it – *ibid*, p 510.

122   This rule is centuries old – for example, 2 *Bl Comm* p 508 affirms that the deceased's estate has responsibility for funeral expenses, as a principal liability, and see the 'older' cases noted immediately below. More recent authorities have framed it in restitution law terms – see *Smith v Tamworth City Council* (1997) 41 NSWLR 680, 694 (noting a "restitutionary action to recover … reasonable costs and expenses") and *Chernichan v Chernichan (Estate)* 2001 ABQB 913, [14] ("the obligation to reimburse arises in restitution, not in contract"). In certain circumstances. the deceased's estate may also be able to recover funeral expenses where the deceased's death was caused by a third party act or omission, as part of an action for damages – see Smale (n 119), pp 84–86 and the authorities cited therein.

person with primary responsibility for arranging the funeral;[123] where the arrangements were made by someone else, he/she can recover expenses from the deceased's estate via the personal representative.[124] This might be someone within the wider family circle, or (reverting to the legal hierarchy noted earlier[125]) a householder, hospital trust or local authority that incurred costs in discharging the duty to dispose of the deceased's remains.[126] Any social fund payments advanced towards funeral expenses are also recoverable.[127]

Of course, this assumes that there are adequate funds available in the estate – or that the deceased actually left an estate to administer.[128] The latter scenario is less common in adult deaths, but often occurs where the deceased was an infant or minor. Here, the child's parents are legally responsible for funeral costs.[129]

2.   'Reasonable' expenses only

Recovery from the deceased's estate is limited to 'reasonable' funeral expenses;[130] the personal representative (or anyone else who orders the

---

123   The personal representative can recover any costs incurred or reimbursed personally, as a first charge on the estate – *Rees v Hughes* [1946] 1 KB 517.

124   Kerridge and Brierly (n 119), p 510 citing *Green v Salmon* (1838) 8 Ad & E 348. This is subject to two important caveats, the first being where the person who discharged the funeral costs made it clear that they did not want to be reimbursed – *MJ Dixon Construction Ltd v Dixon* 2011 ONSC 2430 (son paid his father's funeral expenses from the family company which the son had taken over, but changed his mind when he discovered that he had not received his share of a family cottage when his mother died; court refused to allow him to claim these expenses from the mother's estate, because the son had assured his mother and his estranged siblings that the company would absorb the costs). Second, recovery is not permitted where a third party makes alternative funeral arrangements to those already in place, and claims for any outlay – *Williams v Williams* (1882) 20 Ch D 659.

125   See Pt I.

126   Recovery of local authority costs is specifically included in the Public Health (Control of Disease) Act 1986, s 46(5) (as amended).

127   Social Security Administration Act 1992, s 78(4).

128   Costs are not recoverable if the deceased left no estate – *Davey v Rur Mun Cornwallis* [1931] 2 DLR 80.

129   *Clark v London General Omnibus Co Ltd* (1906) 2 KB 648, and the rules around different sets of parents presumably apply here as well (see pp 64–65). At common law, husbands were legally responsible for their wives' funeral expenses before the latter became entitled to their own estates – see *Jenkins v Tucker* (1788) 1 H Bl 90, as well as *Ambrose v Kerrison* (1851) 10 CB 777 and *Bradshaw v Beard* (1862) 12 CB (NS) 344. *Chernichan v Chernichan (Estate)* 2001 ABQB 913 suggests that this rule still applies, with reciprocal obligations imposed on both spouses.

130   See *Hancock v Podmore* (1830) 1 B & Adol 260, *Edwards v Edwards* (1834) 2 C & M 613 and *Green v Salmon* (1838) 8 Ad & E 348. For more recent confirmation, see the comments in *Smith v Tamworth City Council* (1997) 41 NSWLR 680, 694 (reproduced at n 140) and *Chernichan v Chernichan (Estate)* 2001 ABQB 913, [11] ("[t]he responsibility at law for funeral expenses is not unlimited, and only extends to 'reasonable' expenses"), as well as the cases noted immediately below.

funeral) is liable for any excess costs. Presumably, the underlying intent is to prevent unwarranted and extravagant expenses from depleting the deceased's estate to the detriment of beneficiaries.[131]

There are no set financial limits for what constitutes reasonable expenditure, nor is it restricted to a percentage value of the net estate.[132] However, certain qualifications apply. The deceased's 'position in life' is a material factor,[133] as is his/her religious and cultural beliefs (if, for example, these mandate more elaborate funeral requirements and mourning rituals).[134] Where the deceased has stipulated certain funeral arrangements in a will or pre-paid funeral plan, these can also act as benchmark for recovering costs.[135] Concerns around the potential solvency of the deceased's estate will also dictate what is reasonable, to prevent assets being diminished at the expense of the deceased's creditors.[136] Assuming the personal representative knows the estate is insolvent or has reason to believe it might be,[137] any funeral costs

131   While extravagant expenses are not allowed (*Stacpoole v Stacpoole* (1816) 4 Dow 209 and *Mullick v Mullick* (1829) 1 Knapp 245), expenses viewed as 'luxuries' might be – *Chernichan v Chernichan (Estate)* 2001 ABQB 913, [24].

132   This is a "question of fact to be decided having regard to the circumstances of the particular case" – Kerridge and Brierly (n 119), p 510 and citing various authorities.

133   *Re Walter (Deceased)* [1929] 1 Ch 647, 655. This presumably includes an individual's social standing, successes and achievements; and while we might regard it as an anachronism, the test still applies – see *Chernichan v Chernichan (Estate)* 2001 ABQB 913, [24]. In *Zannetides v Spence* [2013] NSWSC 2032 a son claimed almost AUS$18,000 for his father's funeral expenses (out of an estate worth almost AUS$282,000); the court held that this sum was reasonable, the son having arranged a funeral and memorial which he felt were 'worthy' of his father.

134   *Gammell v Wilson* [1982] AC 27, 43, and *Chernichan, ibid.*

135   In *Szedgedi v Horvath* 2005 CanLII 26603 (ON SC) reasonable funeral expenses included transporting the deceased's ashes from Ontario to Hungary to be placed in a mausoleum purchased by the deceased.

136   The old adage that "[d]ead debtors must not feast to make their living creditors fast" – R Atherton, "Claims on the Deceased: The Corpse as Property" (2000) 7 *Journal of Law and Medicine* 361, p 370. Reasonable funeral expenses take priority over the deceased's preferential debts on insolvency – Kerridge and Brierly (n 119), p 540. However, the timing of the bankruptcy adjudication (whether it occurs before or after the deceased's death) may be relevant – *Re Walter* [1929] 1 Ch 647 and *Re Decleva* (2008) 40 ETR (3d) 144.

137   *Stag v Punter* (1744) 3 Atk 119 and *Bissett v Antrobus* (1831) 4 Sim 512. In *National Westminster Bank plc v Lucas* [2013] EWHC 770 (Ch), the Court of Appeal held that £70,000 spent on Jimmy Savile's funeral by his executors was properly incurred expenditure; the funeral arrangements were in accordance with Savile's wishes and made almost a year before he was exposed as a serial child abuser and sex offender prompting multiple financial claims against his estate.

are limited to "the minimum expenses that will accord a dignified interment [or cremation]".[138]

Expenditure on headstones marking the deceased's grave is one of the most contentious areas. This falls within the rubric of funeral expenses,[139] subject to the reasonableness test;[140] but when the estate is (or is likely to be) insolvent, this particular cost will be closely scrutinised so that "only the simplest and most modest tombstones can be charged against the creditors".[141]

III.   Disposal of the dead and civil law causes of action

Mistreatment of the dead attracts criminal law offences.[142] From a civil law perspective, however, much debate has centred on whether (and to what extent) third parties with temporary custody of the deceased's corpse or ashes, at various stages between death and final disposal, are liable for mishandling the remains. For example, what if a funeral director embalms the deceased's body contrary to the bereaved family's wishes,[143] or places the wrong body in the wrong casket,[144] or allows the deceased's body to be

---

138   *Chernichan v Chernichan (Estate)* 2001 ABQB 913, [23]. Here the court allowed charges for the clergyman, burial plot and interment, but disallowed an expensive funeral lunch (questioning whether this would ever be permissible with insolvent estates) – *ibid*, [26]. Older cases suggested that, where a person died insolvent, only 'necessary' expenses would be allowed – *Stag v Punter* (1744) 3 Atk 119 and *Hancock v Podmore* (1830) 1 B & Adol 260.

139   See *Goldstein v Salvation Army Assurance Society* [1917] 2 KB 291, *Gammell v Wilson* [1982] AC 27, *Booth Estate v Sault Ste Marie (City)* [1994] OJ No 3105 and *Starosielski v Starosielski (Estate)* 2000 AQBD 324 – as well as *Chernichan v Chernichan (Estate)* 2001 ABQB 913, [27]–[31] and the various cites cited therein. The underlying rationale is that, in most societies, "a grave marker is a customary part of burial" – *Chernichan*, [31].

140   *Smith v Tamworth City Council* (1997) 41 NSWLR 680, 694 ("reasonable cost of a reasonable headstone is recoverable from the deceased's estate"). Again, the amount spent is case specific – for example, the respective amounts were permissible in both *Gammell v Wilson* [1982] AC 27 and *Booth Estate v Sault Ste Marie (City)* [1994] OJ No 3105, but not in *Hart v Griffiths-Jones* [1948] 2 All ER 729. In *Starosielski v Starosielski (Estate)* 2000 AQBD 324 expenditure of almost Can$11,400 on a headstone (out of funeral expenses totalling around Can$22,000), was unreasonable – despite the deceased leaving a substantial estate.

141   *Chernichan v Chernichan (Estate)* 2001 ABQB 913, [31].

142   See Ch 1, Pt VI.

143   As in *Scheuer v William F Howard Funeral Home* 385 So 2d 1076 (1980) (funeral home embalmed the deceased's body even though she was Jewish and permission to embalm was refused on religious grounds).

144   A frequently cited example is *Lott v State of New York* 225 NYS 2d 434 (1962) where the original mistake was made by a hospital, which mixed up the bodies of two dead patients, one of whom was Jewish and the other Catholic. The corpse of the Orthodox Jewish woman was embalmed and displayed in an open coffin with a crucifix and rosary; the Catholic family received the corpse of a complete stranger wrapped in a shroud.

cremated before the family have had a chance to view it;[145] what if a crematorium returns the wrong set of ashes, or misplaces them;[146] what if the hospital where the deceased died carries out an unauthorised autopsy?[147] While we might instinctively assume that these scenarios generate a cause of action, invoking established rules of contract, torts and property law is complicated by two things. First, while the alleged wrong is tangible enough, the real issue here is not physical 'damage' to the corpse, but the mental anguish that such actions inflict on surviving relatives. Second, there is the question of who can recover in a civil law action, given that the deceased's immediate family will have suffered the relevant harm – regardless of who has the legal duty of disposal and right to possession of the deceased's remains.

Where the third party has control of the deceased's remains by virtue of a contractual arrangement, basic liability is governed accordingly. The best example is the contract with the funeral director; failure to provide specific services will permit recovery based on the express or implied terms of the contract – for example, where an expensive coffin selected by the deceased's family is surreptitiously replaced with a cheaper one by the funeral director.[148] However, inflicting some sort of 'injury' to the deceased's remains, or conducting the funeral in an inappropriate or irreverent manner, opens up the possibility of damages for non-pecuniary loss such as distress and mental anguish caused by the breach of contract.[149] For example, in the Canadian case of *Kressin v Memorial Gardens*[150] the deceased's widow claimed that the defendant had breached its contractual arrangements for her husband's funeral by interrupting the interment (the coffin lowering mechanism jammed and had to be replaced, resulting in the deceased being buried when his wife was not there to witness it), and failing to record the ceremony because an empty videocassette could not be found. The court awarded damages for mental distress arising out of the breach of contract since the widow was:

> ... by virtue of the defendant's breaches, deprived of much of the peace of mind that she bought and paid for. Rather than relieving the [widow]

---

145   As in *McNeill v Forest Lawn Memorial Services Ltd* (1976) 72 DLR (3d) 556 and noted below.

146   See *Mason v Westside Cemeteries* (1996) 135 DLR (4th) 384 and noted below.

147   As in *Davidson v Garrett* (1899) 5 CCC 200.

148   In *Vigers v Cook* [1919] 2 KB 475 the Court of Appeal held that a contract with a funeral director is one entire contract for an entire price; and where taking the coffin into the church for the service was an essential part of the contract that could not be carried out because the coffin had burst and was leaking, the onus was on the funeral director to prove that he was not responsible for defaulting on the term.

149   See the discussion in H Beale, *Chitty on Contracts, Volume I: General Principles* (Sweet & Maxwell, 31st edn, 2012), pp 1845–1847 citing cases such as *Cook v Swinfen* and *Jarvis v Swan Tours* [1973] QB 233.

150   2004 BCPC 413.

... of some of the emotional distress commonly associated with burying a loved one..., the defendant by its conduct significantly aggravated it.[151]

Contracts with funeral directors have what we might term an 'emotional context', and mental distress would almost certainly be contemplated by the respective parties as a "not unlikely consequence"[152] of any breach.[153] Recovery of damages is, therefore, a distinct possibility.[154] However, there is also the question of who has actually suffered the loss, given that a particular family member will usually have contracted with the funeral director, but other close relatives of the deceased also suffer mental anguish as a result of the funeral director's actions. The general rule is that only the contracting party can recover in contract;[155] and while the point could be made that the contracting party (for example, one of the deceased's children) is effectively contracting on behalf of the immediate family as well (for example, the deceased's other children, spouse, parents, etc), the court might be reluctant to accept this because it risks exposing funeral directors to too much

---

151   *Ibid*, [29]. See also the English case of *Loach & Son v Kennedy* (1952) 103 LJ 76 in which the defendant had instructed the plaintiff funeral directors to take his dead wife's body straight from the hospital mortuary to the cemetery; to make things more convenient for them, the plaintiff removed the body from the mortuary on a Saturday and left it in their machine shop over the weekend before taking it to the cemetery on Monday. The court held that the plaintiff had broken its contract with the defendant, and allowed damages for the anxiety and distress caused by the plaintiff's actions.

152   *Chitty on Contracts* (n 149), p 1847.

153   The same would apply to any mutilation or mistreatment of the dead sustained in the course of bodies being transported from one place to another (as in *Louisville & NR Co v Wilson* 123 Ga 62 (1905) where the deceased's coffin was left on an open railway platform in the rain while being transferred between trains; the body was "soaked and otherwise mutilated" as a result), or where the corpse is misplaced and the funeral delayed as a result. While basic liability would be determined by the contract of carriage, emotional damage for any breach is a distinct possibility.

154   Cases like *Farley v Skinner (No 2)* [2002] 2 AC 732 show that damages for mental distress arising from a contractual breach have come a long way in English law (and elsewhere) since change was instigated in cases such as *Jackson v Horizon Holidays* [1975] 1 WLR 1468. The Canadian case of *McNeill v Forest Lawn Memorial Services Ltd* (1976) 72 DLR (3d) 556 would almost certainly be decided differently today. The body of the plaintiffs' daughter (who had died in violent and suspicious circumstances) was mistakenly cremated before the plaintiffs had a chance to view the body. Although the judge held that the duty to exhibit the body arose from contract and that mental anguish was a foreseeable consequence of the breach, compensation for mental distress was refused; being unable to view the body was only a small part of the plaintiffs' overall distress (given the circumstances surrounding their daughter's death) and there was no evidence that their health had been affected. The result seems harsh, and the law has moved on since this decision.

155   As in the American case of *Wells Fargo & Co Express v Fuller* 13 Tex Civ App 610 (1896) where a husband had contracted with the relevant company to ship his child's remains, but without telling his wife (the child's mother); the wife was unable to recover for injury resulting from the breach of contract.

liability.[156] In contrast, if the contract is made by a non-relative (for example, a family friend or 'independent' executor appointed under the deceased's will), that person will be contracting as the surviving relatives' agent,[157] enabling them to recover for mental distress.[158]

The same broad scenarios could also generate a cause of action in negligence. Once a funeral director assumes responsibility for a corpse, a duty of care arises and actions like careless handling of the deceased's remains or substandard preparation of the body,[159] as well as improper conduct of the funeral proceedings, would breach this. And while the actual duty would be limited on *Caparo Industries plc v Dickman*[160] grounds – in other words, that the harm was reasonably foreseeable, and that there was a relationship of proximity between claimant and defendant – this should not be difficult to establish here. If what is alleged is loss caused by damage to property, the only person with a cause of action would be the person with the legal right to possession of the deceased's remains for disposal purposes – assuming that this entitlement constituted a sufficient "possessory title to the property, at the time when the loss or damage occurred".[161] Recovery for mental illness or injury caused by the acts or omissions of the funeral director (or anyone else with temporary custody of the deceased's remains),[162] would seem a more obvious head of damage in negligence. The traditional view in English law was that negligently inflicted mental injury was not recoverable unless accompanied by some sort of physical injury.[163] Although this is no longer the case, mere emotional distress or grief does not constitute a recognised

---

156   Though, if courts were persuaded to take a more holistic view of who the contract was effectively being made on behalf of, liability could be limited by rules of remoteness and, for example, applying a 'close personal relationship test' similar to that which dictates recovery for negligent infliction of psychiatric harm in tort – see n 165.

157   *Chitty on Contracts* (n 149), pp 1141–1142.

158   In certain circumstances, family members could bring an action against the agent, if that person has exceeded their authority in making the funeral arrangements – *Ayache Estate v Muslim Association of Calgary* 2000 ABPC 101.

159   Causing it, for example, to decay too quickly and prevent family viewing or having an open-casket wake (or funeral).

160   [1990] 2 AC 605.

161   *Leigh and Sillivan Ltd v Aliakmon Shipping Co* [1986] AC 785, 809 (Lord Brandon). Again, both the 'no-property' rule and the legal content of the duty of disposal exception raise questions about the ability to recover here.

162   What was once categorised as 'nervous shock' and analysed in cases such as *Alcock v Chief Constable of South Yorkshire Police* [1991] 4 All ER 907. See generally J Goudkamp and E Peel, *Winfield and Jolowicz on Tort* (Sweet & Maxwell, 19th edn, 2014), pp 129-141 and NJ Mullaney and P Handford, *Tort Liability for Psychiatric Damage* (Law Book Company of Australia, 2nd edn, 2006).

163   *Winfield and Jolowicz on Tort* (n 162), p 130. The same rationale applied elsewhere, and see *Miner v CPR* (1911) 3 Alta LR 408 as a direct example of this in the bodily disposal context (corpse mistakenly delivered to a different location by the defendant railway company, delaying the funeral which the plaintiff had arranged for her son; damages for mental suffering alone not recoverable in negligence).

psychiatric illness,[164] and would not give rise to a cause of action for mistreatment of human remains. Another obstacle is the fact that recovery for any proven mental illness would have to be based on the 'secondary victim' rule, whereby the plaintiff's condition results from witnessing danger to the primary victim, or being proximate (both temporally and spatially).[165] In the scenarios identified here, the deceased's family would not usually have observed or perceived any improper interference with their loved one's remains.[166]

Looking beyond the parameters of negligence, the tort of intentional infliction of emotional distress might be another option.[167] Its scope is

164   *Winfield and Jolowicz on Tort* (n 162), p 130.
165   English law imposes a number of basic control mechanisms to limit liability for negligent infliction of psychiatric injury – see the respective decisions in *Alcock v Chief Constable of South Yorkshire Police* [1991] 4 All ER 907 and *White Chief Constable of South Yorkshire* [1999] 2 AC 455. The plaintiff would also have to establish the requisite tie of 'love and affection' to the deceased, as mandated by *Alcock*. Most close family relationships would fall within this category, so that recovery would (once again) extend beyond those individuals with possessory rights to the deceased's remains for the disposal purposes.
166   The decision in *Owens v Liverpool Corporation* [1939] 1 KB 394 is an esoteric example, involving a tramcar being driven negligently by one of the defendant's employees, which collided violently with a hearse in a funeral procession, overturning the coffin; the deceased's relatives (who were following behind) were entitled to recover damages for what was then described as nervous shock (though the decision has subsequently been questioned – see *Winfield and Jolowicz* (n 162), p 141). Nwabueze suggests a number of circumstances that might constitute negligent infliction of psychiatric harm in relation to the dead, including bodily mutilation caused by an unauthorised autopsy or carelessly allowing a corpse to decompose, as well as failure to notify family members about the deceased's death – see RN Nwabueze, *Biotechnology and the Challenge of Property: Property Rights in Dead Bodies, Body Parts and Genetic Information* (Ashgate, 2007), pp 199–202. In *Bastien v Ottawa Hospital* [2001] 56 OR (3d) 397 the court held that there was a genuine issue for trial where a hospital (which had been authorised by the parents of stillborn twins to bury their remains, but placed them in a communal casket with other stillborns) had broken their obligation to carry out burial in a dignified manner and inflicted psychiatric damage on the twins' parents.
167   Pawlowski (n 8), citing the rule in *Wilkinson v Downton* [1897] 2 QB 57 and assuming the existence of a recognised psychiatric injury. Once again, the constituent elements of the tort (including proof of deliberate mishandling of the deceased's remains, with the intent of harming the plaintiff) would have to be satisfied.

unclear in English law,[168] though American courts have used this particular tort as a basis for recovery where a funeral director has knowingly released the deceased's remains to a former spouse instead of another family member[169] or deliberately retained a body to secure payment of funeral charges.[170] However, it should be pointed out that these courts have been much more proactive than their English counterparts in allowing surviving relatives (provided they are closely enough related) to recover for mental anguish resulting from improper treatment of the dead, with a number of distinct causes of action in tort.[171] While these used to be confined to the individual who had the legal right to control disposal and possession of the

168   However, Nwabueze suggests that it might allow civil law recovery where a grave is disturbed with the intent of provoking the deceased's surviving relatives or causing them acute emotional trauma – Nwabueze (n 166), p 200, citing the example of animal rights activists disinterring the body of 82-year-old Gladys Hammond from a churchyard in Staffordshire in October 2004, as a protest against her son-in-law who was breeding guinea pigs for medical research (something which is also a criminal law offence – see Ch 7, Pt VI). The factual scenario in *R v Moyer* [1994] 2 SCR 899 could be another example of intentional infliction of emotional distress (photographs taken at identifiable graves in a Jewish cemetery, featuring a 'skinhead' wearing neo-Nazi symbols simulating urinating on monuments), while the same tort could conceivably be invoked where the deceased's remains are deliberately taken from the person with the legal right to possession and associated decision-making powers over the funeral, by someone else who disagrees with the proposed arrangements.

169   *Rekosh v Parks* 735 NE 2d 765b (Ill Dist C App 2000).

170   *Levite Undertakers Co v Griggs* 495 So 2d 63 (Ala 1986). Contrast these decisions with the Canadian case of *Sopinka v Sopinka* (2001) 55 OR (3d) 529, where the deceased's parents had disposed of their son's ashes without informing his ex-wife who had a history of violence towards the deceased; the court rejected the ex-wife's claim for intentional infliction of emotional distress. In *Bradley v Wingnut Films* [1994] EMLR 195 the plaintiff's family tombstone had been included in a 'comedy horror' film shot at the same cemetery, appearing for a total of 14 seconds in one scene and never in its entirety. While accepting that the plaintiff's resultant condition met the requirements for a recognisable mental injury, the court rejected a claim for intentional infliction of emotional distress because it was not reasonably foreseeable that the plaintiff would sustain such damage and there was no evidence to suggest the defendant had intended to cause any distress to the plaintiff.

171   Nwabueze (who has written extensively in this area) has identified five separate causes of action: (i) intentional infliction of emotional distress; (ii) negligent infliction of emotional or mental distress; (iii) intentional mishandling of a dead body; (iv) abuse of a corpse; and (v) negligent or wrongful interference with a dead body – RN Nwabueze, "Biotechnology and the New Property Regime in Human Bodies and Body Parts" (2002) 24 *Loyola of Los Angeles International & Comparative Law Review* 19, p 29. For a more in-depth analysis of these various causes of action, as well as their origins and implications, see PE Jackson, *The Law of Cadavers* (Prentice Hall, 2nd edn, 1950), ch VI; comment on "Damages for Mental Suffering Resulting from Mistreatment of a Cadaver" (1960) 9 *Duke Law Journal* 135; J Leavitt, "The Funeral Director's Liability for Mental Anguish" (1964) 15 *Hastings Law Journal* 464; and AW Craigie, "Burial of a Tort: The California Supreme Court's Treatment of Tortious Mishandling of Remains in *Christensen v Superior Court*" (1993) 26 *Loyola of Los Angeles Law Review* 909.

deceased's remains,[172] courts now recognise the basis for liability as the surviving relatives' personal interests in disposing of their loved one's remains in a dignified manner.[173]

However, proprietary remedies may still have a role to play here, given that the body is a tangible physical object with discrete possessory entitlements. While English law does not recognise a tort of wrongful interference with the body,[174] there has been some suggestion that the individual with the legal right to possession of a dead body (usually the personal representative, for the majority of adult deaths) should be able to maintain a corresponding action for mutilation or mishandling of the dead.[175] Most arguments centre on Canadian and Scottish authority,[176] though the fundamental difficulty is ascribing a cause of action. Trespass to the person appears to be one option;[177] conversion might be another, assuming the plaintiff could establish interference with identifiable property rights – though Gage J in *AB v Leeds Teaching Hospitals NHS*[178] suggested that English law does not recognise an

---

172   In the US, the deceased's legally recognised next-of-kin (in descending order) have a quasi-property right in the deceased's remains – see Ch 5, Pt III. Although frequently characterised as a species of property interest, it is not significantly different from the English law position outlined earlier, given that it generates a transient and purposive right to possession of the dead for disposal purposes.

173   See Horan (n 22), pp 434–436 (though noting that US courts, to date, have been reluctant to extend this to non-family) and Hardcastle (n 20), pp 55–56. Other non-proprietary actions might include things like breach of privacy and informed consent stemming from mistreatment of the dead. However, both are problematic. Dead bodies cannot claim protection under privacy laws, and surviving family members cannot claim relational rights to sue for infringement – Nwabueze (n 166), pp 204–211. And in *Bradley v Wingnut Films* [1994] EMLR 195 (see n 170), the court rejected the plaintiff's claim for breach of privacy in relation to the offending film scene, noting that there were few things less private than a tombstone in a public cemetery. Informed consent has been largely overlooked by courts, presumably on the basis that the dead lack capacity to consent to the types of mistreatment being contemplated here – again, see Nwabueze (n 166), pp 211–212.

174   See the comments of Gage J in *AB v Leeds Teaching Hospitals NHS* [2005] 2 WLR 358, 389, and the comment by RN Nwabueze, "Interference with Dead Bodies and Body Parts: A Separate Cause of Action in Tort" (2005) 15 *Tort Law Review* 63.

175   Since, except for natural decay, the body should be in the condition it was in "the moment death intervenes" – Stueve (n 1), p 42.

176   The leading Canadian authority is *Edmonds v Armstrong Funeral Home Ltd* [1931] 1 DLR 676, but see also *Davidson v Garrett* (1899) 30 OR 653 and *Phillips v Montreal General Hospital* (1908) 33 Que SC 483. All three cases involved the performance of an unauthorised autopsy, though the plaintiff failed in *Davidson* because the post-mortem had been authorised by the coroner. Scottish authorities include *Pollok v Workman* 1900 2F 354 and *Hughes v Robertson* 1930 SC 394, as well as *HM Advocate v Dewar* 1945 SC 114.

177   See Matthews (n 79), p 218 taking the view these cases "appear to be cases of trespass" (and citing the American case of *Larson v Chase* 47 Minn 307 (1891) as another example).

178   [2005] 2 WLR 358, 397.

action for conversion in corpses.[179] In contrast, possessory entitlements to the dead could underpin an action in bailment according to the Canadian case of *Mason v Westside Cemeteries*[180] where a son was allowed to recover on this basis after a cemetery authority lost the urns containing his dead parents' ashes – despite not having visited or memorialised the ashes at any time during the 23-year period since his parents had died.[181] While the son was awarded only Can$1,000 for mental distress (bearing in mind that he had effectively ignored his parent's ashes for so many years), Molloy J struggled with the assessment of damages for loss of property; given that the ashes had "no monetary value whatsoever", it would be inappropriate "to fix a completely arbitrary monetary value to the ashes based solely on the plaintiff's emotional attachment to them".[182] In the end, the son was awarded a nominal sum of Can$1 for the ashes and Can$2 for each urn, an outcome which is hard to dispute on the facts.

Overall, emotional injury caused to the living by mishandling or mistreating their dead is an interest that the law seeks to protect. Courts have experimented with different theories of liability; and while contract law is an obvious option, both torts and associated proprietary-based actions can also assist, either as concurrent actions, or in situations that cannot be analysed in contract.

Conclusion

English law sets out a clear order of entitlement when it comes to disposal of the dead and who has decision-making authority, even if the applicable rules have their origin in the administration of estates and liability for funeral expenses. In most instances, a funeral will be arranged by the deceased's immediate family, without much regard being paid to the laws governing this area. However, where surviving relatives cannot agree, the legal duty of disposal and right to possession of the deceased's remains becomes the focal point of any subsequent litigation, revealing flaws in the existing regime and uncertainties around its strict application as the following chapter illustrates.

179　Several US cases have reached the same conclusion, including *Culpepper v Pearl Street Building Inc* 877 P2d 877 (Colo 1994) (court rejected the "fictional theory that a property right exists in a dead body" to facilitate a claim in conversion in respect of a mistakenly cremated corpse – *ibid*, 880) and *Bauer v North Fulton Medical Center Inc* 527 SE2d 240 (Ga Ct App 1999), though contrast these with *Spates v Dameron Hospital Association* 7 Cal Rptr 3d 597 (Ct App 2003) (next-of-kin's quasi-property interest in deceased's corpse supported a cause of action for conversion). The same appears to be true of living body parts, according to the well-known American case of *Moore v Regents of University of California* 51 Cal 3d 120 (Sup Ct Cal 1990) (no claim in conversion in respect of Moore's excised cells).

180　(1996) 135 DLR (4th) 361.

181　Losing the ashes also constituted negligence on the facts.

182　(1996) 135 DLR (4th) 361, [41].

4 Resolving Funeral Disputes

[I]n this area ... difficulties sometimes arise that would test the wisdom of Solomon.[1]

Introduction

Most people probably assume that their funeral will be arranged by close family members who, united in grief, will ensure that this takes place in a dignified and appropriate manner. However, consensus is not always possible as surviving relatives clash over a range of things: whether the deceased should be buried or cremated, adherence to religious or cultural values, the proposed burial location and what happens to the ashes following cremation to list but a few.[2] While most occur within the private sphere, contests involving dead celebrities and high-profile public figures also illustrate death's divisive and destructive powers. For example, in February 2007 the former Playboy model Anna Nicole Smith lay decaying in her coffin during a protracted legal battle between her estranged mother, former lover and infant daughter (represented by the child's guardian ad litem),[3] while disagreement over Nelson Mandela's final resting place also engulfed his

---

1   JA Everhard, "Whose Body?" (1964) 6 *United States Air Force JAG Bulletin* 17, p 17. The early part of this chapter draws on material previously published as H Conway and J Stannard, "The Honours of Hades: Death, Emotion and the Law of Burial Disputes" (2011) 34 *University of New South Wales Law Journal* 860, and reproduced with kind permission of the journal.

2   More trivial points of contention (e.g. the type of coffin, and who should participate in ceremony by, for example, carrying the deceased's coffin or delivering the eulogy) tend to be resolved without litigation.

3   *Arthur v Milstein* 949 So2d 1163 (2007). A similar fate befell James Brown, the self-styled 'Godfather of Soul', whose body was refrigerated at an undisclosed location for almost three months as his family fought over his final resting place – see FH Foster, "Individualized Justice in Disputes over Dead Bodies" (2008) 61 *Vanderbilt Law Review* 1352, pp 1352–1354.

family in the months before the former South African president's death in December 2013.[4]

While the death of a loved one often brings families together, its emotional impact "can also tear survivors apart".[5] Anger, hostility and resentment are basic functional responses to death,[6] and 'funeral disputes' are a classic example of death fracturing family bonds. If consensus or compromise cannot be reached, the law must intervene to ensure that disposal of the dead takes place.

I.   The underlying motives

Case law reveals a number of recurring themes, when surviving relatives disagree over the fate of their dead.[7]

1.   'Doing right' by the deceased

Here, the emphasis is on doing what the deceased wanted.[8] A good example is *Burrows v HM Coroner for Preston*,[9] which involved the remains of a 15-year-old boy who committed suicide while detained in a young offenders' institution. The boy's mother (a recovering heroin addict who had little contact with her son) insisted on burial, despite acknowledging that this was contrary to his wishes; the boy's uncle, with whom he had lived for eight years, wanted to fulfil the boy's request to be cremated. [10]

2.   Religious and cultural dimensions

Funeral rites are often shaped by religious and cultural beliefs,[11] generating tensions within families. Disputes based on religious convictions typically

---

4   "Nelson Mandela Day: How Mandela Family Feud Overshadowed Madiba's Final Year and Death", *International Business Times* (18 July 2014) www.ibtimes.co.uk/nelson-mandela-day-how-mandela-family-feud-overshadowed-madibas-final-year-death-145711 7 (accessed 30 September 2015).

5   EC Rodriguez-Dod, "Ashes to Ashes: Comparative Law Regarding Survivors' Disputes Concerning Cremation and Cremated Remains" (2008) 17 *Transnational Law and Contemporary Legal Problems* 311, p 312.

6   See J Bowlby, "The Process of Mourning" (1961) 42 *International Journal of Psycho-Analysis* 331 and J Archer, *The Nature of Grief: The Evolution and Psychology of Reactions to Loss* (Routledge, 1999).

7   Some of these map onto other areas of intra-familial conflict – for example, disputes over the deceased's memorial (see Ch 8).

8   See for example, *Robinson v Pinegrove Memorial Park* (1986) 7 BPR 15,097, *Grandison v Nembhard* [1989] 4 BMLR 140 and *Keller v Keller* [2007] VSC 118.

9   [2008] EWHC 1387 (Admin).

10   While custody of a dead body would normally be awarded to the mother as closest ranking kin, the court ruled in favour of the uncle – see pp 103–104.

11   See generally D Rees, *Death and Bereavement: The Psychological, Religious and Cultural Interfaces* (Whurr, 2nd edn, 2001).

take two forms, the first occurring where the proposed funeral arrangements are not in keeping with the deceased's own beliefs. For example, the deceased in *Hunter v Hunter*[12] was a staunch Protestant who had converted to Catholicism one month before his death, apparently to enable him to be buried beside his devoutly Catholic wife. The deceased's son opposed this, given his father's strong religious convictions and conflicting evidence surrounding the deceased's mental state before he died.[13] The second and more problematic scenario arises where certain relatives insist on imposing their own religious preferences, regardless of the deceased's ambivalence or non-adherence. A good illustration is *Saleh v Reichert*[14] in which a husband's decision to honour his wife's wishes by cremating her remains was challenged by the deceased's father who wanted to inter his daughter's body in accordance with the Muslim faith in which she had been raised.[15]

Cultural values can generate similar types of disputes. Again, problems arise where the deceased discarded cultural traditions in life, yet family members insist on specific funeral rites as a means of re-establishing those links on death – an issue that has arisen in numerous Australian cases involving members of the Aboriginal community. For example, in *Meier v Bell*[16] one of the deceased's siblings wanted to bury her brother in a cemetery alongside his dead parents and other family members according to Aboriginal burial beliefs, while his de facto partner (who also claimed some 'distant' Aboriginal blood) insisted on burying the deceased in another cemetery where it would be easier for her and her children to visit the grave. A similar scenario arose in *Jones v Dodd*,[17] the deceased's father wanting his son to be buried with other relatives in his geographical and spiritual homeland, while the deceased's former partner claimed that he should be buried in a cemetery close to where she resided so that the couple's children could visit their father's grave. Custody of the deceased's body was awarded to the partner in *Meier*, while the opposite conclusion was reached in *Jones*

---

12   (1930) 65 OLR 586.
13   The court ruled in favour of the son as executor under the deceased's will. See also *Re Lochowiak (Deceased)* [1997] SASC 6301 (deceased's adult son challenged the deceased's partner, the former arguing that his Catholic father would not have wanted to be cremated).
14   (1993) 104 DLR (4th) 384.
15   The court ruled for the husband as presumptive administrator, observing that his conversion to Islam when he married the deceased did not transmute into a legal obligation to dispose of her remains accordingly. See also *Stewart v Schwartz Brothers-Jeffer Memorial Chapel Inc and Scott* 606 NYS 2d 965 (1993) (dispute between the deceased's same-sex partner and his mother; the partner wanted to cremate the deceased in accordance with his wishes while the mother insisted on burial in accordance with the Jewish faith in which her son had been raised), as well as *Abeziz v Harris Estate* [1992] OJ No 1271 and *Privet v Vovk* [2003] NSWSC 1038.
16   (Supreme Court of Victoria, 3 March 1997).
17   [1999] SASC 125.

though the significance attached to cultural imperatives was very different in each.[18]

3.   Resuming old hostilities

Where families are already divided and prone to conflict, death can resurrect old grievances. For example, disputes between adult siblings following the loss of a parent are often childhood problems in contemporary guise, something which the judge acknowledged in *Leeburn v Derndorfer*[19] where a brother and his two sisters quarrelled over the fate of their father's ashes.[20] Disagreements between children from multiple marriages or relationships also reignite simmering tensions, as do those between adult children and a step-parent or surviving partner.[21]

One of the most acrimonious examples involves separated parents fighting over the remains of a dead child, as death prompts one final and decisive posthumous custody dispute.[22] Consumed by feelings of bitterness towards each other, one parent may oppose the other's choice of funeral arrangements just to be vindictive or spiteful.[23] Disputes of this nature invariably involve claims as to who was the better parent in life, suggesting that this confers the right to decide what happens to the child in death.[24]

18   See pp 104–107. Other examples include *Dow v Hoskins* [2003] VSC 206, *Calma v Sesar* (1992) 106 FLR 446 and *Reece v Little* [2009] WASC 30. For similar disputes involving the Maori peoples, see *Doherty v Doherty* [2006] QSC 257 and *Takamore v Clarke* [2012] NZSC 116.

19   [2004] VSC 172.

20   "I suspect ... that the division between them on this matter represents a manifestation of some more deep-seated hostility which I cannot resolve" – *ibid*, [9]. The sisters had interred the ashes in a cemetery against their brother's wishes; he argued that the ashes should be exhumed and divided between the siblings.

21   See for example, *Burnes v Richards* (1993) 7 BPR 15,104, *Re Lochowiak (Deceased)* [1997] SASC 6301 and *Ugle v Bowra & O'Dea* [2007] WASC 82.

22   See for example, *Joseph v Dunn* [2007] WASC 238 and *AB v CD* [2007] NSWSC 1474 (separated parents divided over the funeral arrangements for their young children), as well as *Fessi v Whitmore* [1999] 1 FLR 767 and *Derwen v Ling* [2008] FamCA 644 (dispute over where to scatter children's ashes). See also *Calma v Sesar* (1992) 106 FLR 446, *Burrows v Cramley* [2002] WASC 47 and *Gilliott v Woodlands* [2006] VSCA 46 (separated parents fighting over where to bury their adult child).

23   See for example, *Tully v Pate* 372 F Supp 1064, 1066 (DSC 1973) (parents fighting over the remains of their children – who had died in a tragic accident – were more interested in "vindictive pursuit" than seeking justice).

24   However, judges have been at pains to stress that their decisions are based on the legal merits of the case. See for example, the comments of Harrison J in *AB v CD* [2007] NSWSC 1474, [66]: "It is very important to emphasise that the result in this case is not, and should not appear to be, a prize for who was the better parent".

4. The punitive element

Specific funeral arrangements can punish the deceased for physical or emotional harm which they inflicted on others in life. In *Holtham v Arnold*[25] the deceased's partner wanted to bury him in accordance with his wishes; the deceased's wife insisted on cremation, despite the fact that the deceased had left her and his six children years earlier.[26] To similar effect is *Betty Brannam v Edward Robeson Funeral Home*,[27] in which the deceased's estranged wife wanted to bury her husband's remains, contrary to a stipulation in the deceased's will that he be cremated and the ashes given to his long-term partner with whom he had three children.[28]

5. Posthumous protection from harm

Here the aim is to protect the deceased from the person who caused their death, or repeatedly harmed them in life. A perfect illustration *Scotching v Birch*[29] where a mother had pleaded guilty to the unlawful killing of her son yet wanted to bury the child in a cemetery close to where he was killed, and the father insisted on burial elsewhere.[30] Domestic violence cases are another example. In *Burnes v Richards*,[31] where the deceased had endured numerous assaults during a 17-year relationship with her partner and had left him two months before her death, the deceased's adult children fought for custody of their mother's remains.[32] More sinister motives featured in *Maurer v Thibeault*[33] in which the deceased's estranged husband who had repeatedly threatened to kill her wanted to cremate his wife's remains contrary to her wishes, and those of her family.[34]

---

25   [1986] 2 BMLR 123.
26   The wife claimed not to be acting out of malice, insisting on cremation (and supported by the couple's children) so that the deceased's ashes could be interred in the same grave as the deceased's parents. As the deceased's legal next-of-kin (the couple had never divorced), the wife's wishes were final.
27   No 43141/96 (NY Sup Ct Nov 14, 1996).
28   In this case, however, the deceased's wishes prevailed. See also *Theodore v Theodore* 259 P2d 795 (1953) (deceased's wife wanted to exhume his remains and move them else-where after discovering that she had received nothing under her late husband's will).
29   [2008] EWHC 844 (Ch).
30   Under the forfeiture rule, the mother had no right to possession of her sons' remains – see Ch 3, n 24.
31   (1993) 7 BPR 15,104.
32   See also *Reid v Love and North Western Adelaide Health* [2003] SASC 214 (dispute between the deceased's son and her de facto partner, the former arguing that the deceased and the defendant were alcoholics and had an extremely volatile relationship) and *Sopinka v Sopinka* (2001) 55 OR (3d) 529 (dispute between deceased's ex-wife and his parents, the former with a history of violence towards the deceased).
33   860 NYS2d 895 (2008).
34   Custody of the body was granted to the wife's family. For a similar outcome, see *Spanich v Reichelderfer* 628 NE2d 102 (Ohio Ct App 1993).

6.   Fixing social identities and 'reclaiming' the deceased

Funerals and their attendant rites fix the deceased's social identity,[35] providing an opportunity for certain individuals to publicly affirm (or recast) their relationship with the deceased. This particular classification cuts across many of the areas already discussed; for example, both *Holtham v Arnold*[36] and *Betty Brannam v Edward Robeson Funeral Home*[37] could just as easily be viewed as estranged spouses attempting to reinstate themselves publicly as the deceased's next-of-kin, despite past events. Disputes between a new spouse or partner and the deceased's children from a previous relationship are ways of openly reclaiming the dead, as are conflicts between separated parents over the remains of a dead child.[38]

II.   Resolving competing claims: The common law rules

In the absence of a statutory framework, judges must apply common law rules around the duty of disposal and associated possessory rights to the dead. This results in the vast majority of funeral disputes being resolved by succession law rankings, which prioritise the personal representatives.[39]

Where the deceased made a will, the executor takes precedence over the deceased's immediate family. For example, in *Murdoch v Rhind*[40] the executor prevented the deceased's wife from cremating her husband's remains (fulfilling the deceased's wish to be interred in the family burial plot), while in *Grandison v Nembhard*[41] the executor could repatriate the deceased's remains to his native Jamaica (again, in accordance with the deceased's wishes) despite objections from the deceased's only daughter who wanted to bury her father in England. The executor's authority is not well-known, and being entitled to possession of the deceased's remains may be difficult for family members to accept if the executor is unrelated to the

---

35   PC Jupp, "Virtue Ethics and Death: The Final Arrangements" in K Flanagan and PC Jupp (eds), *Virtue Ethics and Sociology: Issues of Modernity and Ethics* (Palgrave, 2001) 217, pp 217–218.

36   [1986] 2 BMLR 123.

37   No 43141/96 (NY Sup Ct Nov 14, 1996). Both cases were discussed immediately above.

38   And between different sets of parents – see Ch 3, pp 64–65.

39   See Ch 3, Pt I.

40   [1945] NZLR 425.

41   [1989] 4 BMLR 140.

deceased.[42] Of course, the choice of executor may have been based on this person fulfilling the deceased's funeral instructions,[43] in which case the executor should have the final say (regardless of any familial opposition).[44] However, an executor is not obliged to follow the deceased's instructions;[45] as the person legally entitled to possession of the body, it is for them to decide the manner and place of disposal.[46]

Where the deceased died intestate,[47] the highest ranking presumptive administrator is legally entitled to dispose of the deceased's body. Courts will rule in favour of a surviving spouse or civil partner in the event of a dispute with the deceased's children, parents or siblings;[48] where there is no spouse or civil partner, the strength of a particular claim depends on where the

42   The executor can be a close relative (for example, the deceased's son and brother respectively in *Hunter v Hunter* (1930) 65 OLR 586 and *Murdoch v Rhind* [1945] NZLR 425), but can also be a friend (as in *Abeziz v Harris Estate* [1992] OJ No 1271) or an independent executor chosen for their professional or business expertise (see *Schara v Tzedeck v Royal Trust Co* [1951] 2 DLR 228 and *Keller v Keller* [2007] VSC 118). In *C v Advocate General for Scotland* 2012 SLT 103 Lord Brodie questioned whether an independent executor could impose their choice of funeral arrangements on an objecting family. However, see *Laing v John Poyser Solicitors* (High Court (Chancery Division), Manchester Civil Justice Centre (unrep extemporary judgment), 31 July 2012) in which the deceased's solicitor (acting as executor) decided to cremate the deceased in accordance with her wishes; the decision was upheld despite strong objections from the deceased's son (both the deceased's daughter and a long-standing friend confirmed that the deceased wanted to be cremated). Of course, this will only be an issue where there is some sort of conflict within the deceased's family; in most instances, an independent executor will confine their role to estate administration and allow the family to make their own funeral arrangements.

43   As in *Abeziz v Harris Estate* [1992] OJ No 127 where the deceased appointed his friend as executor, because this friend would follow the deceased's request to be cremated even if the deceased's mother insisted on a 'proper' Jewish burial for her son. See the discussion in MZ Zwicker and MJ Sweatman, "Who Has the Right to Choose the Deceased's Final Resting Place" (2002) 22 *Estates, Trusts & Pensions Journal* 43, pp 46–47.

44   According to William Young J in *Takamore v Clarke* [2012] NZSC 116, [204] this factor will be of "considerable and probably decisive significance".

45   These are not legally binding – see Ch 5.

46   The executor can opt for completely different funeral arrangements (as in *Hunter v Hunter* (1930) 65 OLR 586) and can seek an injunction to prevent alternative funeral arrangements from going ahead (see *Waldron v Howick Funeral Home* [2010] NZHC 1467, though courts will not allow an 'eleventh hour' application by an absentee executor to prevent perfectly orthodox funeral arrangements being carried out).

47   Or an executor declines to act, usually because they are aware of conflicting views about the deceased's funeral arrangements – see *Keller v Keller* [2007] VSC 118.

48   See for example, *Saleh v Reichert* (1993) 104 DLR (4th) 384 (deceased's husband ranked above her father); *Mouaga v Mouaga* [2003] OJ No 2030 (deceased's wife ranked above his parents and siblings – even though the spouses were living under separate roofs at the time); *Threlfall v Threlfall* [2009] VSC 283 (coroner had erred in releasing deceased's body to his brother instead of the deceased's widow).

individual falls within the relational hierarchy.[49] Unlike executors, presumptive administrators will have close family connections to the deceased. However, the emphasis on traditional family forms means that not all filial relationships are recognised, generating problems when the intestacy framework is mapped onto funeral disputes.[50] For example, cohabitants are currently excluded from English intestacy laws and cannot claim their dead partner's remains if spouses or other relatives assert preferential rights, and illustrated in *Holtham v Arnold*[51] where the court ruled for the deceased's estranged wife despite objections from his long-term partner.[52]

Succession law rules also determine who controls deceased's the ashes. Groves has noted that their physical structure, and the fact that ashes can be moved around, has "enabled people to bring disputes ... that would be inconceivable if the deceased was still in body form".[53] However, these are every bit as contentious as. As a general rule, the rights of the deceased's executor are paramount on testate deaths.[54] If there is no will, the presumptive administrator takes priority, as illustrated in *Doherty v Doherty*[55] where the deceased's wife was entitled to her husband's ashes and to delay their interment for a short time, despite objections from the deceased's

---

49   For example, the deceased's children would usually outrank parents and siblings in the event of a dispute over the funeral – see for example, *Meier v Bell* (Supreme Court of Victoria, 3 March 1997) and *Roma v Kethcup* [2009] QSC 442.

50   See Ch 3, p 63, noting the consequences for cohabitants (reiterated below), step-families and kinship designations within different ethnic and minority groups.

51   [1986] 2 BMLR 123 (see p 91).

52   The only way of avoiding this would be to ensure that cohabiting partners (especially those in long-term, committed relationships) appoint each other as executor under their respective wills. In contrast, all Australian states place de facto partners in the same position as spouses and will favour cohabitants in disputes with the deceased's parents, children or siblings (see for example, *Brown v Tullock* (1992) 7 BPR 15,101, *Ugle v Brown & O'Dea* [2007] WASC 82, *Reece v Little* [2009] WASC 30 as well as *Spratt v Hayden* [2010] WASC 340) unless the cohabiting relationship had broken down prior to the deceased's death (see *Burnes v Richards* (1993) 7 BPR 15,104 and *Reid v Love and North Western Adelaide Health* [2003] SASC 214; though contrast these with *Dow v Hoskins* [2003] VSC 206 where the deceased's partner was still entitled to his remains despite the couple's temporary separation while the deceased battled alcohol problems). The position is broadly the same in Canada where cohabitants have the same standing as spouses for inheritance purposes – see for example, *Lajhner v Banoub* [2009] OJ No 1327 (though the court favoured the deceased's parents because the deceased and his common law spouse were no longer living in a conjugal relationship; the surviving partner's belief that they would have eventually reconciled was irrelevant).

53   M Groves, "The Disposal of Human Ashes" (2005) 12 *Journal of Law and Medicine* 267, p 270. See also Rodriguez-Dod (n 5).

54   See *Robinson v Pinegrove Memorial Park* (1986) 7 BPR 15,097. The executor can, of course, give the ashes to someone (else) in the deceased's family – see *Szedgedi v Horvath* 2005 CanLII 26603 (ON SC).

55   [2006] QSC 257.

mother and siblings who wanted immediate burial in New Zealand in accordance with Maori tradition.[56]

III. Separating equal claims

The common law framework is ineffective where two or more people have equal rights to possession of the deceased's remains but conflicting views on what should happen to them. This typically occurs on intestate deaths,[57] between relatives within the same kinship tier – for example, where siblings diverge on the funeral arrangements for a dead parent,[58] or where parents are fighting over the remains of a dead child.[59]

The issue has arisen in a number of English and Australian cases, with courts devising a range of tests to separate equal claims.[60] One of the most pressing concerns is the need for burial or cremation "without unreasonable delay",[61] making judges more likely to rule in favour of the individual whose

56   See also *Milenkovic v McConnell* [2013] WASC 421 (deceased's de facto partner entitled to custody of his ashes in preference to the deceased's mother) and *Re Jaworenko Estate* 2013 ABQB 517 (deceased's sister entitled to custody of the ashes as administrator of the estate).

57   Though it can also be an issue where joint executors under a will have joint decision-making powers, but disagree over funeral arrangements – see *Leeburn v Derndorfer* [2004] VSC 172.

58   For a recent illustration, see "Siblings in High Court Battle Over Whether Mother Should Be Buried in Jewish or Christian Ceremony", *The Telegraph* (London, 6 November 2015) www.telegraph.co.uk/news/uknews/law-and-order/11980060/ Siblings-in-High-court-battle-over-whether-mother-should-be-buried-in-Jewish-or-Christian-ceremony.html (accessed 1 December 2015). Sibling feuds can also occur over the remains of another brother or sister who dies without legally closer kin; the surviving siblings assume joint responsibility for disposal.

59   Often involving infants and minors, but also adult children where the parents are still the designated next-of-kin (because, for example, the deceased was unmarried or childless at the time of death). For a practical overview, see M O'Rourke and G Williams, "Burial of a Child's Remains – Resolving Parental Disputes", *Family Law Week* (28 February 2013) www.familylawweek.co.uk/site.aspx?i=ed112114 (accessed 30 September 2015).

60   For an overview, see I Freckleton, "Disputed Family Claims to Bury or Cremate the Dead" (2009) 17 *Journal of Law and Medicine* 178. Succession law rules for administering estates in the event of a dispute between personal representatives offer little in terms of comparable solutions. For example, in England and Wales, reg 27 of the Non-Contentious Probate Rules 1987, SI 1987/2024 favours a person of full age in preference to a guardian of a minor and a living person in preference to the personal representative of a deceased person (reg 27(5)), but otherwise leaves the matter to the court (regs 27(6)–(8)). However, the usual practice is to select the individual who is best placed to administer the estate (*Warwick v Greville* (1809) 1 Phill 123 and *Re Loveday* [1900] P 154), or the person with the largest interest in it (*Dampier v Colson* (1812) 2 Phill 54, as well as *Re Slattery* (1909) 9 SR (NSW) 577).

61   *Calma v Sesar* (1992) 106 FLR 446, 452. See also *Smith v Tamworth City Council* (1997) 41 NSWLR 680, 694 (where two or more persons have equal rights, "the practicalities of burial without unreasonable delay will decide the issue") and cited with approval in *Manktelow v The Public Trustee* [2001] WASC 290.

arrangements are well advanced, and who is not intending to transport the corpse elsewhere.[62] Other factors include the deceased's association with a particular place,[63] majority views within the relevant kinship tier,[64] and the wishes of the custodial parent in disputes involving infant or minor children.[65] And while decisions have occasionally been based on which person had the closest relationship with the deceased,[66] both English and Australian courts appear reluctant to adopt this test.[67]

In contrast, other jurisdictions have specific legislation for resolving equal kinship disputes. For example, several Canadian provinces favour the elder or eldest person within the same category, so that disputes between parents, adult siblings or adult children of the deceased are determined solely on age.[68] Turning to the US, the state of New Jersey looks for majority preference in disputes between the deceased's surviving adult children or the

---

62   See *Calma v Sesar* (1992) 106 FLR 446 (custody of the son's body awarded to the mother as arrangements for burial in Darwin had already been made and there was no legal reason for preferring Port Hedland, Western Australia – over 2,000km away); *Burrows v Cramley* [2002] WASC 47 (mother succeeded because the son had died in Western Australia and burial in Perth would be more practical); *Joseph v Dunn* [2007] WASC 238 (child's remains given to his father, who had made funeral arrangements for the following day); and *Mourish v Wynne* [2009] WASC 85 (paternal grandmother, who had raised the deceased, had already made the necessary arrangements).

63   As in *Fessi v Whitmore* [1999] 1 FLR 767 (12-year-old boy's ashes should be scattered in Nuneaton where his family had always lived, not interred in Wales where he had recently moved with his father and been killed in an accident) and *Hartshorne v Gardner* [2008] EWHC B3 (Ch) (father could inter his 44-year-old son's body in Kington where the deceased had been living and where the deceased's fiancée and brother also wanted him to be buried, despite objections from the deceased's mother who favoured cremation and interment of the ashes elsewhere). See also *Scotching v Birch* [2008] EWHC 844 (Ch) and *Minister for Families and Communities v Brown* [2009] SASC 86.

64   As in *Leeburn v Derndorfer* [2004] VSC 172 (deceased's final resting place had been chosen by two of his three children).

65   Though this practice is by no means universal – contrast the decision in *AB v CD* [2007] NSWSC 1474 (see immediately below) with both *Fessi v Whitmore* [1999] 1 FLR 767 and *Burrows v Cramley* [2002] WASC 47 in which the respective fathers lost despite being the custodial parent. The trend in some US cases is to rule in favour of the custodial parent in burial disputes involving minor children – see n 218.

66   *Keller v Keller* [2007] VSC 118 (deceased's daughter was closer to her mother, and preferred over the deceased's son) and *AB v CD* [2007] NSWSC 1474 (mother allowed to make the funeral arrangements for her 14-month-old son as custodial parent, and because his father had had virtually no contact with his son in almost a year)

67   See the comments of deputy judge Sonia Proudman QC in *Hartshorne v Gardner* [2008] EWHC B3 (Ch), [2]–[3]: "[A] decision between the earnest wishes of two grieving parents requires the wisdom of Solomon, which I do not profess to have. ... [T]he court should be slow to make findings as to the details of the deceased's family relationships".

68   For example, s 5(3) of the Cremation, Interment and Funeral Services Act 2004 in British Columbia; s 36(3) of the General Regulation (Funeral Services Act) Alta Reg 226/1998 and ss 11(2)–(3) of the General Regulation (Cemeteries Act) Alta Reg 249/1998 in Alberta; and s 91(2)(b) of the Funeral Services Act 1999 in Saskatchewan.

deceased's surviving brothers and sisters.[69] This is also the starting point in the District of Columbia and Minnesota,[70] though courts can apply other factors if majority consensus cannot be reached – including the "degree of the personal relationship between the decedent and each of the persons in the same degree of relationship to the decedent".[71] However, in the state of Pennsylvania, this is the *only* factor; in a dispute between equally ranked kin, courts should look to the "person who had the closest relationship with the deceased".[72]

IV.   Flexibility versus finality?

Equal kinship disputes aside, funeral disputes are resolved by fixed rules of entitlement, which are relatively easy to apply and promote certainty of outcome. This has a number of benefits. Decisions can be reached within a compressed timeframe,[73] since public health concerns combined with respect for the dead mandate prompt disposal.[74] Meanwhile, judges do not have to unravel complex family histories or make difficult subjective value judgments on the merits of competing claims. For example, Hale J (as she then was) in *Buchanan v Milton*[75] stressed that "the law cannot establish a hierarchy in which one sort of feeling is accorded more respect than other equally deep and sincere feelings", sentiments echoed by Lord Brodie in *C v Advocate General for Scotland*:[76]

> Determining what are appropriate funeral arrangements by reference to the quality of relationships within a family appears to me a task for which the court is quite unsuited.[77]

---

69   See NJ Stat Ann 45:27–22(a)(2) and NJ Stat Ann 45:27–22(a)(4). The same legislation simply states that a surviving parent or parents are in charge of funeral arrangements for a dead child – NJ Stat Ann 45:27–22(a)(3).

70   See respectively DC Code § 3-413, and Minn Stat § 1489A.80 and Minn Stat § 1489A.80, subd 2.

71   DC Code § 3-413.01 and Minn Stat § 1489A.80, subd 5.

72   20 PA Const Stat Ann § 305(d)(2), and see the respective decisions in *Estate of NP* 22 Fid Rep 2d 473 (Berks Cty OC 2002) and *In re Estate of Weiss* 2009 Philia CtCom Pl LEXIS 236 (Oct 1, 2009).

73   "Time acts as a secret third party in all of the issues that develop over the disposal of bodies" – BL Josias, "Burying the Hatchet in Burial Disputes: Applying Alternative Dispute Resolution Techniques to Disputes Concerning the Interment of Bodies" (2004) 79 *Notre Dame Law Review* 1141, p 1145. Constraints of time, however, contrast sharply with the finality of the decision.

74   Though the public health argument is overstated – see below.

75   [1999] 2 FLR 844, 855.

76   2012 SLT 103, [67].

77   See also the comments of Ashley J in *Meier v Bell* (Supreme Court of Victoria, 3 March 1997), [8] (courts should "identify the person with the best claim in law", which is "likely to be very much easier than attempting to resolve … 'the merits'").

One could also argue that the existing framework actually benefits the deceased's family by curbing the amount of (additional) damage being inflicted on an emotionally vulnerable yet extremely volatile family. As McKechnie J remarked in *Ugle v Bowra & O'Dea*:[78]

> There has to be a balance between the need for prompt expedition of a matter that involves grief and loss to many people, together with the need to secure burial of a person reasonably promptly. ... Pressures of time, stress and pain add to an already emotional situation where there are no winners and losers, only deeply held and legitimate feelings that are exacerbated by uncertainty.

Families will invariably dredge up past histories, as the emotional ante of the conflict is upped significantly by the raw, consuming emotions of death.[79] In these circumstances, a quick, pragmatic solution operates as a vital form of damage limitation.

There is also a sense in which judges do not like having to deal with such cases, and not just because there is little scope for conciliation or finding a middle ground.[80] Post-cremation ashes can be divided to allow feuding relatives to inter, scatter or keep their portion as they see fit;[81] and while a court can sanction such an agreement[82] it cannot force a split if one person objects.[83] Where the dispute is over a dead body, there is no prospect of carving it up in this manner – public health issues concerns alongside the notion of respect for the dead would prevail, even if the protagonists

---

78   [2007] WASC 82, [1].

79   See Conway and Stannard (n 1).

80   Unlike succession law disputes where courts can appease warring factions of the deceased's family by dividing the estate between them.

81   Rodriguez-Dod also posits the idea of a shared custody arrangement whereby each protagonist retains the ashes for a set time, effectively passing them backwards and forwards – Rodriguez-Dod (n 5), p 329.

82   There is no general prohibition on dividing ashes, as long as family members agree to this – *Leeburn v Derndorfer* [2004] VSC 172, [31].

83   See for example, *Fessi v Whitmore* [1999] 1 FLR 767 (unlike division of the living child in the judgment of Solomon, it was "wholly inappropriate" to divide a child's ashes between its parents where the father did not want this – *ibid*, 770) and *Doherty v Doherty* [2006] QSC 257 (division of the deceased's ashes and separate burials in New Zealand, possibly years apart, would be distressing to both the wife and children). In *Kulp v Kulp* 920 A2d 867 (Pa Super Ct 2007) the court held that the trial judge had abused his discretion by dividing a child's ashes between feuding parents where the father had opposed this, though contrast with *In re Estate of KA* 807 NE2d 748 (Ind Ct App 2004) where the deceased's ashes were divided between her parents, despite opposition from the father (the court apparently influenced by the deceased having told her family that she would like her ashes to be scattered in three different places).

favoured such a macabre compromise.[84] At a human level, judicial attitudes towards families fighting over their dead have ranged from expressive empathy[85] to feelings of discomfort and embarrassment.[86] More revealing, perhaps, is the sense of judicial revulsion that permeates numerous decisions. For example, Hargrave J in *Keller v Keller*[87] described the dispute between the deceased's adult children "as disrespectful of the deceased and offensive to ordinary standards of common decency". Repeated references have been also made to contests such as these being "unseemly",[88] while one judge described a fight for custody of the deceased's ashes as "all very sordid and unpleasant".[89] In expressing such negative sentiments, judges may also be reflecting societal distaste towards families fighting over their dead; such disputes infringe the basic concept of respect for the dead, and are the real reason why courts are inclined towards swift outcomes.

Public health concerns are also important, yet the emphasis placed on this factor is misleading. Where the contest centres on the deceased's ashes, there is no associated risk, though courts have still stressed the need for a quick decision.[90] In disputes over the fate of the body itself, there is no legal or scientific impairment to a corpse being placed in cold storage to halt the inevitable process of decay and allow protracted litigation.[91] However, the spectre of human remains being refrigerated (or of ashes being held *in specie*) for prolonged periods while relatives engage in bitter public squabbles over

84 Described as "unthinkable" by Byrne J in *Leeburn v Derndorfer* [2004] VSC 172, [31]. Historically however, division of human remains did occur throughout Europe, with multiple relics carved from the corpses of kings and saints – NL Cantor, *After We Die: The Life and Times of the Human Cadaver* (Georgetown University Press, 2010), p 112.

85 For example, in *Hunter v Hunter* (1930) 65 OLR 586, 587 McEvoy J described the proceedings as "one of those unfortunate cases where honest sentiment and honest religious conviction have brought about an intensely sad struggle between two factions of a family".

86 For example, Byrne J in *Leeburn v Derndorfer* [2004] VSC 172, [10] described feelings of "embarrassment at having to deal … with bitter conflicts within families over the remains of a recently deceased relative or friend", sentiments that were echoed by Hargrave J in *Keller v Keller* [2007] VSC 238, [15]. However, this was given short shrift by Elias CJ in *Takamore v Clarke* [2012] NZSC 116, [11]: "Although the position of a court asked to resolve such differences is not a comfortable one, there is nothing particularly unusual in that".

87 [2007] VSC 118, [1].

88 See for example, *Murdoch v Rhind* [1945] NZLR 425, 426, *Buchanan v Milton* [1999] 2 FLR 844, 854 and *Keller v Keller* [2007] VSC 118, [15].

89 Vaisey J in *Re Korda* (*The Times*, 23 April 1958).

90 See the comments of Byrne J in *Leeburn v Derndorfer* [2004] VSC 172, [27]: "[A]shes are, after all, the remains of a human being and … should be treated with appropriate respect and reverence".

91 Usually several weeks or months, as in *Mourish v Wynne* [2009] WASC 85 (delay of two months). However, compare this with *C v M* [2014] SCRFOR 22 (remains stored for three years – see n 165) and the dispute that began as *Tkaczyk v Gallagher* 222 A2d 226 (1965) in which the deceased's body remained frozen for 5 years while her family embarked on litigation that eventually ended up before the US Supreme Court.

who should arrange the funeral is not something that the law should encourage, and not just because of the potential impact on the deceased's family.[92] Societal attitudes also play an important role, as Martin J explained in *Calma v Sesar*:[93]

> The conscience of the community would regard fights over the disposal of human remains such as this as unseemly. It requires that the court resolves the argument in a practical way paying due regard to the need to have a dead body disposed of without reasonable delay, but with all proper respect and decency.[94]

Any unnecessary delay in the disposal of the dead is "repugnant to the sentiment of humanity" and should be avoided where possible.[95]

The existing common law framework, with its emphasis on mechanistic rules of entitlement, has obvious benefits but suffers from two main drawbacks. The first is that courts cannot insist that the deceased's remains are dealt with in a particular way;[96] the person with the strongest legal claim is free to make whatever funeral arrangements they wish.[97] Second, in applying this framework, judges can legitimately ignore the wishes of the dead as well as the opinions or sentiments of family members who do not qualify as personal representatives – including those based on religious and cultural beliefs. Treating these factors as extraneous to the decision-making process ignores their centrality to the actual dispute, though there are signs that this is changing.

V.   Deviating from the common law framework

Courts in this jurisdiction (and elsewhere[98]) are increasingly willing to evaluate the merits of competing claims in funeral disputes, rather than

---

92   Though this is obviously important. As Hale J explained in *Buchanan v Milton* [1999] 2 FLR 844, 854: "Modern methods of refrigeration may make [these proceedings] ... possible but they are certainly unseemly. They delay the proper disposal of the body and the normal processes of grieving while bringing further grief in themselves".

93   (1992) 106 FLR 446, 452.

94   See also the comments of Bryson J in *Privet v Vovk* [2003] NSWSC 1038, [6] (prompt disposal is in the "public interest ... to avoid or minimise scandal and indecency associated with delay").

95   *Burnett v Surratt* 67 SW2d 1041, 1041 (1934).

96   Since it is "not within the power of the court to control the means of disposition" – *Privet v Vovk* [2003] NSWSC 1038, [17] (Bryson J). See also the comments of Patten J in *Scotching v Birch* [2008] EWHC 844 (Ch), [7] ("[t]he court has no power to direct what form anybody's funeral should take").

97   Aside from the basic requirement to notify and consult with family members (see Ch 3, pp 67–68), the only obligation is to ensure that disposal occurs in a dignified and appropriate fashion – see *Sopinka v Sopinka* (2001) 55 OR (3d) 529 and the various cases cited therein.

98   Especially Australia and New Zealand.

simply applying fixed rules.[99] The final outcome may not be all that different; in many cases the executor or presumptive administrator's right will still prevail, though this is no longer an absolute rule. At the very least, decisions made by personal representatives are now subject to greater scrutiny than before.

1. Reviewing the personal representative's decision

Though usually viewed as absolute,[100] a number of cases have suggested that the personal representative's right to possession of the deceased's remains is a prima facie or refutable one.[101] Equally relevant is the court's ability to review funeral arrangements where the decision-maker is acting unreasonably or capriciously.[102] The nature and scope of this power has been described as "formless and ill-defined".[103] However, a refusal to disclose the deceased's final resting place might be capricious,[104] while unreasonable behaviour can be alleged if the personal representative fails to consult other interested

---

99 As Elias CJ pointed out in *Takamore v Clarke* [2012] NZSC 116, [57], "inflexible rules as to entitlement to bury are impractical and may not accord with community expectations of how the disposal of the dead should be arranged in particular cases".

100 See the respective comments in *Re Belotti v Public Trustee* (Supreme Court of Western Australia, 11 November 1993) and *Meier v Bell* (Supreme Court of Victoria, 3 March 1997). See also *Re Boothman, ex parte Trigg* (Supreme Court of Western Australia, 27 January 1999) (executor rule cannot be defeated by a competing claim, even if it seems more meritorious).

101 See for example, *Jones v Dodd* [1999] SASC 125, [37] (presumptive administrator's right is not a "rigid proposition or principle of law"); *Dow v Hoskins* [2003] VSC 206, [43] (any such entitlement is a "prima facie" one); *University Hospital Lewisham Trust v Hamuth* [2006] EWHC 1609 (Ch), [13] (executor "in general" has the right to arrange disposal, but is one of a range of views to consider); and *Threlfall v Threlfall* [2009] VSC 283, [9] (the right of the executor followed by next-of-kin is "not an inflexible rule ... [and] will yield to the circumstances of the case"). However, the common law hierarchy is still important – see for example, *Burrows v Cramley* [2002] WASC 47, [27] (while the administrator rule is not a fixed principle of law, it would be "extremely rare" to depart from this approach) and *University Hospital Lewisham Trust v Hamuth* [2006] EWHC 1609 (Ch), [16] (executor's view has "high priority").

102 Vinelott J in *Grandison v Nembhard* [1989] 4 BMLR 140 suggested that courts could interfere where an executor's decision was "wholly unreasonable", while HHJ Hodge QC in *Laing v John Poyser Solicitors* (High Court (Chancery Division), Manchester Civil Justice Centre (unrep extemporary judgment), 31 July 2012), [23] stated that "the court should not interfere ... unless it can be shown that the [executor's] discretion ... has been exercised dishonestly, capriciously or unreasonably".

103 R Croucher, "Disposing of the Dead: Objectivity, Subjectivity and Identity" in I Freckleton and K Peterson (eds), *Disputes and Dilemmas in Health Law* (Federation Press, 2006), p 335.

104 As in *Re Popp Estate* 2001 BCSC 183 (husband's refusal to tell his dead wife's family what he intended to do with her ashes, 5 years after her death, was capricious).

parties about the funeral arrangements or to take their views on the deceased's funeral into account.[105]

The 2012 decision of the Supreme Court of New Zealand in *Takamore v Clarke*[106] is one of the few substantive appellate judgments on the issue of funeral disputes, and contained (amongst other things) a substantive review of the role of the executor or presumptive administrator. In deciding the funeral arrangements, personal representatives should take account of "the views of those closest to the deceased", including those based on "customary, cultural or religious practices, which a member of the deceased's family ... considers should be observed".[107] However, a majority of the court went further, suggesting that the personal representative's decision could be challenged if "was not appropriate".[108] Once again, the nature and scope of this judicial power is unclear, though the threshold for "re-making" a particular decision is "still likely to be high".[109] There is also the broader question of whether courts outside New Zealand will follow suit, or simply confine themselves to reviewing decisions that are unreasonable or capricious.[110]

In England and Wales, s 116 of the Senior Courts Act 1981[111] offers a more direct route for challenging the personal representative's choice of funeral arrangements. This provision allows the High Court to appoint a different executor or administrator if "it appears ... necessary or expedient" to do so because of "special circumstances",[112] and extends to funeral

---

105   The executor's decision in *Grandison v Nembhard* [1989] 4 BMLR 140 could not be challenged on this basis, nor in *Laing v John Poyser Solicitors* (High Court (Chancery Division), Manchester Civil Justice Centre (unrep extemporary judgment), 31 July 2012) (and upheld on appeal, [2012] EWCA Civ 1240). Note that failure to follow the deceased's own funeral directions may not be unreasonable, especially where there are legal or practical obstacles to doing what the deceased wanted – see *Sullivan v Public Trustee* [2002] NTSC 107.

106   [2012] NZSC 116.

107   *Ibid*, [156], Tipping, McGrath and Blanchard JJ also suggesting that the deceased's own views were another "important consideration". Broadly similar comments were made by Elias CJ (*ibid*, [82]).

108   *Ibid*, [162].

109   New Zealand Law Commission, *The Legal Framework for Burial and Cremation in New Zealand: A First Principles Review* (Issues Paper 34, October 2013), p 196.

110   A view apparently shared by William Young J in one of the dissenting judgments, when suggesting that a personal representative's decision could only be reviewed if it was "unreasonable, capricious or perhaps otherwise erroneous" – [2012] NZSC 116, [171].

111   Formerly the Supreme Court Act 1981.

112   1981 Act, s 116(1). Any such appointment is at the court's discretion – s 116(2).

arrangements.[113] In *Buchanan v Milton*[114] the deceased, one of a generation of 'stolen children',[115] lived in England with his adoptive family but re-established contact with his birth mother and birth family in Australia as a young adult. Following his death in a road accident, both the deceased's adoptive mother and the mother of his infant daughter were intent on cremation but were informed by the deceased's birth mother that this was contrary to Aboriginal beliefs; she asked for the deceased's remains to be returned to Brisbane for burial in his birthplace. Having initially agreed to this, the deceased's adoptive mother changed her mind.[116] Faced with a claim by the birth mother that she should be appointed as her son's administrator under s 116 (and therefore entitled to decide the funeral arrangements), Hale J (as she then was) listed a number of relevant factors – including the deceased's Aboriginal heritage and the importance attached to correct burial procedures, alongside the interests of the deceased's Australian family and those of his infant daughter. While these constituted 'special circumstances' under s 116, it was neither 'necessary' nor 'expedient' to replace the deceased's daughter (represented by her mother and the deceased's adoptive mother) as administrator. A different decision was reached in *Burrows v HM Coroner for Preston*[117] where the deceased's uncle successfully invoked s 116 to oust the deceased's mother as presumptive administrator. The court accepted that there were "powerful arguments" in favour of the deceased's natural mother, as the highest ranking next-of-kin.[118] However, in granting custody of the deceased's remains to his uncle, Cranston J held that the deceased's desire to be cremated (which the uncle was intent on fulfilling) combined with the fact that the uncle and his wife had been the deceased's "psychological parents"[119] for many years constituted 'special

---

113   The decision in *Holtham v Arnold* [1986] 2 BMLR 123 (noted at p 91) must be regarded as wrong on this point. In attempting to replace the deceased's estranged wife as administrator of his estate and take control of the funeral, the deceased's long-term cohabiting partner invoked s 116, arguing that 'special circumstances' included the deceased's unhappy marriage and the fact that the wife had previously filed for divorce. Hoffman J (as he then was) rejected the application on the basis that s 116 only dealt with the administration of estates.

114   [1999] 2 FLR 844.

115   The description used in the media at the time, to describe Aboriginal children who were forcibly removed and integrated into white Australian society

116   The deceased's birth family had claimed that an evil spirit (possibly that of the deceased's late adoptive father) had caused the fatal accident, and that the deceased's daughter should be raised in Australia.

117   [2008] EWHC 1387 (Admin) and noted at p 88.

118   [2008] EWHC 1387 (Admin), [22] citing *R v Gwynedd County Council, ex parte B* [1992] 3 All ER 317.

119   Citing the comments of Baroness Hale in *R (G) Children* [2006] 1 WLR 2305.

circumstances' under s 116, and necessitated the uncle's appointment as presumptive administrator.[120]

2.   Religious and Cultural Imperatives

Religious and cultural imperatives shape attitudes towards death and disposal, yet can be a source of conflict within families.[121] Different factions may have fundamentally different beliefs, resulting in divergent funeral preferences;[122] even shared beliefs can generate disagreements around the extent to which these should be reflected in the funeral arrangements.[123] The impact on the common law priority rankings is unclear, as is the influence of the deceased's own religious and cultural beliefs when determining who gets possession of the body. Neither issue has been comprehensively addressed in English law, beyond the litigation in *Ghai v Newcastle City Council*,[124] which raised a slightly different question: whether Mr Ghai could compel Newcastle City Council to facilitate his wish to be cremated on an open-air funeral pyre, in accordance with his orthodox Hindu beliefs. However, funeral disputes fuelled by competing religious and cultural preferences may arise more frequently, as multi-faith or multi-ethnic families have different views about what happens to the body on death. Again, English courts could look to Australia for guidance, where the issue has driven numerous disputes involving indigenous peoples.

Initially, Australian courts were of the opinion that spiritual and cultural factors need not influence the deceased's funeral arrangements. For example, in *Calma v Sesar*[125] the deceased and his parents were of Aboriginal descent; the mother had arranged for a Catholic burial in Darwin, while the father had organised burial in the family plot at Port Hedland where the deceased

---

120   In reaching a final decision on s 116, Cranston J was also influenced by Article 8 of the European Convention on Human Rights – see p 109 and Ch 5, pp 139–140. See also *Ibuna v Arroyo* [2012] EWHC 428 (Ch), which involved the body of a Filipino congressman who died while receiving medical treatment in London; the dispute was between the deceased's estranged wife, and his long-term partner (supported by one of the deceased's daughters, who was also executrix under his last will). Although both sides intended to repatriate the remains to the Philippines, they disagreed on where the deceased's wake would be held and who should make the funeral arrangements. The court accepted that there was no evidence that the deceased wanted his body to be disposed of by his estranged wife, and that the deceased's partner was intent on giving effect to his expressed wishes. This latter point combined with the fact that the deceased's executrix was willing to allow his partner to make the funeral arrangements constituted 'special circumstances' under s 116 and resulted in Smith J appointing the partner as a joint personal representative of the estate.

121   See pp 88–90.

122   As in *Hunter v Hunter* (1930) 65 OLR 586.

123   As in *Doherty v Doherty* [2006] QSC 257 and *Takamore v Clarke* [2012] NZSC 116.

124   [2009] EWCH 978 and *Ghai v Newcastle County Council* [2010] EWCA Civ 59, discussed in Ch 2, pp 44–45.

125   (1992) 106 FLR 446.

had been born. Relatives on both sides of the family were deeply divided over the issue, the deceased's paternal grandfather, arguing that the deceased should be buried in his Aboriginal homeland. Refusing to be swayed by these and other "imponderables",[126] Martin J concluded:

> [I]ssues such as these could take a long time to resolve if they were to be properly tested by evidence in an adversary situation. A legal solution must be found; not one based on competing emotions and the wishes of the living, except insofar as they reflected a legal duty of right. That solution will not embrace the resolution of possibly competing spiritual or cultural values.[127]

Agreeing with this statement, Ashley J in *Meier v Bell*[128] stressed the need to adhere to established legal principles when resolving funeral disputes; courts could not depart from these simply to accommodate competing claims, whether "founded on matters religious, cultural, or of some other description".[129]

Both judgments were heavily criticised, not just for their refusal to take such matters into account but for their rigid adherence to a common law framework that did not reflect the fundamentally different kinship networks that exist within indigenous communities, where the emphasis is on collateral (as opposed to lineal) relationships and where extended notions of kinship do not necessarily match conventional Western constructs of family.[130]

---

126 *Ibid*, 452.
127 *Ibid*.
128 (Supreme Court of Victoria, 3 March 1997). The facts were noted at p 89.
129 At p 6 judgment. Similar views have been expressed in a number of Canadian cases. For example, in *Saleh v Reichert* (1993) 104 DLR (4th) 384, 391 the court stressed that "religious law had no bearing on the case" and that there were "only legal obligations" (citing virtually identical comments in *Abeziz v Harris Estate* [1992] OJ No 1271). Whereas the position has now shifted in Australia (see below), Canadian courts have apparently maintained their original stance – see for example, *Lajhner v Banoub* [2009] OJ No 1327, [29] (religious views or beliefs "are not a factor that the court may take into consideration") and *Buswa v Canzoneri* 2010 ONSC 7137 (deceased's daughter could cremate her father's remains, despite the deceased's sisters alleging that cremation was inconsistent with First Nations teachings).
130 The fact that mapping a succession law framework onto funeral disputes may not match the reality of the deceased's kinship networks has already been noted (see p 94), but is even more apparent in indigenous communities where there are much lower rates of will-making, and patrilineal and blood-based definitions of 'family' under intestacy statutes do not reflect kinship patterns – see P Vines "Consequence of Intestacy for Indigenous People in Australia: The Passing of Property and Burial Rights" (2004) 8 *Australian Indigenous Legal Reporter* 1. See also the comments of Cummins J in *Dow v Hoskins* [2003] VSC 206, [45]: "[T]he statutory regimes for intestacy are all based on a non-Aboriginal view of family and kinship. This creates a serious mismatch between the legislative scheme and Aboriginal cultural expectations".

However, the decision in *Jones v Dodd*[131] – involving a dispute between the deceased's former de facto partner[132] (and mother of his two young children), and the deceased's father – signalled a different approach. The initial decision[133] favoured the father as the deceased's next-of-kin,[134] who was consequently free to bury his son within the geographical area of his family's tribe, according to Aboriginal law and custom. Affirming this decision, Perry J (who delivered the judgment of the Full Court) refuted claims that the deceased's partner could claim letters of administration on behalf of her children because the presumptive administrator rule was not "a rigid proposition or principle of law".[135] Perry J also criticised the fixation with strict legal rights in both *Meir* and *Calma*:

> [T]he proper approach in cases such as this is to have regard to the practical circumstances, which will vary considerably between cases, and the need to have regard to the sensitivity of the feelings of the various relatives and others who might have a claim to bury the deceased, bearing in mind also any religious, cultural or spiritual matters which might touch upon the question.[136]

Judges should not be influenced by "expressions of pure emotion or arbitrary expressions of preference";[137] however, "proper respect and decency"[138] required them to take account of religious and cultural values in resolving funeral disputes.

Since *Jones*, Australian courts have been more cognisant of religious and cultural imperatives, even if the final outcome has not been significantly different. For example, in *Dow v Hoskins*[139] the court accepted the importance that Aboriginal custom and culture placed on burying the deceased beside his dead father; and while cultural imperatives were of "primary importance",[140] these did not displace the common law right of the deceased's partner as next-of-kin and mother of two of his children to bury

131   [1999] SASC 125. The facts were noted briefly at p 89.
132   The couple had not lived together for 10 years.
133   [1998] SASC 6769.
134   According to common law degrees of kinship as opposed to degrees of kinship under Aboriginal customary law, which emphasise the father's role as head of the family. Both would have produced the same result here.
135   [1999] SASC 125, [37].
136   *Ibid*, [51].
137   *Ibid*, [54].
138   *Ibid*, [53].
139   [2003] VSC 206.
140   *Ibid*, [38], citing similar comments in *Sullivan v Public Trustee for Northern Territory* [2002] NTSC 107.

him elsewhere.[141] The deceased's own beliefs have also featured in a number of cases, with courts dismissing claims from other family members where the deceased did not espouse the same spiritual values or maintain a traditional lifestyle, and the person entitled to custody of the deceased's remains intends to organise the funeral based on what the deceased wanted.[142]

While these cases suggest a more nuanced approach, judges still tend to apply the common law framework in legalistic matter; religious and cultural values have much greater prominence, yet often fail to displace the rights of the executor or presumptive administrator when it comes to deciding the fate of the deceased's remains.[143] However, human rights arguments may strengthen the case for courts to deviate from the existing framework.[144]

3. Human rights arguments

Human rights are engaged in funeral disputes, because "the disposal of human remains touches on matters of human identity, dignity, family, religion and culture".[145] In England and Wales, these basic protections are enshrined in the European Convention on Human Rights, now part of domestic law throughout the UK since October 2000.[146] There is growing evidence in both the jurisprudence of the European Court of Human Rights ('ECtHR')[147] and domestic courts that Convention rights can influence the

---

141 Identical outcomes were reached in *Ugle v Bowra & O'Dea* [2007] WASC 82 (deceased's de facto partner entitled to decide the funeral arrangements, despite objections from the deceased's son on the basis of cultural values; while these are "all important, or highly important at the least" (*ibid*, [10]), the coroner was entitled to release the deceased's remains to his partner) and in *Garlett v Jones* [2008] WASC 292 (deceased's long-term partner entitled to possession of the remains as presumptive administrator, despite objections from the deceased's father and other relatives; cultural traditions were important and "might ... be decisive" in another case (*ibid*, [55]), but did not displace the partner's entitlement here).

142 For example, in *Re Dempsey* (Supreme Court of Queensland, 7 August 1987) the court granted custody of the deceased's remains to her de facto partner, who intended to have a religious ceremony in the Methodist church that the couple had attended for years before burying the deceased in a cemetery close to where the couple had lived, according to her wishes. The deceased's mother (supported by other family members) had insisted her daughter be returned to her birthplace for burial according to Aboriginal beliefs. See also *Minister for Families and Communities v Brown* [2009] SASC 86 and *Spratt v Hayden* [2010] WASC 340.

143 Different sides of a family can observe distinct cultural traditions, which point towards fundamentally different outcomes in funeral disputes. Here, courts can simply resort to established legal principles – or occasionally devise an alternative solution, depending on the facts before them (see *State of South Australia v Smith* [2014] SASC 64).

144 See pp 112–114.

145 *Takamore v Clarke* [2012] NZSC 116, [82] (Elias CJ).

146 Under the Human Rights Act 1998.

147 Section 2(1)(a) of the 1998 Act instructs domestic courts to take ECtHR jurisprudence into account when determining a question that has arisen in connection with any Convention right.

fate of the recently dead.[148] For example, specific Convention rights may strengthen the case for upholding the deceased's funeral directions, something that is analysed in the next chapter.[149] For now, the focus is on using human rights to challenge the personal representative's choice of funeral arrangements.

(a)   Family preferences and Article 8

Harris has argued that the right to private and family life under Article 8(1)[150] includes the process of bereavement and execution of funeral rites,[151] something that has been reflected in ECtHR jurisprudence to date. Although not dealing with funeral disputes, the decisions in *Pannullo and Forte v France*[152] and *Girard v France*[153] both suggest a legitimate family interest in the disposal of a loved one. In the former, damages were awarded to the parents of a 4-year-old child, after a delay of seven months between French authorities performing an autopsy and returning the child's body to her parents. The court held that the authorities had not struck the right balance between the applicants' right to private and family life and the need to conduct an effective investigation into the child's death, in violation of Article 8. In *Girard*, the applicants' daughter had died in suspicious circumstances. Following a post-mortem, the body was returned to the applicants for burial; it was exhumed three months later and new investigative samples taken,[154] which eventually led to a man being convicted of the deceased's murder. Following repeated requests from the applicants, a formal decision to return the samples was issued in March 2004 but the samples were not actually returned until late July. While taking DNA and other material from a corpse for forensic purposes does not infringe Article 8,[155] the court held that the right to bury one's relatives is protected by Article 8(1). Like *Pannullo*, the four-month delay in returning the samples after the formal

---

148   See the various cases discussed below, as well as the specific comments of Cranston J in *Burrows v HM Coroner for Preston* [2008] EWHC 1387 (Admin), [1]: when determining who has the right to make funeral arrangements, the "common law provides a bright line rule although the issue which has arisen ... is whether that bright line has been blurred through the lens of ... the European Convention on Human Rights".

149   See Ch 5, pp 139–144.

150   See generally P van Dijk, F van Hoof, A van Rijn and L Zwaak (eds), *Theory and Practice of the European Convention on Human Rights* (Intersentia, 4th edn, 2006), ch 12 and C Grabenwarter, *European Convention on Human Rights – Commentary* (CH Beck, 2014), pp 183–233.

151   J Harris, "Law and Regulation of Retained Organs: The Ethical Issues" (2002), 22 *Legal Studies* 527, p 545.

152   (Application no. 37794/97, ECHR, 30 January 2002).

153   (Application no. 22590/04, ECHR, 30 June 2011).

154   The body was reburied immediately.

155   Citing *Estate of Kresten Filtenborg Mortensen v Denmark* (Application no. 1338/03, ECHR, 15 May 2006).

decision had been made infringed the applicants' rights to private and family life.[156]

Domestic courts have also acknowledged that Article 8(1) is engaged in cases involving disposal of the dead. For example, the Article 8(1) rights of the deceased's mother and uncle (as well as the deceased himself) featured strongly in *Burrows v HM Coroner for Preston*,[157] while the same judge in *Ghai v Newcastle City Council*[158] accepted that "the respect accorded to private (and indeed family) life ... can extend to aspects of funeral arrangements ... because they are so closely related to a person's physical, psychological or familial identity".[159] These (and other judicial comments) were endorsed by Lord Brodie in *C v Advocate General for Scotland*.[160] In this case, the deceased was a soldier who had died while on a training mission in Germany; his widow wanted to bury him with full military honours in Forfar, while the deceased's mother (who was also executrix under her son's will) wanted to bury him some 45 miles away in the Wemyss, Fife.[161] The widow's petition for judicial review of the decision to release the deceased's remains to his mother as executrix was dismissed as incompetent by the Court of Session,[162] though the underlying arguments were discussed at length. A key

---

156 See also *Ploski v Poland* (Application no. 26761/95, ECHR, 12 December 2002) and *Giszczak v Poland* (Application no. 40191/08, ECHR, 29 November 2012) (granting compassionate leave to a prisoner to attend a close relative's funeral engaged Article 8, though the latter case stressed that this was not an unconditional right). The funeral arrangements for the applicant's child were central to the decision in *Hadri-Vionnet v Switzerland* (Application no. 55525/00, ECHR, 14 May 2008) (applicant had given birth to a stillborn baby while in a centre for asylum seekers, and was unable to attend the funeral; the fact that the body was taken away, and buried without her knowledge in a communal grave violated the applicant's Article 8 rights), and in *Maric v Croatia* (2015) EHRR 2 (Croatian hospital had breached the complainant's Article 8 rights by disposing of the body of his stillborn child as clinical waste). Meanwhile the applicants' inability to pay their last respects at a loved one's funeral was a key feature of *Maskhadova and Others v Russia* and *Sabanchiyeva and Others v Russia* (Application nos. 18071/05 and 38450/05, ECHR, 6 June 2013) (Russian authorities' refusal to return the bodies of the Chechen separatist President and insurgents to their families, and a ban on disclosing the location of the graves, breached Article 8).

157 [2008] EWHC 1387 (Admin).

158 *Ghai v Newcastle City Council* [2009] EWCH 978, [141].

159 However, the claimant ultimately failed to establish a breach of Article 8 for reasons noted in Ch 5, n 107. See also the decision of the English Court of Appeal in *Esfandiari v Secretary of State for Work and Pensions* [2006] HRLR 26 (the need for a decent funeral was a basic requirement of human dignity, whether from the point of view of the individual or from that of the family) and the comments of Patten J in *Scotching v Birch* [2008] EWHC 844 (Ch), [13] ("it seems to me that the right to organise a funeral and to stipulate the place of burial is probably sufficient to engage Art.8").

160 2012 SLT 103.

161 The deceased had appointed his mother as joint executor with his brother, following erroneous advice from the army that he could not appoint his wife as both executrix and sole beneficiary. The deceased's brother gave his mother sole authority to act.

162 The decision to transfer the deceased's remains to his mother was not amenable to judicial review.

feature was an alleged interference with the widow's Article 8(1) rights, in being deprived of the opportunity to arrange her husband's funeral and to determine where he was interred. Lord Brodie was in no doubt that these rights were engaged:

> [Article 8] may be engaged by an act of the state which touches on a family's freedom to determine what may be described as the place and modalities of burial of a deceased member of that family, to have custody of the body for the purpose of burial and to participate in any funeral ceremony.[163]

On the facts, any interference would be justified in accordance with Article 8(2),[164] though the deceased's widow eventually succeeded in civil proceedings against his mother and a full military burial was held in Forfar.[165]

While most of these cases concerned alleged infringements by the State or its various bodies, the doctrine of horizontal effect means that these and other Convention rights could be raised in private law disputes around the fate of the dead. Creative use of Article 8(1) and its guarantee of a right to respect for private and family life might enable the deceased's immediate family or those in a close personal relationship with the deceased to contest the personal representative's decision.[166] For example, where a spouse (acting as executor or presumptive administrator) is refusing to inter the body in the family burial plot, this could constitute an interference with the Article 8(1) rights of the parents and siblings, as well as other family members.[167] The fact that Article 8(1) extends beyond traditional domestic law classifications of family[168] could also benefit those omitted by intestacy-based kinship designations. For example, a cohabiting partner might oppose the common law entitlement of the deceased's parents (as joint presumptive administrators) to make their son's funeral arrangements, especially where the parties have

---

163   2012 SLT 103, [13].
164   This aspect of the case is discussed below.
165   *C v M* [2014] SCRFOR 22, the presiding sheriff granting custody of the deceased's remains to his widow on the basis that an executor does not have priority over close family members in Scottish law, and also because of the mother's failure (as executor) to take account of the wishes of her late son (who had apparently wanted to be buried in Forfar) and his widow. The deceased's remains were stored in a London mortuary for three years, while attempts were made to solve the dispute through mediation and court proceedings. However, trouble flared at the funeral – "Police Called to Soldier's Funeral After Family Feud", *BBC News Online* (9 January 2015) www.bbc.co.uk/news/uk-scotland-tayside-central-30750749 (accessed 30 September 2015).
166   Although if opposition to the funeral arrangements is based on religious or cultural beliefs, the applicant could invoke Article 9.
167   While more distant family ties may not attract the protection of Article 8(1), the term 'family' covers a range of relationship – Grabenwarter (n 150), p 195.
168   See generally van Dijk *et al* (n 150), pp 690–695 and the cases cited therein.

different views on disposal of the body.[169] A step-child of the deceased (especially one who lived with their step-parent) could also rely Article 8(1) in the event of a funeral dispute,[170] as could foster parents.[171]

However, even if such rights are engaged, they are subject to certain limitations. As with other Convention rights, those guaranteed by Article 8(1) are defeasible and legitimate interference is permissible under Article 8(2) where this is "in accordance with the law" and deemed to be necessary in the interests of "public safety" or "for the protection of health or morals". For example, in *Scotching v Birch*[172] the mother's Article 8(1) rights did not override the common law forfeiture rule and its underlying public policy rationale that a person should not benefit from their own crime. Other restrictions may also come into play. For example, a husband's desire to have his wife's remains preserved and permanently displayed in their living room (arguably within the scope of Article 8(1)) might be defeated by his wife's family on the grounds of public health and human dignity,[173] despite the husband's entitlement as personal representative.[174] Article 8(1) rights are also subject to practical limitations; for example, a desire to bury the deceased in family grave (again, a possible aspect of private and family life) might be impossible to fulfil because the plot is full.

Before leaving Article 8, two other points must be addressed. First, because funeral disputes inevitably involve warring family factions, there will be valid Article 8 arguments on both sides. This has been acknowledged in a number of cases, including *C v Advocate General for Scotland*.[175] In deciding that any interference with the widow's Article 8(1) rights would be both proportionate under Article 8(2) and necessary to protect the rights of another family member (the deceased's mother), Lord Brodie remarked:

> The [widow's] complaint is that the transfer of custody and therefore control of the body to the [mother] ... adversely affects the [widow's] rights to family life. However, were the transfer to be ... to the [widow], the [mother] ... could complain that her rights to family life have been adversely affected. Compromise appears impossible. A state of impasse

169 Cohabitants constitute a 'family' for the purposes of Article 8 (see *Johnston and others v Ireland* (1987) 9 EHRR 203).
170 Since 'belonging to the same household' often denotes family ties – van Dijk *et al* (n 150), p 694.
171 See the reference to *Re LL (Application for Judicial Review)* [2005] NIQB 83 at n 80.
172 [2008] EWHC 844 (Ch) and noted at p 91.
173 Concerns for human dignity, while not expressly mentioned in Article 8(1), are implicit in this provision and throughout the Convention.
174 Although there is no real risk to public health from a properly preserved body, and the fact that a corpse is on display might not be regarded as immoral or contrary to human dignity (though there is a difference between agreeing to be put on display, and simply having one's remains posed in this manner – whether in a public or private place).
175 2012 SLT 103.

has been reached. The first respondent[176] has been put into a position where he has to exercise a choice as between near relatives each of whom has rights protected by art.8.1. To respect the rights of both he cannot avoid favouring one.[177]

Cranston J had been placed in a similar position in *Burrows v HM Coroner for Preston*[178] when assessing the respective Article 8(1) rights of the deceased's mother and paternal uncle.[179] In these circumstances, the court must "focus intensely on the comparative importance of the different rights being claimed, and ... balance those competing rights so as to minimise the interference with each to the least possible extent".[180] Second, it seems inconceivable that judges would ignore the deceased's own funeral preferences when evaluating rival family claims under Article 8. For example, would a court really allow a personal representative to proceed with cremation where the deceased opposed this method of disposal and wanted to be buried, or *vice versa?* As in *Burrows*, courts may look more favourably on the Article 8(1) rights of those family members who are willing to do what the deceased wanted.

(b)   Religious and cultural imperatives and Article 9

Religious and cultural values could also have greater weight under the Convention, as courts balance these against the prevailing legal norms around disposal of the dead. Article 9(1) guarantees a right to freedom of thought, conscience and religion, and to manifest a particular religion or belief in public or in private.[181] The provision is quite broad in scope, given that the words 'religion or belief' have been interpreted widely by both the

---

176   The Advocate General for Scotland, representing the Secretary of State for Defence (since the decision to release the deceased's body to his mother was made by the Ministry of Defence).

177   2012 SLT 103, [68]. Article 8(2) contemplates legitimate interference where necessary for the "the protection of the rights and freedoms of others".

178   [2008] EWHC 1387 (Admin).

179   The uncle's Article 8(1) rights were based on *Boyle v UK* (1994) EHRR 179, while *Berrehab v The Netherlands* (1988) 11 EHRR 322 established that the mother's family life with her son had not been broken by long periods of physical separation and limited contact.

180   [2008] EWHC 1387 (Admin), [21]. See also *Scotching v Birch* [2008] EWHC 844 (Ch) (while the mother had Article 8 rights, these had to be balanced against the father's equivalent rights) and *Re LL (Application for Judicial Review)* [2005] NIQB 83 discussed in Ch 3, p 65 (even if the mother's Article 8 rights survived the freeing order for adoption (which the court doubted), any alleged interference with those rights was proportionate under Article 8(2); and granting the mother's request would infringe the Article 8(1) rights of the child's foster parents, which were also engaged here).

181   See generally van Dijk *et al* (n 150), ch 13 and Grabenwarter (n 150), pp 234–250, as well as C Evans, *Freedom of Religion under the European Convention on Human Rights* (OUP, 2001).

ECtHR and domestic courts; as such, they are not limited to traditional religions, but embrace a wide range of convictions and philosophies, as well as non-religious beliefs such as pacifism.[182] Like Article 8(1), Article 9(1) rights are engaged in issues around death and disposal of the dead. A small number of exhumation cases have addressed this point,[183] while Article 9(1) was also central to the applicant's case in *Ghai v Newcastle City Council*[184] when asserting a right to open-air cremation in accordance with Hindu beliefs and traditions.

As before, creative use of Article 9(1) might enable the deceased's relatives to contest the personal representative's choice of funeral arrangements where the former insist that a particular method of disposal or funeral rite is essential to their religious or cultural values – for example, where the deceased is to be cremated but the family argue that spiritual beliefs mandate burial, or there are issues around the proposed grave site where the deceased was a humanist yet is to be interred in consecrated ground, or where the personal representative is not intending to return the body to the deceased's spiritual homeland. However, even if raised in funeral disputes, alleged Article 9(1) violations are subject to two caveats. First, while generous in determining what constitutes a religion or belief, Convention jurisprudence is not so liberal in determining whether a particular activity manifests a religion or belief.[185] Certain funeral practices may not be classed as an expression of spiritual or cultural beliefs under Article 9(1).[186] Second, Article 9(2) permits state interference with the right to freedom of thought, conscience and religion where such limitations are "prescribed by law" and deemed to be necessary in the interests of "public safety", or for "the protection of public order, health or morals". For example, cremating the deceased on a funeral pyre would be impermissible under Article 9(2), even

182   *Re Crawley Green Road Cemetery* [2001] 2 WLR 1175 citing *Arrowsmith v UK* (1978) 3 EHRR 118 and *Kokkinakis v Greece* (1993) 17 EHRR 397. See also the views of the House of Lords in *R (Williamson) v Secretary of State for Employment and Education* [2005] 2 AC 246.

183   See *Re Durrington Cemetery* [2000] 3 WLR 1322 and *Re Crawley Green Road Cemetery* [2001] 2 WLR 1175, both discussed at Ch 7.

184   [2009] EWCH 978 and see in Ch 5, pp 141–142.

185   See van Dijk *et al* (150), pp 761–764, as well as the respective judgments of the House of Lords in *R (Williamson) v Secretary of State for Employment and Education* [2005] 2 AC 246 and *R (on the application of Begum) v Denbigh High School* [2007] 1 AC 100 and the various authorities cited therein.

186   For example, an individual wanting to have his ashes scattered on his own land (*X v Federal Republic of Germany* (1981) 24 D & R 137 and discussed in Ch 5, p 141) or wanting to place a photograph on his daughter's memorial stone (*Jones v UK* (Application no. 42639/04, ECHR, 13 September 2005) and discussed in Ch 8, pp 220–221).

if motivated by a particular religion or belief,[187] while practical considerations would place further restrictions on the exercise of Article 9(1) rights.[188]

Finally, as with Article 8, funeral disputes involving different members of the deceased's family may give rise to conflicting Article 9(1) arguments where there are different religions or beliefs on both sides – a distinct possibility given increasing rates of inter-marriage within multi-faith, multi-ethnic societies. In these circumstances, courts would have to balance the competing claims, and interference with one person's rights under Article 9(1) might be necessary to protect the rights of another family member.[189] More importantly, the deceased's own beliefs are likely to be a major factor when judges are addressing Article 9(1) arguments. For example, it is unlikely that courts would allow a personal representative to proceed with cremation where the deceased actively opposed this method of disposal on religious or cultural grounds, or that they would uphold the Article 9(1) rights of a family member (perhaps where that person insists on a religious funeral ceremony) where the deceased did not espouse or had actively rejected the same teachings during his/her lifetime.

VI.   Promoting statutory reform

Should the legal resolution of funeral disputes still be based on an outdated succession law paradigm, which also fails to take account of the wishes of the deceased? The Canadian provinces of Alberta,[190] Saskatchewan[191] and British Columbia[192] have introduced detailed statutory frameworks setting out who has the legal right to dispose of the deceased's remains (including post-cremation ashes), though the most comprehensive provisions exist in the

---

187   As decided at first instance in *Ghai v Newcastle City Council* [2009] EWCH 978.

188   Similar to those noted in respect of Article 8(1). For example, a family's wish to bury the deceased in a particular churchyard (arguably within the scope of Article 9(1)) being defeated because the ground is full, or there being insufficient funds in the deceased's estate to repatriate the body to the deceased's spiritual homeland.

189   Article 9(2) contemplates legitimate interference where necessary for the "the protection of the rights and freedoms of others". In *H v Norway* (1992) 73 D & R 155 the applicant had failed to prevent his pregnant partner from undergoing a termination, and subsequently requested the hospital to hand over the remains of the aborted foetus to him for interment in accordance with the Jewish faith. When this was refused, the applicant claimed a violation of Article 9(1). The Commission affirmed that the right to manifest one's religion or beliefs is not unlimited when it violates the rights of others. Granting custody of the foetal remains to the applicant would have been degrading to his partner as the person primarily concerned with the pregnancy and its continuation or termination.

190   General Regulation (Funeral Services Act) Alta Reg 226/98, s 36 and General Regulation (Cemeteries Act) Alta Reg 249/1998, s 11.

191   Funeral Services Act 1999, s 91.

192   Cremation, Interment and Funeral Services Act 2004, s 5.

latter jurisdiction.[193] Similar legislative schemes are currently being considered in both Queensland[194] and in New Zealand,[195] strengthening the case for analogous reforms in England and Wales as increasing numbers of funeral disputes expose the inherent weaknesses in the common law framework.

These statutory regimes only apply where a dispute arises over disposal of the deceased's remains;[196] and while the common law rules are replaced entirely, the legislation mimics them in certain ways. Again, the central theme is a ranking of entitlement, which favours a specific individual or group of individuals within the same class (the 'designated decision-maker(s)') in descending order.[197] Decision-making powers are extensive, going beyond the form of disposal to attendant funeral rites and the type of ceremony as well as the deceased's final resting place.[198] The result is a clear and accessible legal framework, which allows disputes to be resolved without protracted litigation in keeping with the basic principle of respect for the dead while minimising the amount of emotional damage being inflicted on the living.[199] However, the British Columbia statute deviates significantly

193 Similar statutory provisions exist in many US states – see AM Murphy, "Please Don't Bury Me Down in that Cold Cold Ground: The Need for Uniform Laws in the Disposition of Human Remains" (2007) 15 *Elder Law Journal* 400 (especially Appendix A).

194 Queensland Law Reform Commission, *A Review of the Law in Relation to the Final Disposal of a Dead Body* (Report No 69, December 2011), Ch 6.

195 New Zealand Law Commission (n 109), Ch 16. The Scottish Government is also contemplating a different priority ranking for adult deaths, which would move completely away from the personal representative and focus on the deceased's 'nearest relative' as defined in s 50 of the Human Tissue (Scotland) Act 2006 Scottish Government, *Consultation on a Proposed Bill Relating to Burial and Cremation and Other Related Matters in Scotland* (January 2015), [35]–[40].

196 Surviving relatives are free to make their own arrangements, where they are all in agreement.

197 The basic expectation to consult with the deceased's wider family (see Ch 3, pp 67–68) has not been transmuted into a legal obligation under any of these statutory frameworks, and the Queensland Law Reform Commission has adopted a similar stance. According to the Commission, it would be "difficult to define the nature and extent of consultation required and the range of persons who should be consulted", while also increasing the "time and complexity involved in the decision-making process, and open[ing] up additional points of dispute" – Queensland Law Reform Commission (n 194), p 200. In contrast, the New Zealand Law Commission supports a statutory duty to consult – New Zealand Law Commission (n 109), pp 215–216.

198 Though not commemoration of the dead – see *Wiebe v Bronstein* [2013] BCSC 1041 (s 5 of the Cremation, Interment and Funeral Services Act 2004 only deals with the disposition of human remains or ashes; it does not deal with the placement of memorials). Likewise, issues such as exhumation and organ donation are still governed by their own distinct statutory provisions.

199 This also makes the task easier for courts, as well as funeral directors who are equally sensitive to the potential for disputes (and likely to be involved from the outset).

from the common law approach when it comes to the deceased's written funeral preferences, by making these binding on the designated decision-maker.[200]

Devising a list of designated decision-makers, and in what order, is important here. Each of the Canadian statutes listed above prioritises the executor named in the deceased's will – apparently because, having been chosen by the deceased, this individual knows (or should know) the deceased's funeral preferences.[201] If there is no executor,[202] decision-making powers devolve on specified adult relatives. Each Canadian statute favours the deceased's spouse or partner, followed by a child or children of the deceased. After that however, they all differ slightly in their ranking of relationships and the respective positions of grandchildren, parents, siblings as well as nieces and nephews and other relatives of the deceased vis-à-vis each other.[203] Of course, individual priority rankings are a matter for each jurisdiction. However, an advantage of abandoning the common law approach in favour of a statutory scheme is that law-makers can reflect more contemporary notions of 'family' and 'kinship' when prioritising relationships with the deceased – breaking free of an intestacy law framework that clings to a nuclear family model and has traditionally been reluctant to embrace change.[204] If English law were to follow suit, for example, an obvious adjustment would be to give cohabitants the same standing as spouses and civil partners in funeral disputes since the former do not currently have any rights on intestacy.[205] Other contemporary family relationships could also be included – for example, adult step-children could be granted certain decision-making powers over the fate of a step-parent's remains and *vice versa*.[206] Yet closer inspection of the different priority rankings listed in the Canadian statutes suggests that, beyond marriage and

---

200   Cremation, Interment and Funeral Services Act 2004, s 6. This is an important policy issue and one which is discussed at length in Ch 5.

201   Cremation, Interment and Funeral Services Act 2004, s 5(1)(a) in British Columbia; General Regulation (Funeral Services Act) Alta Reg 226/98, s 36(2)(a) and General Regulation (Cemeteries Act) Alta Reg 249/1998, s 11(2)(a) in Alberta; and Funeral Services Act 1999, s 91(1)(a) in Saskatchewan. The executor would also top any new priority ranking in Queensland (Queensland Law Reform Commission (n 194), pp 143–144) though possibly not in New Zealand where there are doubts about retaining the executor rule (New Zealand Law Commission (n 109), chs 14–15).

202   Or the executor declines to act.

203   Likewise, s 4E(3) of the draft Cremations and Other Legislation Amendment Bill 2011 in Queensland proposes a slightly different pecking order than the various Canadian statutes.

204   See for example, FH Foster, "The Family Paradigm of Inheritance Law" (2001) 80 *North Carolina Law Review* 199.

205   Change has been mooted (see Law Commission for England and Wales, *Intestacy and Family Provision Claims on Death* (Law Com No 331, 2011), Pt 8), but is not imminent.

206   Though presumably with subordinate decision-making powers of the deceased's natural/adoptive children or natural/adoptive parents.

marriage-like relationships, there is still a fixation with ties of consanguinity and lineal descent within what are still fairly traditional constructs of family.[207]

An ordering of entitlement that embraces all of the relationships that a person valued in life would be virtually impossible to achieve. Questions would also be asked about where a particular relationship should be positioned within any statutory ranking; even if 'family' was defined more broadly under a new bodily disposal statute, it would still be conceptualised as a hierarchy of individuals based on proximity of relationship to the deceased. Faced with these difficulties, perhaps the only workable solution is to confine the statutory hierarchy to a discrete number of fairly conventional and easily identifiable family relationships, which reflect the dominant social model. A more expansive approach could be achieved by following the Canadian example and enacting a residual provision that confers decision-making powers on an adult person who does not fall within the other categories yet had a "personal or kinship relationship with the deceased".[208] This contemplates a range of possibilities, from foster parents on the death of a minor, to members of the deceased's step-family, a fiancé(e), close friends and possibly even an unpaid carer – albeit that these individuals are located at the bottom of the pecking order, and consequently well down the order of entitlement.[209] Alternative family forms aside, the fact that existing bodily disposal statutes are premised on normative constructs of family may not sit easily with certain ethnic groups where there are more collective and extended notions of kinship, as well as different views on where certain individuals rank within a family hierarchy. Again, incorporating all the potential variants within a generic legislative framework would be virtually impossible to achieve.

Each of the Canadian statutes also provides that, if the designated decision-maker is unable or unwilling to act, the legal right to decide the deceased's funeral arrangements passes to the person who is next in priority. Disinclination is straightforward enough, but other factors can render someone incapable of assuming this role – for example, where the designated

---

207   See Foster (n 3) where the author questions the extent to which legal frameworks (both common law and statutory) incorporate relationships outside the traditional family paradigm. Similar issues were raised in TK Hernández, "The Property of Death" (1999) 60 *University of Pittsburgh Law Review* 971.

208   This is the exact wording used in s 5(1)(k) of the Cremation, Interment and Funeral Services Act 2004 in British Columbia and is also being proposed in Queensland under s 4F(3)(k) of the draft Cremations and Other Legislation Amendment Bill 2011. The Alberta provisions refer to a person "having some relationship with the deceased not based on blood ties or affinity" (General Regulation (Funeral Services Act) Alta Reg 226/98, s 36(2)(k) and General Regulation (Cemeteries Act) Alta Reg 249/1998, s 11(2)(k)), while broadly similar wording can be found in the Funeral Services Act 1999, s 91(1)(k) in Saskatchewan.

209   However, there may still be scope for challenging the designed decision-maker's entitlement – see below.

decision-maker is away and cannot be contacted in time to organise the funeral. Being responsible for the deceased's death would also be a disqualifying event, the decision in *Scotching v Birch*[210] suggesting that the common law forfeiture rule, which results in disinheritance, maps onto the legal right to possession of the deceased's remains as well.[211] Any bodily disposal statute could include a specific provision to this effect when dealing with family killings,[212] even if it is rare for the issue of criminal responsibility to be resolved before disposal of the deceased's remains.[213]

A more fundamental issue is whether someone lower down (or even outside) the statutory priority ranking should be able to challenge the designated decision-maker's authority. Once again, the British Columbia statute leads the way, with a specific provision allowing a "person claiming that he or she should be given the sole right to control the disposition" of the deceased's remains to apply to court "for an order regarding that right".[214] A default discretionary provision of this nature has obvious benefits, allowing courts to deviate from the statutory ranking if a convincing case is made (beyond the applicant simply preferring different funeral arrangements). Applications could be considered where the designated decision-maker is ignoring the deceased's own disposal preferences, or is proposing something that is at variance with the deceased's cultural and

---

210   [2008] EWHC 844 (Ch) and discussed at p 91.

211   As a result, the mother (and joint highest ranking next-of-kin) in this case forfeited any potential right to inherit from her son's estate (had one existed), and could not have applied for letters of administration; her estranged husband and father of the boy was the presumptive administrator and entitled to make the funeral arrangements. However, a mere suspicion of having been in some way responsible for or contributed to the deceased's death is not enough – see *Joseph v Dunn* [2007] WASC 238 and *AB v CD* [2007] NSWSC 1474.

212   "Forfeiture is essential to protect not only a decedent from being victimized a second time ... but also the victim's grieving survivors from being rendered powerless and unable to control the final resting place of their loved one" – ME Bremenstul, "Victims in Life, Victims in Death: Keeping Burial Rights Out of the Hands of Slayers" (2013) 74 *Louisiana Law Review* 213, p 215. The article provides an in-depth analysis of so-called 'slayer statutes' in the US, increasingly extended to prevent slayers from arranging their victims' funerals, though with varying levels of protection between states.

213   The mother in *Scotching* had pleaded guilty to the manslaughter of her son on grounds of diminished responsibility and was on remand in prison awaiting trial for murder (the Crown having rejected her plea); the boy's body had been under the coroner's control in the seven months since his death.

214   Cremation, Interment and Funeral Services Act 2004, s 5(4). The relevant statutory provisions in Alberta are not as explicit, but (unlike Saskatchewan) appear to contemplate the possibility of court intervention by making the statutory priority ranking "subject to" a court order – General Regulation (Funeral Services Act) Alta Reg 226/98, s 36(2) and General Regulation (Cemeteries Act) Alta Reg 249/1998, s 11(2). In contrast, s 4F of the draft Cremations and Other Legislation Amendment Bill 2011 in Queensland contains detailed provisions for courts to make orders following an application concerning the exercise of the right to control the disposal of a deceased person's remains, notwithstanding the statutory priority ranking.

spiritual beliefs (and those of the immediate family). Different constructs of 'family' could also encourage someone lower down the statutory pecking order to challenge a higher ranking claim on the basis that their familial tie to the deceased was less conventional yet much stronger, or because the applicant believes that the statute is not reflective of alternative kinship structures within the deceased's own cultural or ethnic group.

In deciding whether to deviate from the statutory order of entitlement,[215] judges could be guided by specific criteria. For example, as well as the rights of interested parties, courts in British Columbia must consider a combination of factors when hearing these applications – namely (i) the "feelings of those related to, or associated with, the deceased" (especially a spouse); (ii) the bodily disposal practices and beliefs "followed or held by people of the religious faith of the deceased"; (iii) any "reasonable" funeral directions given by the deceased; and (iv) whether the funeral dispute driving the application "involves family hostility or a capricious change of mind" around disposal of the deceased's remains.[216] The deceased's own funeral instructions are an obvious reference point, as are that person's spiritual and cultural beliefs around disposal of the dead; whether the family's views on the latter are a material factor is an entirely different issue, especially where the two conflict, though courts might look at the interests of family members more generally. Another consideration might be the practicalities around disposal of the deceased's remains and ensuring that all those connected with the deceased can pay their last respects, alongside some sort of marker for breaking the deadlock in equal kinship disputes – not something as arbitrary as age,[217] but possibly favouring the custodial parent in parental disputes over a dead child[218] or majority rule where feuding siblings fail to agree on the funeral arrangements for a dead parent.[219] An alternative might be to look for

---

215 Or, in equal kinship disputes (see Pt III), whether to strike-out others within the same class of entitlement and vest sole decision-making powers in the applicant.

216 Cremation, Interment and Funeral Services Act 2004, s 5(5). Section 5(6) of the same statute states that, if the application is successful, that particular individual is then deemed to be at the top of the statutory order of priority. Queensland has proposed a broadly similar approach, with s 4F(2) of the draft Cremations and Other Legislation Amendment Bill 2011 instructing courts to have regard to (i) the need to dispose of human remains "in a dignified, respectful and timely way"; (ii) the deceased's written funeral instructions; (iii) any other directions given by the deceased in relation to his/her funeral; (iv) the deceased's cultural and spiritual beliefs or practices in relation to bodily disposal; and (v) the interests of any person falling under the statutory priority ranking.

217 The test applied in British Columbia, Alberta and Saskatchewan – see p 96.

218 A number of US cases have favoured the custodial parent in disputes involving minor children – see for example, *Rader v Davis* 134 NW 849 (Iowa 1912), *Robinson v Robinson* 237 SW2d (Ark 1951) and *Tully v Pate* 372 F Supp 1064 (DSC 1973). For a general analysis, see TE Ellis, "Loved and Lost: Breathing Life Into the Rights of Noncustodial Parents" (2005) 40 *Valparaiso University Law Review* 267, pp 293–296.

219 As in the US state of New Jersey – see pp 96–97.

the individual who had the 'closest relationship' to the deceased in life, as a means of separating equally ranked relatives.[220]

The latter test could apply to all funeral disputes where someone challenges the designated decision-maker's authority, raising the possibility that a stronger ante-mortem connection with the deceased generates an equivalent post-mortem entitlement. Affirming real-life, actual relationships between the living and the dead has a certain intuitive appeal, since it prevents those who were on less favourable (or actively hostile) terms with the deceased from having the final say on funeral arrangements.[221] When deciding who had the closest relationship, the legislation could include emotional estrangement from the deceased as a potential ground for displacing the designated decision-maker, something that addresses separated spouses or couples living apart but not formally separated as well as family scenarios in which the deceased and their surviving son, daughter or sibling have not been on speaking terms for years but the latter is still the legally designated kin. Closely aligned with this is the concept of 'misconduct' as a disqualifying factor. This is something that has been prevalent in the US, with courts using evidence of past transgressions as a means of negating an otherwise valid legal entitlement to possession of the deceased's remains.[222] The outcome in *Maurer v Thibeault*[223] is a good example, the Supreme Court of the State of New York deciding that an estranged husband's cruel and, at times, physically aggressive treatment of his wife denied him any right

220   Again, drawing on the US approach where this is one of a range of factors (as in the District of Columbia and Minnesota) or the sole factor (as in Pennsylvania) – see p 97.
221   Writing in the US context, Foster has noted a "long tradition of looking beyond family status" to consider the "actual relationship" between the deceased and his/her survivors – see Foster (n 3), pp 1393–1395. For example, in *Felipe v Vega* 570 A2d 1028 (NJ Super Ct Ch Div 1989) the court awarded custody of the deceased's remains to his cohabiting partner of seven years following a dispute between the partner and the deceased's father; although the partner did not qualify under New Jersey intestacy laws, the judge ruled in her favour because of a strong emotional tie to the deceased.
222   This idea has been mooted in succession law, especially in the US, as a means of preventing those who are guilty of certain types of behaviour towards the deceased from inheriting – see for example, FH Foster, "Towards a Behavior-Based Model of Inheritance?: The Chinese Experiment" (1998) 32 *University of California Davis Law Review* 77 and LC Dummond, "The Undeserving Heir: Domestic Elder Abuser's Right to Inherit" (2010) 23 *Quinnipiac Probate Law Journal* 214.
223   860 NYS2d 895 (NY Sup Ct 2008) and noted at p 91.

to determine the fate of her remains.[224] Again, a specific provision to this effect could be included in any bodily disposal statute, directing courts to look at evidence of prior misconduct on the part of the designated decision-maker towards the deceased – for example, violent behaviour or domestic abuse,[225] threats of violence, sexual abuse and perhaps even allegations of emotional cruelty.

Replacing common law rules with a discrete statutory framework for adjudicating funeral disputes would not prevent families from resorting to litigation on the death of a loved one, but would at least provide a modern and authoritative reference point for judges. Any new statutory regime must have well-defined legal rights and obligations at its core. However, a one-dimensional hierarchy of decision-making entitlement is not always appropriate; instead, a principled discretion to confer decision-making powers on someone else, regardless of the statutory order, guarantees some measure of flexibility in a scheme that cannot encapsulate the increasingly complex scenarios that end up before the courts.

VII. Alternative dispute resolution?

Whether governed by common law rules or customised legislation, family disagreements over the fate of their dead will end up before the courts in the event of an impasse. This has obvious drawbacks. Set against a highly charged emotional backdrop, adversarial legal proceedings inflict additional (and often irrevocable) damage on everyone involved, with the decision typically viewed as a "winner-take-all outcome",[226] which polarises the warring camps even further. The court is also a very public (and expensive) forum to engage in such intensely private disputes, and, unlike other litigation where those involved can avoid future contact, this is not always possible within families.

---

224 American courts have traditionally taken this approach. In *Scott v Riley* 16 Phila 106 (Pa 1883) the court granted custody of the deceased's remains to friends with whom she had been living as opposed to the deceased's father who had abandoned his daughter even as she lay dying. In *Boyd v Gwyn* 6 Pa D & C 275 (1925), which involved a dispute between a sister and two brothers over the fate of their father's corpse, the court ruled in favour of the daughter who had been a dutiful and loving child for the last 17 years of her father's life and cared for him in her own home. The daughter was also intending to bury her father in a family grave in his boyhood home of Nashville; in contrast, the deceased's two sons had had little contact with their father for almost 20 years and, driven by feelings of hostility and animosity, intended to bury his body in Philadelphia where the deceased had no family or contacts. For an overview of *Boyd* and other US cases applying a 'behaviour-based' approach to funeral disputes, see Foster (n 3), pp 1395–1398.

225 Including any related criminal convictions, as well as protective court orders secured by the deceased.

226 Josias (n 73), p 1166.

Against this backdrop, there is scope for more widespread use of alternative dispute resolution ('ADR') techniques – especially mediation – in funeral disputes. One of the strongest advocates is Josias, who sees ADR as a way of "dealing with the highly emotional demands of ... a sensitive subject" while allowing the "disputants to come together to find their own solution to the problem, which is essential to restoring fractured relationships".[227] Mediation would shift the emphasis from a hierarchical to a consensus-based resolution, allowing the parties to negotiate a mutually acceptable outcome instead of being presented with a court ruling based on a mechanistic ordering of entitlement. As a more expansive and less formal process, it could also take a broader range of family interests into account while avoiding the high costs associated with litigation. Some of the additional benefits associated with the use of mediation in wills and trusts contests would also translate across to funeral disputes.[228] For example, it would provide a forum for those involved to articulate the non-legal issues, which are such a key aspect of the conflict; there may also be greater scope for honouring the deceased's wishes, where these have been clearly set out.[229]

However, several factors inhibit the use of mediation, beyond basic service provision. While offering the potential for more innovative and individualistic solutions, there will be a limited number of viable outcomes where families are divided over the fate of their dead, unlike other situations which use ADR.[230] In disputes involving a corpse, mediation would also have to be time-limited to ensure swift disposal, unless the remains were frozen and stored. Perhaps the biggest obstacle though, is that English law has not yet adopted mandatory mediation; judges can only encourage feuding families to consider it as an alternative. Even where both sides agree to mediation, there is no guarantee that an acceptable outcome will be reached. Situations may also arise in which the "disallowed party"[231] rejects the negotiated settlement and goes to court anyway.

---

227   *Ibid*, p 1178 and see Pt III of the same article for a detailed overview. Josias aside, there is surprisingly little commentary on the use of ADR here. For example, Ellis (n 218), p 299 also advocates mediation, but only in disputes between divorced or separated parents over the remains of a dead child.

228   See for example, LP Love, "Mediation of Probate Matters: Leaving a Valuable Legacy" (2001) 1 *Pepperdine Dispute Resolution Law Journal* 255 and T Mayersak, "Examining the Use of Arbitration and Dealing with Decedent's Wishes in Wills, Trusts and Estates" (2010) 12 *European Journal of Law Reform* 404.

229   Though Josias suggests that arbitration may be more appropriate here, since the deceased's wishes should act as a "guiding hand" for the dispute – Josias (n 73), p 1179.

230   For example, if a family is divided over burial or cremation, alternative methods of disposal are unlikely to be serious options.

231   Josias (n 73), p 1179.

VIII.   Pre-emptive rulings?

Should families be able to litigate over a loved one's remains where the individual in question is still alive, but death is imminent and the likelihood of conflict over funeral arrangements considerable? Courts have been asked to make pre-emptive rulings on at least two occasions, in cases involving children. In the Northern Ireland case of *Re LL (Application for Judicial Review)*,[232] Deeny J was willing to determine whether foster parents or the child's mother should have the legal right to bury the body of an 11-year-old boy dying from cancer.[233] However, Heath J refused to make any such determination in the New Zealand case of *Re JSB (A Child); Chief Executive, Ministry of Social Development v TS*.[234] Here, a 2-month-old baby boy had been left severely brain-damaged as a result of injuries inflicted by his mother, who was subsequently sentenced to six years in prison. Aged five, the boy suffered a serious medical episode and doctors believed he would die within a short time; a court-appointed lawyer, fearing a dispute over funeral arrangements between the paternal grandmother and the boy's parents who were still together (at this stage, the mother had been released on parole, on condition that she had no contact with her child), sought advance directions from the High Court.[235] Heath J accepted that it would probably be in the child's best interests to have a mechanism in place to avoid conflict after his death, but regarded any order as premature. Circumstances could change (the child's parents and paternal grandmother might reconcile, or any one of them might predecease the child), and predetermining the child's funeral arrangements would not be appropriate.[236]

Both *Re LL* and *Re JSB* are context-specific decisions, and it may not be prudent to read too much into either outcome. Although dealing with children, it is not difficult to imagine similar situations involving adults who are close to death yet incapacitated in some way – for example, where someone is on life-support, has suffered brain damage in a serious accident or has a severe mental disability, and that person's family are already at

---

232   [2005] NIQB 83 and discussed in Ch 3, p 65.

233   The court ruled in favour of the foster parents on the facts before it.

234   [2009] NZHC 2054. For an overview, see M Henaghan, "Family Law After Death: Control of the Body of a Child Killed by the Actions for a Parent" (2010) 6 *New Zealand Family Law Journal* 263.

235   Both sides favoured cremation; however, the paternal grandmother wanted the ashes to be buried with her in a family plot; the child's parents wanted the ashes to be interred in the father's family plot.

236   The parties (aided by the child's court-appointed lawyer) agreed a compromise solution when the child died. The child's mother did not attend the funeral, although the child's father did, along with his own mother (the paternal grandmother). The ashes were given to the child's parents and placed in their chosen cemetery, though in a separate plot beside the father's family plot – Henaghan (n 234), p 265.

loggerheads.[237] At a basic human level, it seems distasteful for families to be disputing the funeral arrangements for someone who is not yet dead, and most legal systems would not want to encourage this. The court's jurisdiction to determine the post-mortem fate of a living person is also debatable,[238] while asking it to make what are effectively hypothetical declarations raises obvious policy concerns. The forthcoming event would have to be inevitable; the fact that the individual in question was likely to die would not suffice. Where death is imminent *and* litigation seems unavoidable, then there may be a case for making some sort of pre-emptive ruling on who should have custody of the deceased's remains. Of course, this may be a complete waste of time where death results in some sort of rapprochement between the various protagonists. Faced with a dispute of this nature, a prudent judge might hear the arguments on both sides, declare any ruling to be premature, but indicate the likely outcome if the parties came back to court (assuming there has been no material change of circumstances).

Conclusion

Funeral disputes following the death of a loved one are, in some ways, inevitable. Difficult decisions must be made within a short space of time, and emotions are running high as the bereaved struggle to cope with their grief and family rifts are exposed. The law also confronts its own limitations if the parties resort to litigation; judges do not like having to determine conflicts that delay the final disposal of the dead and cannot always be settled by discrete legal principles.

Rulings based on who has the best legal claim to make the deceased's funeral arrangements may facilitate disposal, but frequently overlook the merits of competing claims. More importantly, the deceased's own views about the fate of their mortal remains need not be respected. The following chapter argues that this approach is inherently wrong, and that funeral instructions should be determinative when courts arbitrate disputes over the dead.

237   Though in some of these scenarios, the individual's own preferences may be important, if specified at an earlier stage.
238   Heath J specifically questioned whether the court had jurisdiction to make any such order in *Re JSB* [2009] NZHC 2054.

5   Funeral instructions

The case for ante-mortem planning

It is a fact that a man's dying is more the survivors' affair than his own.[1]

Introduction

An individual's ability to dictate the post-mortem fate of their body is limited. Donation of organs and other material for transplant or research purposes is permissible,[2] but constitutes little more than a statutorily sanctioned expression of preference.[3] Specific funeral directions also carry no weight in English law, and in many other countries with derivative legal systems.[4] The person with the legal duty of disposal is not obliged to implement the deceased's instructions; possessory entitlements to the dead include full decision-making powers over what form the funeral should take.

Yet, this legal stance ignores practical realities. Many people have strong opinions on what form their funeral should take, from the method of disposal (usually burial or cremation) and attendant rites, to the final resting place for their corpse or ashes. This chapter argues that the deceased's own preferences should be paramount, and that the person entrusted with disposal should be obliged to carry out the deceased's funeral instructions whenever possible. As well as facilitating ante-mortem planning, allowing an individual to make legally binding directions would ensure that these were the first point of reference for judges faced with conflicting claims to the dead.

1   T Mann, *The Magic Mountain* (1924, translated by HT Lowe-Porter), p 532. This chapter draws on material previously published as H Conway, "Burial Instructions and the Governance of Death (2012) 12 *Oxford University Commonwealth Law Journal* 59 and reproduced with kind permission of Hart Publishing. Substantively similar arguments were presented by Muinzer (then a PhD candidate at Queen's University School of Law) in an article published two years later – T Muinzer, "The Law of the Dead: A Critical Review of Burial Law, with a View to its Development" (2014) 34 *Oxford Journal of Legal Studies* 791.

2   Under the Human Tissue Act 2004, discussed in Ch 6, Pt I.

3   Since the deceased's next-of-kin have an effective power of veto – see p 148.

4   Though exceptions do exist – see Pts III and IV.

I.   Funeral instructions: Why do they matter?

Funeral instructions can be driven by reasons of sentiment or personal preference. For example, someone may want to be cremated rather than buried (or *vice versa*),[5] or to have their ashes scattered in a particular place for sentimental reasons.[6] Opting for a specific gravesite may also indicate a desire to be reunited with loved ones after death, such as being buried in the family plot,[7] or may be aligned with notions of citizenship and belonging where the individual insists on having their corpse or ashes returned to their native homeland.[8] Religious or cultural beliefs may also play an important role, with funeral rites reflecting traditions that the individual adhered to in life, and facilitating the transition to the spiritual afterlife.[9]

Funerals and their attendant rites can also be seen as posthumous representations of the deceased's life and character, which create a shared social memory for everyone who attends the funeral.[10] The fact that modern funerals with their emphasis on the life and attributes of the deceased are increasingly viewed as a "strategy for self-expression"[11] can be another motivation for leaving specific instructions.[12] These typically reflect some aspect of the deceased's personality – for example, Hollywood icon Elizabeth Taylor insisted that she be 15 minutes late for her funeral as was her tradition in life.[13] Of course, such requests are not exclusive to the rich and famous; more typical illustrations include someone who favoured an eco-friendly lifestyle insisting on green burial, or an individual requesting a personalised coffin, which reflects some particular aspect of their character.[14]

Despite the personal importance attached to them, however, setting out detailed funeral directions is no guarantee that they will be followed.

---

5   See for example, *Saleh v Reichert* (1993) 104 DLR (4th) 384 and *Holtham v Arnold* [1986] 2 BMLR 123.

6   See for example, *Robinson v Pinegrove Memorial Park* (1986) 7 BPR 15,097.

7   However, in some instances, the deceased may want to sever certain ties on death – see *Maurer v Thibeault* 860 NYS2d 895 (2008) (wife insisted on being buried separately from her husband due to his persistently violent conduct and alleged threats to kill her).

8   See for example, *Grandison v Nembhard* [1989] 4 BMLR 140.

9   This is increasingly important in modern multi-cultural societies with increasing numbers of non-adherents, second marriages and inter-marriages; in these circumstances, the individual may want to ensure that their wishes are known and subsequently respected.

10   See generally E Hallam and J Hockey, *Death, Memory and Material Culture* (Berg, 2001).

11   C Schafer, "Corpses, Conflict and Insignificance? A Critical Analysis of Post-Morten Practices" (2012) 17 *Mortality* 305, p 305.

12   The growing trend towards personalised funerals was discussed in Ch 1, pp 21–22.

13   J Brown, "As a Final Act, Elizabeth Taylor Is Late For Her Own Funeral", *The Independent* (London, 26 March 2011) www.independent.co.uk/news/world/americas/as-a-final-act-elizabeth-taylor-is-late-for-her-own-funeral-2253475.html (accessed 30 September 2015).

14   See the example mentioned at n 199.

II.   The 'no property' rule and the common law legacy of Williams v Williams

Funeral instructions are ineffective because of the conceptual difficulties posed by the common law rule that there is no property in a dead body[15] and its application in a well-known nineteenth-century case. In *Williams v Williams*,[16] the deceased's will stipulated that his body be given to his friend, Eliza Williams, to be dealt with according to his instructions. A private letter directed her to burn the body; any expenses incurred were repayable from the estate. However, things did not go to plan when the deceased died and his wife buried him in Brompton cemetery. Three months later, Miss Williams (without notifying the deceased's executors or his family) obtained a licence to disinter the remains and shipped them to Milan for cremation.[17] In an action against the executors for expenses, Kay J held that these were not recoverable since the deceased's directions could not be enforced:

> [A] man cannot by will dispose of his dead body. If there be no property in a dead body it is impossible that by will or any other instrument the body can be disposed of.[18]

Nwabueze has rightly criticised the decision on the basis that the "critical question" of whether funeral directions are enforceable "was left almost completely unanswered".[19] However, like the 'no property' rule itself, Kay J's dictum in *Williams* is now part of legal folklore: a will disposes of property on death, but if a corpse is not property it cannot be dealt with in this way. As a result, funeral instructions are simply precatory statements, which do not impose any legal obligation on those tasked with the funeral.[20]

---

15   See Introduction, pp 2–3.

16   (1882) 20 Ch D 659 and discussed in S White, "A Burial Ahead of its Time? The Crookenden Burial Case and the Sanctioning of Cremation in England and Wales" (2002) 7 *Mortality* 171, pp 172–179.

17   Miss Williams lied in order to obtain the licence, informing the Home Secretary that she intended to move the deceased's body from the unconsecrated part of Brompton Cemetery to consecrated ground elsewhere. At the time, there were serious doubts over the legality of cremation in England – see Ch 2, p 40.

18   (1882) 20 Ch D 659, 665.

19   RN Nwabueze, "Legal Control of Burial Rights" (2013) 2 *Cambridge Journal of International and Comparative Law* 196, pp 209-210.

20   This common law position has not been altered by statute in England and Wales. The only stipulation that comes close is s 46(3) of the Public Health (Control of Disease) Act 1984 which prohibits cremation of unclaimed remains by a local authority where the authority has reason to believe it would be contrary to the deceased's wishes. Section 46(3) is an isolated provision, and an anomalous one given that an individual cannot – as general rule – prevent their corpse being cremated by leaving directions to the contrary. The position used to be different; reg 4 of the original Cremation Regulations 1903 (introduced alongside the original Cremation Act 1902) made it unlawful to cremate a dead body if that person had left written directions to the contrary. However, this was repealed by the Cremation Regulations 1965.

Both the 'no property' rule and consequent ruling in *Williams* were adopted in other common law jurisdictions, with numerous Canadian, Australian and New Zealand authorities confirming that funeral instructions are not legally binding.[21] Instead, the ultimate decision lies with the executor appointed by the deceased's will or the presumptive administrator where someone dies intestate.[22] There is a certain irony here; a dead body is not property, yet legal entitlements to dispose of it are allocated by the same succession law rules that oversee the post-mortem transfer of property. Where the deceased made a will it has been suggested that appointing an executor operates as a type of "surrogate autonomy",[23] especially where the choice of individual is influenced by their willingness to undertake specific funeral arrangements. If the deceased's family insist on something else, the executor can assert a stronger legal entitlement to possession of the deceased's remains and associated decision-making powers.[24] However, there is nothing to prevent an executor from disregarding the deceased's instructions when under pressure to implement something different;[25] in these circumstances, the executor is entitled to follow the wishes of the surviving relatives against those of the deceased.[26] The same basic principles apply on intestacy, where the presumptive administrator may comply with any non-testamentary directives, yet can equally ignore them and substitute something else.[27]

---

21   See for example, the Canadian cases of *Hunter v Hunter* (1930) 65 OLR 586, *Re Waldman and City of Melville* (1990) 65 DLR (4th) 154, *Saleh v Reichert* (1993) 104 DLR (4th) 384 and *Sopinka v Sopinka* (2001) 55 OR (3d) 529, and the New Zealand cases of *Murdoch v Rhind* [1945] NZLR 425, *Awa v Independent News Auckland Ltd* [1995] 3 NZLR 701 (affirmed on other grounds at [1997] 3 NZLR 590), and the first instance decision in *Clarke v Takamore* [2009] NZHC 901. There is a wealth of Australian case law on this topic; for a flavour, see *Robinson v Pinegrove Memorial Park* (1986) 7 BPR 15,097, *Burnes v Richards* (1993) 7 BPR 15,104, *Smith v Tamworth City Council* (1997) 41 NSWLR 680, *Jones v Dodd* [1999] SASC 125, *Manktelow v The Public Trustee* [2001] WASC 290, *Keller v Keller* [2007] VSC 118 and *Tufala v Marsden* [2011] QSC 222.

22   See Ch 3, Pt I.

23   R Croucher, "Disposing of the Dead: Objectivity, Subjectivity and Identity" in I Freckelton and K Peterson (eds), *Disputes and Dilemmas in Health Law* (2006, Federation Press), pp 330–331.

24   As in *Grandison v Nembhard* [1989] 4 BMLR 140 (see Ch 4, p 92). Such proceedings are a test of the executor's resolve, especially if they are not related to the deceased.

25   As in *Williams v Williams* (1881) 20 Ch D 659 (executor acceded to the wishes of the deceased's wife), and *Hunter v Hunter* (1930) 65 OLR 586 (deceased's son and executor ignored his father's wish to be buried in a Catholic cemetery). A direction in the deceased's will that his/her body should be handed over to someone other than the executor is void – *Williams*.

26   See the comments of Gallop J in *Sullivan v Public Trustee* [2002] NTSC 107.

27   A good illustration is *Holtham v Arnold* [1986] 2 BMLR 123 (see Ch 4, p 91). This is not to suggest that a presumptive administrator will frequently ignore the deceased's wishes, since there are numerous illustrations to the contrary – see for example, *Meier v Bell* (Supreme Court of Victoria, 3 March 1997) and *Saleh v Reichert* (1993) 104 DLR (4th) 384.

The situation would be the same where a pre-paid funeral plan contains specific disposal preferences.[28] Despite assurances about 'peace of mind', such plans can only guarantee this in a financial sense. The contract itself is between the funeral company and the individual, and is activated on death. Thus the individual has no control over the actual performance of the contract, and whether or not their wishes are complied with depends once again on the actions of the personal representative. If he/she decides to comply with the deceased's wishes, then this is the end of the matter. However, if the executor or presumptive administrator declines to follow the deceased's specific requests (for example, by substituting cremation for burial, or opting for a different type of funeral ceremony under pressure from the deceased's family), no legal steps can be taken to prevent this.[29]

In short, those entitled to possession of the deceased's remains may (and often will) comply with funeral directions but are not legally obliged to do so. As Bryson J acknowledged in *Privet v Vovk*:[30]

> [W]ishes expressed by a deceased person, whether in a testamentary document or otherwise, are not binding on the persons on whom the duty falls of disposing of the remains and will not be enforced by a court.[31]

However, there are ways of getting around this. Financial constraints play a role, since a personal representative is liable for any unreasonable funeral costs, or additional expenditure beyond that covered by a pre-paid funeral plan.[32] And while an executor or presumptive administrator cannot be legally compelled to carry out the deceased's wishes,[33] the prospect of non-compliance may prompt others to challenge the funeral arrangements – for

28   See generally Ch 3, p 75.
29   Opposing members of the deceased's family would not, presumably, be able to sue on the contract – aside from issues of standing, it is questionable whether this specific element of the contract could actually be enforced. Contrast this with the approach in certain parts of Canada where such contracts are regulated by statute – see MW Zwicker and MJ Sweatman, "Who Has the Right to Choose the Deceased's Final Resting Place?" (2002) 22 *Estates, Trusts & Pensions Journal* 43, pp 50–51.
30   [2003] NSWSC 1038, [12].
31   See also *Brown v Tullock* (1992) 7 BPR 15,101 ("[u]nfortunately ... there is no law which enables a court to give directions as to how the body of a deceased person should be disposed of") and *Dow v Hoskins* [2003] VSC 206, [26] ("the law is that the wishes of the deceased are not the determining factor. However cold that may sound, that is the law applicable to this and every other case...").
32   Thus, where the deceased has requested cremation and the personal representative insists on burial, the latter is personally liable for additional costs.
33   If the personal representatives "decide to act differently" and ignore the individual's funeral preferences, "no one can take any legal steps to make them observe [those] directions" – L Skene and B Masters, "What Legal Rights Do You Have Over Your Body After Your Death?" (2002) 81 *Australian Law Reform Commission Reform Journal* 38, p 38.

example, by invoking selected provisions of the European Convention on Human Rights[34] or using s 116 of the Senior Courts Act 1981.[35] In the English context, *Burrows v HM Coroner for Preston*[36] is the strongest indication yet that the deceased's views are an important factor when adjudicating funeral disputes,[37] a sentiment that has been echoed in other recent cases both here and elsewhere. For example, in *Hartshorne v Gardner*[38] the court indicated that "the deceased's wishes are one of the relevant factors to be taken into consideration",[39] while McKechnie J in *Ugle v Bowra & O'Dea*[40] suggested that "[t]he views of the deceased, though not decisive, should nevertheless be accorded considerable weight".[41] Meanwhile Elias CJ in the Supreme Court of New Zealand in *Takamore v Clarke*[42] was highly critical of the existing position, because in "modern conditions" it was "unacceptable to say that the views of the deceased are views that can be ignored".[43] Yet, while the deceased's wishes are an increasingly important reference point, there is no legal obligation to uphold them under the current common law framework.

III.   The common law position in the US, and subsequent statutory developments

American courts have traditionally allowed individuals to direct the post-mortem fate of their remains, taking the view that the wishes of the dead are "paramount to all other considerations".[44] One possible explanation stems

---

34   See pp 139–144.

35   See Ch 4, pp 101–104.

36   [2008] EWHC 1387 (Admin) and discussed in Ch 4, p 88 and pp 103–104.

37   Though note the criticisms in *Ibuna v Arroyo* [2012] EWHC 428 (Ch), [55] (personal representatives ought to take account of the deceased's wishes, but these "should not … have the paramountcy" suggested in *Burrows*).

38   [2008] EWHC B3 (Ch).

39   *Ibid*, [7] citing *Grandison v Nembhard* [1989] 4 BMLR 140.

40   [2007] WASC 82.

41   *Ibid*, [16], sentiments echoed more recently by Nicholson J in *State of South Australia v Smith* [2014] SASC 64, [61] ("the wishes of a deceased person, where they can be ascertained with some confidence, should be given significant weight"). See also *Heafy v McRae* (1999) 5 ETR (3d) 121, [15] ("[a]s much as possible, the wishes of [the deceased] … should be respected and honoured in death").

42   [2012] NZSC 116, [82]. See also the comments of Tipping, McGrath and Blanchard JJ at [168]: "Consideration must also be given to the [deceased's] wishes … since the law has moved on from its early rejection of the significance of this factor".

43   Citing factors such as modern social conditions and the increasing emphasis on human rights.

44   *In re Eichner's Estate*, 18 NYS 2d 573, 573 (1940). See generally, 22A Am Jur 2d, *Dead Bodies*, Part II and PE Jackson, *The Law of Cadavers* (Prentice Hall, 2nd edn, 1950), ch III, as well as KE Naguit, "Letting the Dead Bury the Dead: Missouri's Right of Sepulcher Addresses the Modern Decedent's Wishes" (2010) 75 *Missouri Law Review* 248 and the various cases referenced throughout this section.

from the fact that testamentary freedom is much more entrenched in the United States than elsewhere;[45] funeral instructions could be viewed as a logical extension of this, in a country that jealously safeguards the personal liberties and freedoms of the individual. However, much more important is a long-standing judicial willingness to recognise limited proprietary interests in dead bodies.[46] By the end of the nineteenth century, the English 'no property' rule was regarded as a "juridical enigma"[47] by American courts. In *Pierce v Proprietors of Swan Point Cemetery*,[48] one of the leading authorities, Potter J accepted that "the body is not property in the usually recognized sense of the word" yet labelled it as a "sort of *quasi* property, to which certain persons may have rights, as they have duties to perform towards it arising out of our common humanity".[49] Subsequent cases adopted this reasoning, with the duty of disposal falling on the deceased's spouse followed by next-of-kin (children, then parents, then siblings etc) in order of inheritance entitlement.[50]

In some ways, this is similar to the English position: regardless of whether the right is labelled *quasi* property, the practical outcome is the same since it falls short of full property, is for a limited purpose only[51] and vests in designated kin.[52] However, this relational hierarchy *only* comes into play where the deceased has failed to make their wishes known.[53] As Benedict J acknowledged in *Cooney v English*,[54] American common law "gives great weight, if not controlling force, in such matters to the wishes of the

---

45   See RC Brashier, "Disinheritance and the Modern Family" (1994) 45 *Case Western Reserve Law Review* 83.

46   See Jackson (n 44), pp 25–28, as well as TM Kester, "Uniform Acts: Can the Dead Hand Control the Dead Body? The Case for Uniform Bodily Remains Law" (2007) 29 *Western New England Law Review* 571, pp 573–576. The same sources also note the historical absence of ecclesiastical courts in the United States as another important factor.

47   M Barish, "The Law of Testamentary Disposition: A Legal Barrier to Medical Advance" (1956) 30 *Temple Law Quarterly* 40, p 40.

48   10 RI 227 (1872).

49   *Ibid*, 242–243.

50   See for example, *Larson v Chase* 47 Minn 307 (1891), *Pettigrew v Pettigrew* 56 A 878 (Pa 1904), *Litteral v Litteral* 131 MoApp 306 (1908) and *Floyd v Atlantic Coast Line Ry Co* 83 SE 12 (1914). However, the idea of *quasi* property rights was not universally accepted – RN Nwabueze, "Property Interest in a Burial Plot" [2007] *Conveyancer and Property Lawyer* 517, p 535.

51   Any rights that exist are for disposal purposes only; there is no property in a corpse in any commercial or general sense – see *Goldman v Mollen* 168 Va 354 (1936), *In re Shipley's Estate* 53 Erie CLJ 6 (1970) and *Massey v Duke University* 503 SE2d 155 (1998).

52   This has occasionally been recognised as a constitutionally protected right – *Brotherton v Cleveland* 923 F2d 477 (6th Cir 1991) and *Whaley v County of Tuscola* 58 F 3d 1111 (6 Cir CA 1995).

53   See WH Lorshbough, "The Disposition by Will of One's Body After Death" (1945) 22 *Bar Briefs* 272.

54   86 Misc 292, 293–294 (1914).

deceased".[55] This is still the position today;[56] for example, in *Kasmer v Guardianship of Limner*[57] the court ordered cremation of the deceased's corpse in accordance with his wishes, despite religious objections from surviving relatives,[58] while in *Cottingham v McKee*[59] the executor was allowed to exhume and cremate the deceased's body against the wishes of the next-of-kin where the dead man's will had specified this method of disposal.[60] Courts have occasionally weighed the deceased's preferences against other competing interests, and in particular the views of a surviving spouse or close kin; however, funeral instructions are typically prioritised in conflicts between the living and the dead.[61]

The fact that the body is not property in any strict legal sense means that an individual is not actually bequeathing their remains,[62] and need not comply with testamentary formalities.[63] However, a will is only one option and American courts have upheld funeral instructions contained other written documents,[64] or occasionally based on verifiable oral

---

55   Similar views were expressed in *Wales v Wales* 190 A 109 (1936) and in *In re Eichner's Estate* 18 NYS2d 573 (1940).
56   Kester (n 46), p 576 citing *Newman v Sathyavaglswaran* 287 F3d 786 (9th Cir 2002).
57   697 So2d 220 (1997).
58   See also *In re Eichner's Estate* 18 NYS2d 573 (1940) and *Tkaczyk v Gallagher* 26 ConnSupp 290 (1965).
59   821 So2d 169 (2001).
60   Exhumation raises slightly different issues. While the deceased's preferences are important (see P Zablotsky, "'Curst Be He That Moves My Bones': The Surprisingly Controlling Role of Religion in Equitable Disinterment Decisions" (2007) 83 *North Dakota Law Review* 361), they are not decisive and American courts have refused to sanction disinterments simply because funeral directives were ignored – see *Fischer's Estate v Fischer* 117 NE2d 855 (1954), *Guerin v Cassidy* 119 A2d 780 (1955) and *Estate of Moyer v Moyer* 577 P2d 108 (1978).
61   Despite isolated suggestions that the deceased's wishes should be subordinate to those of a surviving spouse (see *Burnett v Surratt* 67 SW2d 1041 (Tex Civ App 1934)), funeral instructions tend to take precedence over any conflicting family views (see *Cordts v Cordts* 154 Kan 354 (1941) and *Holland v Metalious* 105 NH 290 (1964)).
62   See for example, *Fidelity Union Trust Co v Heller* 16 NJSuper 285 (1951) and *Guerin v Cassidy* 38 NJ Super 454 (1955), as well as *Estate of Moyer v Moyer* 577 P2d 108, 110 (1978) and *In re Estate of Medlen* 286 IllApp3d 860 (1997).
63   Thus funeral directions contained in a will do not need to be probated before being implemented, and need not comply with revocation formalities if the individual changes their mind – see T Hernández, "The Property of Death" (1999) 60 *University of Pittsburgh Law Review* 971, pp 1020–1021. While funeral directions could still be valid where a will fails for non-compliance with technical formalities, the position may be different where it fails on grounds of mental capacity – *Rosenblum v New Mount Sinai Cemetery Association* 481 SW 2D 593 (1972).
64   See for example, *Fidelity Union Trust Co v Heller* 16 NJSuper 285 (1951) (deceased's wishes ascertained from various documents indicating detailed plans to erect a mausoleum in a particular cemetery). A pre-paid funeral contract is another option – see FH Foster, "Individualized Justice in Disputes over Dead Bodies" (2008) 61 *Vanderbilt Law Review* 1351, pp 1377–1378.

statements[65] – even if these contradict prior written directives. A good illustration is *In re Scheck's Estate*[66] where the deceased's children interred their mother's body in a cemetery in New York state (the mother had repeatedly said she wanted to be buried there and purchased a plot), despite her will – executed years earlier – stipulating burial in Palestine where she had been living with her (now estranged) husband.[67] As a general rule, primary responsibility for upholding the deceased's directions falls on the personal representative, though courts have allowed other next-of-kin[68] and those with close emotional ties to the deceased[69] to challenge funeral arrangements if these individuals claim to represent the deceased's wishes.

A significant number of US states have codified the common law position by enacting 'mortal remains statutes', allowing an individual to direct the posthumous disposal of their remains with a default list of authorised decision-makers where the deceased's wishes cannot be ascertained or carried out.[70] Most of these statutes require written instructions, whether contained in a will, pre-paid funeral contract, designated power of attorney or health care directive.[71] Some also allow the deceased to nominate an agent or proxy to take charge of the funeral arrangements, with comprehensive statutory

---

65   See *Wales v Wales* 190 A 109 (1936). The validity of written or oral directions is determined by ordinary rules of evidence – *Leschey v Keschey* 374 Pa 350 (1953).

66   14 NYS2d 946 (1936).

67   However, the fact that the deceased had changed her mind about this did not invalidate the remainder of the will. For more recent illustrations, see *Cohen v Guardianship of Cohen* 896 So2d 950 (2005) (deceased's 1992 will had specified a traditional Jewish burial in the family plot in New York, but after moving to Florida the deceased told his daughter that he wanted to be buried there so his non-Jewish wife could be interred alongside him; oral directions were clear evidence of a change of intent), as well as *Arthur v Milstein* 949 So2d 1163 (2007) (former Playboy model, Anna Nicole Smith buried in the Bahamas beside her recently deceased son, based on Smith's alleged statements to this effect).

68   *Cohen v Guardianship of Cohen* 896 So2d 950 (2005) (wife's decision challenged by deceased's siblings).

69   The deceased's same-sex partner in *Stewart v Schwartz Brothers-Jeffer Memorial Chapel* 606 NYS2d 965 (1993) and the deceased's girlfriend in *Pittman v Magic City Memorial Co Inc* 985 So2d 156 (2008).

70   Though this practice is not universal, with some states still allowing the deceased's next-of-kin to decide the means of disposal – see for example, Ala Code § 34-13-11 (Alabama).

71   See for example, Wash Rev Code § 68.50.160 (Washington), DC Code § 3-413 (District of Columbia), Colo Rev Stat § 12-34-204 (Colorado), Mont Code Ann § 35-21.810 (Montana), Tex Health and Safety Code Ann § 711.002 (Texas) and Minn Stat Ann § 1489.80 (Minnesota). The Texas and Minnesota statutes both state that, where funeral directions are included in a will, these can be implemented without probate (Tex Health and Safety Code Ann § 711.002 (g) and Minn Stat Ann § 1489.80, subd 1). The latter statute also stipulates that, in the event of a dispute over directions in more than one written instrument, a witnessed or notarised document takes priority (Minn Stat Ann § 1489.80, subd 2).

forms authorising the appointment and facilitating specific directives.[72] Although state dependent, the wishes of the deceased are generally paramount.[73]

IV.   Legislative directions elsewhere

Several other common law jurisdictions also recognise certain types of funeral instructions. In Australia, laws in both Queensland and the Australian Capital Territory uphold an individual's signed instructions to be cremated, or not to be cremated.[74] However, the Canadian province of British Columbia leads the way in granting an individual control over their own funeral arrangements. Section 5 of the Cremation, Interment and Funeral Services Act 2004 Act lists certain individuals who are entitled to control the disposition of the deceased's remains, and instructs courts to consider "any reasonable directions given by the deceased respecting the disposition of his or her ... remains or cremated remains" when resolving funeral disputes.[75] However, s 6 goes one step further:

> A written preference by a deceased person respecting the disposition of his or her human remains or cremated remains is binding on the person who under section 5 ... has the right to control the disposition of those remains if

---

72   See 755 Ill Comp Stat 65/1 (Illinois) and Del Code Ann Tit 12 § 265 (Delaware). The respective forms are reproduced in AM Murphy, "Please Don't Bury Me Down in that Cold Cold Ground: The Need for Uniform Laws in the Disposition of Human Remains" (2007) 15 *Elder Law Journal* 400, Appendices B and C. See also DG Fish, "To Avoid Burial Disputes, New Statutory Form is Available" (2006) 235 *New York Law Journal (Elder Law)* 1, discussing equivalent legislative provisions in the state of New York under Public Health Law § 4201 (New York).

73   As in *Bruning v Eckman Funeral Home* 300 NJSuper 424 (1997) (deceased's directive that he be interred in a mausoleum with his girlfriend upheld under the prevailing New Jersey legislation, despite opposition from the deceased's estranged wife). For other illustrations, see *Sherman v Sherman* NJSuperCh 638 (1999), *Caseres v Ferrer* 774 NYS2d 372 (2004) and *Maurer v Thibeault* 860 NYS2d 895 (2008).

74   See respectively s 7 of the Cremations Act 2003 in Queensland and s 20(2) of the Cemeteries and Crematoria Act 2003 in the Australian Capital Territory, as well as the other statutory provisions noted in Queensland Law Reform Commission, *A Review of the Law in Relation to the Final Disposal of a Dead Body* (Report No 69, December 2011), pp 97–101. In Queensland, where the deceased has not requested cremation, s 8 of the 2003 Act enables certain family members to prevent cremation where they object to this method of disposal (see the respective discussions in *Reid v Crimp* [2004] QSC 304 and *Tufala v Marsden* [2011] QSC 222). The Queensland Law Reform Commission has suggested that, because a similar provision was first introduced in 1913, it possibly reflects attitudes at a time when cremation was not lawful yet "nevertheless retained a stigma and might be distressing for family members" – *ibid*, p 108.

75   2004 Act, s 5(5)(c) though these are simply one factor to be taken into account – see Ch 4, p 119.

(a)   the preference is stated in a will or preneed cemetery or funeral services contract, [and] ...

(c)   compliance with the preference would not be unreasonable or impracticable or cause hardship.[76]

In other words, the 2004 Act prioritises funeral instructions contained in a will or pre-paid funeral contract; the person with the legal right of disposal under s 5 is expected to comply with them.[77]

A number of civil law jurisdictions also allow an individual to dictate the fate of their remains. For example, Article 42 of the Civil Code of Quebec 1991 allows "a person of full age ... [to] determine the nature of his funeral and the disposal of his body",[78] though there is no requirement that this is set out in writing.[79] Likewise, both France and Spain also focus on the

---

76   Section 6(b) also insists that compliance is consistent with the provincial legislation on human organ and tissue donation.

77   *Kartsonas v Stamoulos* 2010 BCCA 336, though in this case the deceased's will had failed to specify a non-religious funeral, allowing his estranged children – all of whom favoured a Greek Orthodox burial – to assert decision-making authority under s 5 of the 2004 Act (and displacing the deceased's niece as executor). Looking elsewhere, Queensland is also proposing that signed instructions become binding, and that the person making the funeral arrangements takes reasonable steps to carry them out – Queensland Law Reform Commission (n 74), ch 5 and draft Cremations and Other Legislation Amendment Bill 2011, ss 4A and 4D. A debate on the topic has been instigated in the state of Victoria (Victorian Law Reform Commission, *Funeral and Burial Instructions: Consultation Paper* (November 2015)), while tentative proposals for upholding funeral instructions are also being considered in New Zealand – New Zealand Law Commission, *The Legal Framework for Burial and Cremation in New Zealand: A First Principles Review* (Issues Paper 34, October 2013), ch 16. Both Ontario and Western Australia have proposed similar measures in the past, though these have not been implemented – see the key recommendations noted in Queensland Law Reform Commission (n 74), pp 104–106.

78   Similar rights extend to minors, subject to the written consent of a parent or guardian.

79   However, a casual comment about the relative merits of cremation and burial, made while watching television, does not constitute clear and unequivocal evidence for the purposes of Article 42 – *Dalexis v Kelly* 2011 QCCS 1583 (the court also noting that there is no sanction for non-compliance with the provision). In *Lapolla Longo v Lapolla* 2003 QCCS 731, which involved a dispute between siblings over whether their father's remains should be laid to rest in Montreal or the Italian village of Panni where he and his wife had lived before emigrating to Canada, the court examined a number of factors. These included the fact that the deceased had purchased two adjacent niches at the cemetery in Panni in 1978, following the death of his wife whose remains were placed there; regular visits to Panni; and the fact that the deceased had not remarried following his wife's death. According to the Quebec Superior Court, the deceased's intent under Article 42 was to be buried alongside his wife in Italy.

deceased's funeral preferences if families cannot agree.[80] The origins can probably be traced back to Roman law, where the deceased's heirs had to respect funeral instructions set out in a will or written testament.[81]

V.   Upholding funeral instructions: The arguments for ...

A number of arguments can be made for putting a legal mechanism in place in England and Wales (and elsewhere), which allows people to make binding directions about the disposal of their remains.

1.   Fallacy of the 'no property' rule

Funeral instructions are ineffective because of the doctrinally suspect 'no property' rule – even if its consequences are more apparent than real. The law already recognises limited property rights in dead bodies, with possessory entitlements vesting in certain individuals on death.[82] This is not the problem; it is the fact that the deceased cannot dictate how and by whom such rights should be exercised. More importantly, upholding funeral instructions is not actually dependent on dead bodies being classed as property; all that is needed is an effective legal mechanism for allowing an individual to direct the posthumous fate of their remains.

2.   'It's what the deceased wanted'

The core value of autonomy is important here. Allowing an individual to dictate the posthumous fate of their remains would mirror existing rights to make lifetime decisions that the law recognises on death.

From a succession law perspective, a person enjoys significant control over the post-mortem destiny of their property, and the principle of testamentary freedom (regarded as the hallmark of common law legal systems) ensures that the wishes of the deceased are usually paramount.[83] Hernández places this firmly within an individualistic framework premised on autonomous choices:

---

80   EC Rodriguez-Dod, "Ashes to Ashes: Comparative Law Regarding Survivors' Disputes Concerning Cremation and Cremated Remains" (2008) 17 *Transnational Law and Contemporary Legal Problems* 311, pp 323–328.

81   See Hernández (n 63), pp 981–982 and RC Groll and DJ Kerwin, "The Uniform Anatomical Gift Act: Is the Right to a Decent Burial Obsolete?" (1971) 2 *Loyola University Chicago Law Journal* 275, p 275 (the latter citing Democrates, whose body was embalmed in honey, and Lycurgus, whose corpse was cremated and the ashes thrown into the sea, as two examples).

82   See Ch 3, Pt I.

83   Of course, self-determination in the will-making context can occasionally be tempered by the interests of the deceased's family and dependants – see p 148.

[A]utonomy is the foundation for many rights to control one's life and one's possessions, including the law of wills. ... [F]reedom of testation is a fundamental value ... because it accords with the strong human desire to exert control over one's own property.[84]

Drawing on this general theme, Sperling has argued for a similar right in relation to the body:

> If by enforcing a will what we care deeply about is respecting the decedent's prior wishes and autonomy, then it is not clear why this principle should be defeated in situations where the decedent's wishes are concerned with the disposal of her own body. On the contrary, it seems unambiguous that a person's body is one of the most precious things about which she cares, certainly more than her real property.[85]

An individual has sole decision-making authority over their body while alive.[86] For example, the basic concept of patient autonomy prevails in the medical law context,[87] and specific treatments can be refused even if their effects would be life-saving or prolonging. In the transition from life to death, the law allows competent adults to exert greater control over key decisions. The growing use of advance directives and lasting powers of attorney are both good examples[88] – placing an increased emphasis on self-determination that could equally extend to deciding the fate of the post-mortem body. More importantly, situations already exist in which an individual can make autonomous bodily choices that transcend death, such as posthumous reproduction and organ donation.[89] Why, then, should the law ignore funeral instructions?

---

84 Hernández (n 63), p 976 and see the sources cited therein.

85 D Sperling, *Posthumous Interests: Legal and Ethical Perspectives* (CUP, 2008), pp 152–153. See also Nwabueze (n 19), p 202 ("[a] person's interest in the distribution of their property after their death provides a quintessential example of a persisting or critical post-mortem interest ... However, a person's interest in the time, place and manner of their burial is no less important than testamentary distribution...") as well as the discussion in M Brazier, "Retained Organs: Ethics and Humanity" (2002) 22 *Legal Studies* 550, pp 561–565.

86 As Cardozo J famously stated in *Schloendorff v New York Hospital* 105 NE 92 (1914): "Every human being of sound mind and adult years has the right to determine what shall be done with his own body".

87 See J Herring, *Medical Law and Ethics* (OUP, 5th edn, 2014), pp 198–203 and the various sources cited therein as well as S McLean, *Autonomy, Consent and the Law* (Routledge-Cavendish, 2010). For an alternative view, see C Foster, *Choosing Life, Choosing Death: The Tyranny of Autonomy in Medical Law and Ethics* (Hart, 2009).

88 See for example, A Maclean, "Advance Directives, Future Selves and Decision-Making" (2006) 14 *Medical Law and Ethics* 291 and C Johnston, "Advance Decision Making: Rhetoric or Reality" (2014) 34 *Legal Studies* 497.

89 See Ch 6.

Harris and others have argued that autonomy ceases on death, and that the law should not recognise posthumous rights.[90] Yet, while death results in an inevitable loss of sovereignty, what we are talking about here is a logical and morally intuitive extension of an individual's right to self-determination in life. Upholding funeral instructions validates a conscious choice that the deceased made while alive, in the same way as the law respects ante-mortem directions about the disposal of property or the fate of certain body parts after a person's death. Describing an individual's interest in prearranging certain post-mortem events as a form of "prospective autonomy",[91] Cantor goes on to describe funeral directions as being part of a "person's prerogative to shape his or her memory picture" – something which he also places within the right to self-determination.[92] Closely aligned with this is the concept of personhood. The increased power for individuals to control their lives up to the moment of death could be seen as part of a collective personhood interest,[93] as could the power to control certain aspects of what happens to the body after death. Most people probably expect that their funeral preferences will be respected. According to Sperling, these ante-mortem wishes and desires are "so important to the living person that the meaning attached to them is ... conditioned upon their being fulfilled after the person is dead".[94] And, as a means of self-expression and of cultivating one's own posthumous image, allowing others to frustrate these directions is to inflict harm on that person:

> [O]ne can argue that people define themselves in terms of their physical selves, and so invasion of the body after death, especially through acts performed contrary to a person's prior wishes regarding disposal of her body, injures the personality of this person and the image she would have wanted after death.[95]

Meanwhile, Bray has placed such interests within broader constructs of human dignity and respect for the wishes of others:

> Corpses are ... central to our personhood and identity ... [S]howing respect to the dead and carrying out burial arrangements in accordance

---

90   See generally J Harris, "Law and Regulation of Retained Organs: The Ethical Issues" (2002) 22 *Legal Studies* 527, pp 532–534 and the various authorities cited therein.

91   NL Cantor, *After We Die: The Life and Times of the Human Cadaver* (Georgetown University Press, 2010), p 29.

92   *Ibid*, p 50. For a similar argument see Sperling (n 85), pp 148–149 where the author talks about prospective autonomy interests in a "person's effort to shape, and interest in shaping, other people's recollections of her character and values".

93   See for example, the discussion in M Ford, "The Personhood Paradox and the Right to Die" (2005) 13 *Medical Law Review* 80.

94   Sperling (n 85), p 168.

95   *Ibid*, p 169. The concept of posthumous harm is revisited in pp 146–147.

with the deceased's wishes is a way of according dignity to all human beings, because death is the one thing all individuals have in common. Because interests in corpses are critical to the sense we have of ourselves as individuals, families and communities, under a personhood analysis these issues should be protected ...[96]

3.   Human rights arguments

Specific provisions of the European Convention on Human Rights may influence family funeral disputes.[97] However, they could also bolster arguments for allowing an individual to dictate the posthumous fate of their remains and, in the event of a dispute, for granting possession of the corpse or any post-cremation ashes to the person intent on complying with such directions. While it is unclear whether Convention rights alone would persuade a court to rule decisively on the issue, case law increasingly emphasises the wishes of the dead.

(a)   Precatory preferences and Article 8

Personal autonomy is an important interpretative element of Article 8(1) and its right to respect for private and family life.[98] With this in mind, Article 8(1) could encompass funeral instructions based on sentiment or mere personal preference[99] – for example, directions to 'bury me in the family plot in the local cemetery' or to 'cremate my remains and scatter my ashes in the Lake District'. A direct authority on this is *X v Federal Republic of Germany*[100] in which the European Commission suggested that the applicant's request to have his ashes scattered on his own land (rejected by authorities in Hamburg) was a means of expressing his personality and was so closely related to his private life that it fell within Article 8(1). Thus, the idea of private life as an inner-circle within which individuals may live their personal lives as they choose[101] would appear to encompass death-related directions. However, from a domestic law perspective, perhaps the strongest indication that the deceased's preferences might be determinative under Article 8(1)

---

96   M Bray, "Personalizing Property: Towards a Property Right in Human Bodies" (1990) 69 *Texas Law Review* 209, p 244.

97   See Ch 4, pp 107–114.

98   *Pretty v United Kingdom* (2002) 35 EHRR 1. However, Article 8(1) "does not confer any general rights of autonomy or self-determination" – *Ghai v Newcastle City Council* [2009] EWCH 978, [129].

99   As opposed to funeral directions based on religious or cultural beliefs, which would probably fall within Article 9(1) of the Convention – see below.

100   (1981) 24 D & R 137.

101   See *Niemietz v Germany* (1992) 16 EHRR 97.

comes from *Burrows v HM Coroner for Preston*.[102] The fact that the uncle was intent on carrying out the deceased's wishes in this case was a powerful factor in the court's decision to disregard the mother's common law entitlement to her son's remains. According to Cranston J:

> One thing is clear, that in as much as our domestic law says that the views of a deceased person can be ignored it is no longer good law. ... It is quite clear from the jurisprudence of the European Courts [*sic*] of Human Rights that the views of a deceased person as to funeral arrangements and the disposal of his or her body must be taken into account.[103]

In identifying 'special circumstances' for the purposes of s 116 of the Senior Courts Act 1981,[104] the wishes of the deceased were important "in accordance with Article 8 jurisprudence".[105]

Of course, the rights guaranteed by Article 8(1) are qualified by Article 8(2).[106] For example, an individual's request to be buried in a particular cemetery might be defeated because the cemetery is full, while directions to 'burn my corpse in the middle of a forest in front of my family and friends' might be construed as contrary to public health and morality, as well as being vulnerable to environmental and fire safety concerns.[107] In *X v Federal Republic of Germany*[108] the Commission concluded that, since the legislation which guided the Hamburg authorities' decision was intended to protect the public interest, this was consistent with Article 8(2) and justified any infringement of the applicant's Article 8(1) rights.

---

102   [2008] EWHC 1387 (Admin) and discussed at Ch 4, p 88 and pp 103–104.
103   [2008] EWHC 1387 (Admin), [20], citing *X v Federal Republic of Germany* (1981) 24 D & R 137 and *Dödsbo v Sweden* (2007) 45 EHRR 22. The latter case is discussed in Ch 7, pp 204–205.
104   See Ch 4, pp 102–104.
105   [2008] EWHC 1387 (Admin), [26]. Post-*Burrows*, the latest edition of *Halsbury's Laws of England* sets out the general rule that an individual cannot "by will or otherwise legally dispose of his body after death", but qualifies this by suggesting that Article 8(1) may, in certain circumstances, result in the deceased's expressed wishes being "'special circumstances' justifying the court in overriding the right of the personal representatives to direct the disposal of the body" – "Cremation and Burial", 24 *Halsbury's Laws of England* (5th edn, 2010), para [1102].
106   See Ch 4, p 111.
107   The latter direction would also infringe existing cremation laws – see *Ghai v Newcastle City Council* [2009] EWCH 978, in which refusal of the claimant's request to be cremated on an open-air funeral pyre did not infringe his Article 8(1) rights because the public nature of the activity justified state interference under Article 8(2).
108   (1981) 24 D & R 137.

(b)   Religious and cultural preferences and Article 9

Burial instructions that stem from the deceased's spiritual or cultural beliefs could come within Article 9(1) of the Convention and its right to freedom of thought, conscience and religion – for example, where a devout Catholic insists on being buried with attendant funeral rites, or a person of Aboriginal descent wants to be buried in a manner that respects the spiritual and cultural values of their indigenous community. The fact that Article 9(1) embraces a wide range of convictions and philosophies (including non-religious beliefs)[109] means it could also be invoked where an individual has requested a secular funeral.

The impact of Article 9(1) on the fate of the dead has been raised in a number of English exhumation cases, where courts have been swayed not only by the beliefs of the family but by those of the deceased as well.[110] However, restrictive interpretations on what amounts to 'manifesting' a religion or belief under Article 9(1)[111] may limit the scope for upholding funeral directions. In *X v Federal Republic of Germany*[112] the applicant alleged that preventing him from having his ashes scattered on his own land violated Article 9(1) by denying the right to practise his non-Christian beliefs. Rejecting this claim, the Commission held that scattering ashes on one's own land was not a discernible practice that manifested a belief.[113] Likewise, the fact that the applicant would have to be buried in a public cemetery did not violate Article 9(1) since he was not obliged to have a religious funeral or a tomb decorated with Christian symbols.[114] Article 9(2) could also come into play, whereby state interference is permission when "prescribed by law" or deemed necessary for reasons of "public safety" or "the protection of public order, health or morals". These qualifications were discussed at length in the first instance decision of *Ghai v Newcastle City Council*.[115]

Having ruled that open-air cremations were prohibited under Cremation Act 1902 and accompanying regulations,[116] Cranston J had to consider whether this was inconsistent with Ghai's right to freedom of religion under Article 9(1). The court accepted that the claimant's desire for an outdoor

---

109 See Ch 4, pp 112–113.
110 See *Re Durrington Cemetery* [2000] Fam 33 and *Re Crawley Green Road Cemetery* [2001] 2 WLR 1175 (though *Re Blagdon Cemetery* [2002] 3 WLR 603 was more critical). These cases are discussed in Ch 7, Pt V.
111 See Ch 4, p 113.
112 (1981) 24 D & R 137.
113 Since the term 'practice' does not cover each act which is motivated by a religion or belief – *Arrowsmith v UK* (1978) 3 EHRR 118, and restated in *Pretty v UK* (2002) 35 EHRR 1.
114 See also *Jones v UK* (Application no. 42639/04, ECHR, 13 September 2005) discussed in Ch 8, pp 220–221.
115 [2009] EWHC 978 (Admin) and discussed in Ch 2, pp 44–45.
116 Most notably the Cremation (England and Wales) Regulations 2008, SI 2008/2841.

funeral pyre stemmed from and was a manifestation of his religious beliefs,[117] regardless of conflicting evidence as to whether this was a core tenet of the Hindu faith. Thus prevailing cremation laws constituted an interference with Ghai's right to freedom of religion under Article 9(1). As to whether this was justified under Article 9(2), Cranston J noted that the claimant already had significant freedom to manifest his beliefs; the only restriction was the requirement that cremation took place in a building, and the vast majority of Hindus living in Britain did not consider open-air funeral pyres to be essential from a religious perspective.[118] Safety concerns aside,[119] the most important consideration was the protection of public morals as well as the rights and freedoms of others. The court had to consider how members of the public might react to open-air cremations; even if not witnessed directly by mourners and passers-by, most people would probably be offended by the fact that bodies were being disposed of in this way.[120] Balancing individual rights against the interests of other citizens was a matter for elected representatives who had decided firmly in favour of the public interest under existing cremation laws. The interference with Ghai's funeral wishes was, accordingly, justified under Article 9(2).[121]

(c)   Public statements and Article 10

The right to freedom of expression under Article 10(1) could be invoked here since, when setting out specific funeral directions, the deceased is not only communicating personal preferences but asserting his/her own values and individuality.[122] However, Convention jurisprudence suggests that Article 10(1) is concerned with imparting views to the wider community and society as whole,[123] rather than communicating a personal desire to relatives and close friends. This provision may be of assistance where the deceased prescribes a specific type of funeral or graveside eulogy as a means of making a public statement – for example, the mother of a soldier killed in battle

---

117   Applying the core tests laid down in *R (Williamson) v Secretary of State for Employment and Education* [2005] 2 AC 246.

118   Though this did not prevent the claimant's own beliefs on the matter from falling within the scope of Article 9(1) – [2009] EWHC 978, [100]–[101].

119   The potential risk from open-air fires, as well as pollutants and emissions.

120   [2009] EWHC 978 (Admin), [12]–[14]

121   See generally P Cumper and T Lewis, "Last Rites and Human Rights: Funeral Pyres and Religious Freedom in the United Kingdom" (2010) 12 *Ecclesiastical Law Journal* 131.

122   See generally P van Dijk, F van Hoof, A van Rijn and L Zwaak (eds), *Theory and Practice of the European Convention on Human Rights* (Intersertia, 4th edn, 2006), ch 14 and C Grabenwarter, *European Convention on Human Rights: Commentary* (CH Beck, 2014), pp 251–296. There may also be some overlap with Article 9 here, since "expression of personal beliefs and ideas is for many an integral part of the holding of those beliefs and ideas" – RCA White and C Ovey, *Jacobs, White and Ovey: The European Convention on Human Rights* (OUP, 5th edn , 2010), p 425.

123   van Dijk *et al* (n 122), pp 778–783.

directing that the body be brought in a funeral procession past 10 Downing Street as a protest against the Prime Minister sanctioning that particular conflict.[124] However, most funeral directions would fall short of this high threshold and be outside the scope of Article 10(1).[125]

(d)   Posthumous human rights and competing entitlements?

Invoking Convention rights to safeguard funeral directions raises two fundamental issues. The first is whether the Convention can actually be used a means of upholding the rights of the dead, given its primary function as an instrument that safeguards the rights of living.[126] Judicial statements in a number of cases suggest that a human rights framework is inapplicable to dead bodies, one of the most emphatic rejections coming from the judgment of Peter Smith J in *Ibuna v Arroyo*:[127]

> I confess that I have some difficulty in a post-mortem application of human rights in relation to a body as if it has some independent right to be heard. ... [I]n my view there is no room further [*sic*] for any application of any human rights concepts to protect the right of the body to speak from death ...[128]

However, most of the cases discussed immediately above would seem to dispel this notion in the context of funeral arrangements, and the post-

---

124   The restrictions prescribed in Article 10(2) may also come into play here if, for example, freedom of expression would incite violence or impinge on public safety.

125   However, there may be scope for invoking Article 10 where the deceased prescribes a particular form of commemoration which, by its very nature, is a public statement – see Ch 8, pp 221–222.

126   "Legal orthodoxy ... stipulates that human rights are personal to living human beings and, therefore, expire upon the bearer's death" – Nwabueze (n 19), p 220.

127   [2012] EWHC 428 (Ch), [50].

128   In *R (Pretty) v DPP* [2002] 1 AC 800 the House of Lords suggested that Article 8 rights only apply while the individual is alive. However, these comments dealt with assisted suicide and the manner in which an individual departs from life, while the ECtHR in the subsequent decision in *Pretty v UK* (2002) 35 EHRR 1 did not accept that Article 8 has no relevance to the manner of leaving life. In contrast the ECtHR seemed to take a different view on the facts of *Estate of Kresten Filtenborg Mortensen v Denmark* (Application no. 1338/03, ECHR, 15 May 2006) (court doubted whether the deceased's estate could allege an interference with his corpse's right to private and family life under Article 8(1) following exhumation of the deceased's remains for DNA testing in a disputed paternity claim), while the following views were put forward in *Jones v UK* (Application no. 42639/04, ECHR, 13 September 2005): "[T]he exercise of Article 8 rights of family and private life pertain, predominantly, to relationships between living human beings. While it is not excluded that respect for family and private life extends to certain situations after death ... there is no right as such to obtain any particular mode of funeral or attendant burial features". However, this particular aspect of *Jones* was criticised by Cranston J at first instance in *Ghai v Newcastle City Council* [2009] EWHC 978 (Admin), [140].

mortem rights of the individual may still be operative.[129] At a more basic level, allowing an individual to stipulate the posthumous fate of their remains is arguably in keeping with the ethos of the Convention (and analogous instruments) since the "dignity and autonomy of the individual are fundamental values of legal systems that seek to honour and uphold human rights".[130]

Second, while funeral instructions are predicated on the deceased's wishes, surviving relatives could argue that these directions infringe their Convention rights. Case law suggests a legitimate family interest in what happens to the body of a dead relative,[131] which raises the question of whether courts would be willing to uphold the deceased's preferences where these conflicted with those of the deceased's family. For example, would a judge compel a family member who opposed cremation on religious grounds (and could possibly invoke Article 9(1) in support of this) to organise one in keeping with the deceased's instructions that his/her remains be dealt with in this way (such directions potentially falling within the scope of Article 8(1))? This clash of rights between the living and the dead is something that arises frequently in bodily disposal contests – and not just from a human rights perspective. However, the decision in *Burrows v HM Coroner for Preston*[132] is the strongest indication yet that courts will be swayed by the deceased's wishes when deliberating an alleged conflict of rights under the Convention.

4.   Reducing the potential for funeral disputes

Funeral disputes wreak havoc within families, inflicting extensive and often irreversible emotional damage.[133] While such cases can never be completely avoided, upholding funeral instructions might discourage adversarial litigation if the parties are aware that courts will simply rule in favour of the person who is intent on carrying out the deceased's wishes. Meanwhile, an individual who envisages post-mortem conflict – for example, between first and second families, or on grounds of divergent religious or cultural values –

---

129   Some of these cases have been brought by the living in an attempt to safeguard their interests after death – see for example, *X v Federal Republic of Germany* (1981) 24 D & R 137 and *Ghai v Newcastle City Council* [2009] EWHC 978. While there is no direct authority on whether an application can be made by someone acting on behalf of the deceased to uphold their directions where that person is already dead, such arguments were central features of *Burrows v HM Coroner for Preston* [2008] EWHC 1387 (Admin) and *Re Crawley Green Road Cemetery* [2001] 2 WLR 1175.

130   New Zealand Law Commission (n 77), p 77.

131   See for example, *Pannullo and Forte v France* (Application no. 37794/97, ECHR, 30 January 2002) and *Girard v France* (Application no. 22590/04, ECHR, 30 June 2011) discussed in Ch 4, pp 108–109.

132   [2008] EWHC 1387 (Admin).

133   See generally Ch 4.

can try to avoid this by setting out their preferences in advance.[134] Arguing in favour of a legislative framework that safeguards the wishes of the deceased, Naguit has suggested that:

> Not only [would] such provisions give people more confidence that their intentions will be carried out after death, but they also should help to eliminate … protracted disputes among survivors. … Certainty in the law is vital to ensuring that, when a decedent is finally laid to rest, she truly is able to rest in peace.[135]

Even where families are not divided, funeral directions can act as a guide when numerous decisions have to be made at an emotionally difficult time.

VI. The arguments against …

Several arguments can be made against upholding funeral directions, the first two raising the most substantive issues.

1. 'Dead people don't have rights'

Death results in a loss of autonomy, an inevitable ceding of control as the individual ceases to exist as a living being and a legal entity.[136] As Harris has argued:

> [A]utonomy involves the capacity to make choices, it involves acts of the will and the dead have no capacities – they have no will, no preferences, wants nor desires, the dead cannot be autonomous and so cannot have their autonomy violated.[137]

On this basis, funeral instructions serve no real purpose; the sentiments and preferences of the dead are immaterial because they no longer have rights that can be violated.

However, not everyone subscribes to the view that a person's rights and interests automatically expire when they do. Emphasising the core values of autonomy and human dignity, Smolensky has suggested that the dead, "although unable to make real-time choices, are capable of being legal

---

134 Kester (n 46), p 593.
135 Naguit (n 44), p 269.
136 "Naturally speaking the instant a man ceases to be, he ceases to have any dominion…" – William Blackstone, *Commentaries on the Law of England* (Tucker ed, 1803), pp 10–11, cited in R Chester, *From Here to Eternity* (Vandeplas Publishing, 2007), p 29.
137 Harris (n 90), p 531.

right-holders";[138] recognising certain posthumous legal rights not only adheres to the principle of self-determination but "gives the dead significant moral standing ... [where] lawmakers are driven by a desire to treat the dead with dignity".[139] Meanwhile, others have addressed the issue on the basis of antecedent rights that persist beyond an individual's demise, building on the dual concepts of prospective autonomy and anticipatory choices noted earlier.[140] For example, Young has claimed that the living "have an interest in what happens to their bodies after they die, even though those people will no longer exist at the relevant time",[141] while Sperling highlights the ability to express one's own character and values in the bodily disposal context as an important aspect of the right to self-determination:

> Autonomy is first and foremost the moral privilege of a person to cultivate and nurture her particular vision of herself as a human being. It is the prerogative of shaping the images, conceptions and recollections which other persons have or will have of her regardless of whether she will physically witness those images, conceptions and recollections.[142]

These and similar narratives suggest that the law is not conferring positive rights on the dead; certain posthumous interests have their origin in the ante-mortem individual. Closely aligned with this are notions of harming the dead – a subject of intense philosophical debate. While some have rejected the idea that the dead can suffer harm,[143] others have identified the ante-mortem person as the subject of any damage inflicted by posthumous transgressions.[144] Thus we cannot simply say that frustrating someone's

---

138   KR Smolensky, "Rights of the Dead" (2009) 37 *Hofstra Law Review* 763, p 764. See also MH Kramer, "Do Animals and Dead People Have Legal Rights?" (2001) 14 *Canadian Journal of Law & Jurisprudence* 29.

139   Smolensky (n 138), p 764.

140   See pp 136–139.

141   H Young, "The Right to Posthumous Bodily Integrity and Implications of Whose Right It Is" (2013) 14 *Marquette Elder's Advisor* 197, p 211.

142   Sperling (n 85), p 147. Meanwhile, others have argued that autonomous choices should be upheld after death because an individual is usually best-placed to decide what should happen to his/her body – see generally S McGuinness and M Brazier, "Respecting the Living Means Respecting the Dead Too" (2008) 28 *Oxford Journal of Legal Studies* 297.

143   See for example, JC Callahan, "On Harming the Dead" (1987) 97 *Ethics* 341; E Partridge, "Posthumous Interests and Posthumous Respect" (2001) 91 *Ethics* 243; W Glannon, "Persons, Lives and Posthumous Harms" (2001) 32 *Journal of Social Philosophy* 127; JS Taylor, "The Myth of Posthumous Harm" (2005) 42 *American Philosophical Quarterly* 311; and JS Taylor, *Death, Posthumous Harm and Bioethics* (Routledge, 2012).

144   See for example, G Pitcher, "The Misfortunes of the Dead" (1984) 21 *American Philosophical Quarterly* 183; D Grover, "Posthumous Harm" (1989) 39 *Philosophical Quarterly* Q 334; G Scarre, "Privacy and the Dead" (2012) 19 *Philosophy in the Contemporary World* 1; and G Scarre, "Speaking of the Dead" (2012) Mortality 36. See also D Price, "Property, Harm and the Corpse" in B Brooks-Gordon, F Ebtehaj, J Herring, MH Johnson and M Richards (eds), *Death Rites and Rights* (Hart, 2007).

funeral directions is morally acceptable because the dead cannot experience emotional upset; the real harm is inflicted on the autonomous living person.[145]

Moving beyond a rights-based discourse, it has been suggested that discharging funeral instructions is embedded within the basic social and moral institution of keeping promises. Part of this entails fulfilling our obligations to the dead, based on who they were in life and the existence of relational ties that transcend death.[146] More important, perhaps, are what we might term 'reciprocity' arguments around the practice of promise-keeping. In other words, it is not simply a question of obligations to the dead; upholding funeral instructions has a public interest dimension because it reassures the living that, if they respect the deceased's wishes, there is a stronger likelihood of their own preferences being respected.[147] This, in itself, "reaffirms the social institution of making promises and builds faith among members of the society".[148]

2.   The wishes of the dead versus those of the living

Death is not just about the deceased; it involves the living, and in particular the deceased's family as the social group most affected by death.[149] For them, arranging the funeral goes beyond disposal of the dead; the various rituals are an essential part of the grieving process. As Hernández explains:

> [A] funeral service addresses the emotional needs of a decedent's survivors by providing a socially acceptable outlet for feelings of grief and pain. Planning the funeral service also assists survivors in coming to terms with the loss and their grief ...[150]

However, problems arise where the deceased and their family had different expectations around, for example, the manner of disposal or attendant funeral rites.[151]

---

145   See Nwabueze (n 19), p 203.
146   See the discussion in B Brecher, "Our Obligation to the Dead" (2002) 19 *Journal of Applied Philosophy* 109.
147   See Young (n 141), pp 200–201 ("[t]he reason for respecting people's wishes, even after they are dead, is to give comfort to the living").
148   Sperling (n 85), p 169.
149   M Bowen, "Family Reaction to Death", in F Walsh and M McGoldrick (eds), *Living beyond Loss: Death in the Family* (WW Norton and Co, 1991).
150   Hernández (n 63), p 991.
151   "[F]unerals are important both for the dead and for the survivors, which raises interesting problems ...where the deceased and survivors had different expectations. ... Whose interests should prevail: the deceased's interest in the posthumous identity and memorialisation; or the survivors' interest in obtaining psychological relief through a ritualised mourning of their choosing?" – Nwabueze (n 19), p 202.

Succession law is one area that has always had to mediate between the dual concepts of individualism and familial interests. The principle of testamentary freedom reigns supreme in common law jurisdictions but must yield in most to the family provision system (or its equivalent) whereby certain relatives and dependants can challenge the deceased's estate distribution based on financial ties.[152] Thus death is not simply about respecting the clear and unambiguous wishes of the deceased; the needs of others will vie for priority as a result of financial obligations that transcend death. Of course, the same economic rationale does not apply to funeral instructions. More importantly, family provision does not give courts *carte blanche* to override the deceased's wishes simply because the applicant seeks a different outcome. Judges can only encroach on testamentary freedom where a dissatisfied relative or dependant satisfies the statutory criteria; otherwise, the wishes of the dead prevail, in keeping with succession law norms.

Turning to medical law, family objections cannot defeat a patient's wishes under a valid advance directive. In the post-mortem context, organ donation can also be a source of conflict where the deceased's decision to donate is countered by surviving relatives.[153] The Human Tissue Act 2004 clearly prioritises the wishes of the deceased, as does the accompanying code of practice; family members do not have a legal right of veto, even if the medical profession affords them a de facto one by declining to retrieve organs in the face of familial objections.[154] Nonetheless, the legislative ethos is clear: respect for autonomy, alongside utilitarian arguments for increasing the available supply of organs,[155] ensures that the individual's donative intent is paramount.

With funeral instructions, the core concepts of individual autonomy and family interest values cannot be reconciled in any meaningful way if the two conflict. However, the increased emphasis on self-determination and associated ideas such as personhood, human dignity and the rights of the individual[156] are powerful arguments for upholding the deceased's preferences; a family power of veto would be a significant encroachment on these core ideals.[157] Meanwhile, others have challenged the notion that funerals are predominantly about the bereaved. As Horan has remarked:

---

152   The Inheritance (Provision for Family and Dependants) Act 1975 is the governing legislation in England and Wales – see generally RD Oughton, *Tyler's Family Provision* (Tottel, 3rd edn, 1998). For similar mechanisms elsewhere, see L Englefeld, *Australian Family Provision Law* (Lawbook Company, 2011) and C Harvey and L Vincent, *The Law of Dependants' Relief in Canada* (Carswell, 2nd edn, 2006).

153   Organ donation is discussed in more detail in Ch 6.

154   Ch 6, pp 165–167.

155   Of course, bodily disposal preferences have a less apparent social benefit.

156   "[L]aws granting individuals the right to decide how their corpses will be disposed of exemplify a legal right to posthumous bodily integrity that takes precedence over the wishes of family" – Young (n 141), p 251.

157   This theme is discussed further in pp 157–158.

Today, funerals are not just about the living while the dead go on to a 'better place'. People have expectations about their own autonomy to decide what will happen to their bodies after they die.[158]

Allowing the dead to have the final say, while initially distressing, might benefit the living in the longer-term. Vetoing a loved one's funeral instructions could be a source of regret – in the same way as relatives who disregard the deceased's intent to be an organ donor often regret this decision.[159]

3.   Replacing one form of litigation with another

Upholding the deceased's preferences would not eliminate funeral disputes; instead, the focus would probably shift to arguing over what the deceased's wishes actually were, and whether they should be rejected on any grounds.[160] To some extent this is unavoidable. However, courts would have a clear reference point, and the overall number of legal disputes might be reduced.

VII. Implementing funeral instructions: Options and issues for reform

If funeral instructions are to be upheld in England and Wales (and in other jurisdictions contemplating similar reforms), how can this be done? Changing the current law through a combination of judicial pragmatism and rights-based discourse would take time. A specific legal mechanism would be more effective, and two options are canvassed here.[161]

1.   National register

The government could introduce a national register, allowing individuals to enter their personal details and record basic funeral preferences in a similar way to the existing NHS organ donor register.[162] The potential benefits are obvious. As a centralised, computerised database, the register would be easy to access after a person's death, and would give a clear indication of their wishes. Online entries would also be easy to update, if someone's family circumstances or personal preferences changed, and the person organising

---

158   JE Horan, "'When Sleep at Last Has Come': Controlling the Disposition of Dead Bodies for Same-Sex Couples" (1999) 2 *Journal of Gender, Race and Justice* 423, p 459. The growing trend towards personalised funerals is one example of this.

159   See Ch 6, n 46.

160   The American approach has been criticised on this basis – see L Griggs and K Mackie, "Burial Rights: The Contemporary Australian Position" (2007) 7 *Journal of Law and Medicine* 404, pp 408–409.

161   For several other options see Sperling (n 85), pp 154–165 (though the author acknowledges that some of these would not work in practice).

162   See www.organdonation.nhs.uk (accessed 30 September 2015).

the funeral (as well as funeral service providers) could be legally obliged to consult the register before making the necessary arrangements.

However, there are drawbacks – including the financial costs of setting up and maintaining a state-run register. Any scheme would have to be developed from first principles; and since funeral instructions operate in the private sphere, the register would not have any discernible public benefit (unlike the organ donor register) and might not be classed as a funding priority. Even if a centralised system were established, uptake might be low given that apathy and a general reluctance to contemplate one's mortality have been well documented in other aspects of posthumous choice.[163]

2.   A new statutory framework

Legislative reform offers a simple and cost-effective means of transforming the current law, and one that can draw on numerous precedents from elsewhere. The previous chapter envisaged a new statutory priority ranking for funeral disputes, with a revised list of designated decision-makers.[164] The same statute could prioritise the deceased's funeral instructions by allowing anyone aged 18 or over[165] to stipulate the method of disposal, any attendant funeral rites, and the desired gravesite (or who should retain any ashes that are not being scattered or interred). Legal responsibility for implementing these directions would fall on the designated decision-maker, or an agent or proxy nominated by the deceased if the legislation contemplated such an arrangement.[166] The deceased's instructions would also be the focal point of any subsequent dispute where, for example, another family member wanted to substitute alternative funeral arrangements.[167] If such directions were impermissible,[168] or the deceased simply failed to list any, the law would revert to the new statutory hierarchy.

---

163   Particularly in the will-making context – see A Humphrey, G Morrell, L Mills, G Douglas and H Woodward, *Inheritance and the Family: Attitudes to Will-Making and Intestacy* (National Centre for Social Research, August 2010), p 13.

164   See Ch 4, Pt VI.

165   Children who die before the age of majority lack effective decision-making capacity, and will not normally have made their wishes known. However, isolated examples do exist – see *Burrows v HM Coroner for Preston* [2008] EWHC 1387 (Admin) and *L.A.W. v Children's Aid Society of the District of Rainy River* [2005] OJ No 1446 where the individuals in question (both teenage boys) had made their funeral preferences clear.

166   If the designated decision-maker, agent or proxy was unwilling to implement the deceased's instructions, the individual would be expected to step aside – see below.

167   If the statute is reflecting a conscious policy decision that an individual should be allowed to determine the disposal of their body, then family wishes must be subordinate to what the deceased wants.

168   See pp 153–157.

(a) Formal requirements: Mode of expression

Most bodily disposal statutes[169] insist on written instructions, and any new legislative scheme would include the same basic requirement. A will is an obvious example, where funeral directions are part of a solemn indication of dispositive intent, which has been signed and witnessed accordingly. Of course, certain allowances would have to be made – for example, avoiding the need for probate before the deceased's remains could be disposed of accordingly.[170] More difficult issues would arise if there were question-marks over the validity of the deceased's will due, for example, to non-compliance with the statutory formalities,[171] or lack of capacity.[172] If raised in the immediate post-mortem period, this might persuade courts to sanction alternative funeral arrangements.[173]

Of course, it is a statistical reality that large numbers of people fail to make wills,[174] and even where one has been executed it is often not consulted until after the funeral.[175] Funeral instructions could also be set out in a pre-paid funeral plan;[176] other documentary alternatives might include an advance directive or other legal memorandum, as well as a standardised form (if one were introduced as part of any new legislative framework[177]) or a signed and dated letter of instructions left by the deceased. An issue which might arise here is the existence of conflicting funeral instructions across a number of

---

169   Those found throughout the US and in the Canadian province of British Columbia.

170   Drawing on the US experience where funeral instructions contained in a will are classed as non-testamentary dispositions and not subject to the same technical requirements, including probate – see p 132.

171   Under the Wills Act 1837 (as amended) in England and Wales.

172   According to the tests enshrined in *Banks v Goodfellow* (1870) LR 5 QB 549. Although the higher mental capacity threshold for will-making does not apply to funeral directions (given that these are not testamentary dispositions), the fact that a will fails for lack of capacity may also raise doubts as to whether the deceased had the necessary capacity to set out funeral instructions as well. Other possible grounds of invalidity include duress and undue influence – see pp 157–158.

173   Of course, this assumes both an urgent court application and sufficient evidence to dispute the will's validity. For an illustration see *University Hospital Lewisham Trust v Hamuth* [2006] EWHC 1609 (Ch) discussed at n 214.

174   Recent estimates suggest that around one-third of adults in England and Wales have made a will – Humphrey *et al* (n 163), p 13.

175   See Hernández (n 63), p 1020 and Kester (n 46), p 584. This can be addressed by the individual in question ensuring that surviving relatives are aware of specific funeral instructions, and leaving a copy of the will alongside any personal belongings.

176   Foster has suggested that courts should be even more determined to uphold directions set out in a pre-paid funeral plan because the deceased has paid for everything in advance – see Foster (n 64), pp 1377–1378 and the sources cited therein. A small number of US states only allow funeral instructions where the deceased has already paid for funeral costs – see for example, Idaho Code § 54-1139(1)-(2).

177   Again, drawing on the US example – see pp 133–134.

documents; in these circumstances, the legislation could simply provide that the latest dated document prevailed.[178]

Oral instructions might also be allowed in two situations. The first is where the deceased left no written instructions, but has made clear (and repeated) statements about their funeral preferences in front of others. The second is where the deceased's oral statements clearly contradict a previous written directive, something that American courts have allowed if there is "convincing ... oral evidence of a change in intent".[179] A good illustration is *Sacred Heart of Jesus Polish National Church v Soklowski*,[180] where the deceased was buried in ground consecrated according to the rites of the Catholic church rather than in the graveyard of the schismatic church which he had founded, despite repeatedly having said that he wanted to be buried in the latter.[181] The decision was prompted by the fact that the deceased, on his deathbed, had called for a Catholic priest to administer last rites.[182] Any statutory framework could only sanction oral directions in exceptional circumstances such as these.[183] Additional safeguards could also be implemented – for example, that any verbal statement must have been made in the presence of two or more people – while generic rules about admissibility of oral evidence in civil law proceedings would also apply.

(b)   Enforceability and effective compliance?

Ensuring that funeral instructions are legally binding is more problematic. The fact that the individual is dead and unable to enforce their own directives creates obvious problems, as Smolensky explains:

> The rights of the living are most often enforced because the individual whose right had been violated speaks up ... But the dead are physically incapable of enforcing their posthumous rights and keeping their postmortem affairs in order.[184]

---

178   Another option would be to follow the Minnesota example, and prioritise the witnessed and notarised document – see n 71.
179   *Cohen v Guardianship of Cohen* 896 So2d 950, 955 (2005). See also *In re Scheck's Estate* 14 NYS2d 946 (1936) discussed at p 133.
180   199 NW 81 (Minn 1924).
181   This 'breakaway' church followed certain Catholic doctrines, but did not recognise the Pope as its head.
182   However, a deathbed change of intent may be ineffective due to lack of capacity – as in *Curlin v Curlin* 228 SW 602 (Tex Civ App 1921) where the deceased had been delirious and only occasionally able to recognise members of his immediate family for several days before he died.
183   While oral directions would not allow others to deviate from the distributive contents of the deceased's will, the position is different with funeral instructions since these are not testamentary provisions.
184   Smolensky (n 138), p 799.

However, there are still ways of ensuring that the living abide by the wishes of the dead. Several jurisdictions have legislation that specifically states that the deceased's funeral instructions are binding on the designated decision-maker (or anyone else taking charge of the funeral), who has a duty to fulfil them insofar as this is possible.[185] A similar approach could be adopted in the statutory framework being proposed here, with further measures to guard against the possibility of deliberate non-compliance. Take the scenario where the designated decision-maker is intent on ignoring the deceased's funeral instructions, either because of their own objections or due to pressure from (other) family members. In these circumstances, the legislation might allow someone lower down or outside the statutory hierarchy to usurp the decision-maker's authority, on the basis that the applicant intends to carry out the deceased's instructions.[186] Another option is to impose some sort of penalty for deliberate non-compliance with the deceased's funeral instructions.[187] Of course, if all the parties involved in the funeral arrangements choose to ignore these instructions there is little that can be done;[188] unless someone else contests the decision, the surviving relatives are free to implement their own choice of funeral arrangements, and no statutory regime could prevent this.[189]

3. Legal, practical and public policy constraints

To be permissible, funeral directions must be lawful, practical and not contrary to public policy. The first two limitations are relatively straightforward. What is legal is dictated by existing bodily disposal laws, as well as public health concerns.[190] Thus wanting to be interred in a specific cemetery or to be cremated will cause few problems, unless the individual has died from a notifiable disease[191] or the proposed cremation is on an outdoor funeral

---

185   See s 6 of the Cremation, Interment and Funeral Services Act 2004 in British Columbia, as well as the respective provisions in Ariz Rev Stat Ann § 36-831.01 (Arizona) and Tex Health and Safety Code Ann § 711.002 (Texas).

186   This is effectively what happens in British Columbia under the Cremation, Interment and Funeral Services Act 2004, s 5(6) – see Ch 4, p 118–119.

187   French law imposes criminal law sanctions on those who knowingly contravene the deceased's funeral wishes contained in a will – Rodriguez-Dod (n 80), p 324 citing Art 433-21-1 of the Penal Code (Le Code Pénal). The punishment is significant – namely, a prison sentence of 6 months and a fine of (what is currently) €7,500.

188   "If the family gets the body into the ground quickly enough in the manner they wish, there is no legal recourse, and the family may countermand projected desires of the deceased" – Groll and Kerwin (n 81), p 285.

189   "[A] decedent's wishes regarding cadaver disposal may, in practice if not in law, be dependent on survivors' faithfulness. If survivors unanimously favor breach of instructions, that dereliction may simply become a fait accompli" – Cantor (n 91), p 52.

190   See Ch 2.

191   See Ch 3, n 7.

pyre.[192] Fire safety laws would also influence the latter scenario, in the same way as road safety laws were invoked in 2011 to prevent the family of a Pembrokeshire woman from placing garden gnomes along part of her funeral route in fulfilment of her last wishes.[193] Meanwhile, funeral directions will be impractical where they are unworkable, incapable of being performed or not financially viable – for example, where a chosen cemetery is not accepting new interments,[194] or there are insufficient funds in the deceased's estate to pay for the extravagant send-off and memorialisation that they wanted.[195]

Basic constructs of decency alongside normative behaviours around disposal of the dead bring the third limitation into play. Public policy – a generic term for basic constructs of social welfare and morality that the law regards as paramount to individual preferences – has a significant role to play in curbing certain types of behaviour.[196] Mere whimsical or idiosyncratic directions would probably be permissible – for example, someone who wanted to be buried standing up could stipulate this[197] but not where the individual is 6ft tall and insists on being buried in a grave which is only 5ft

192   Which is unlawful, according to the first instance decision in *Ghai v Newcastle City Council* [2009] EWCH 978 (Admin). Likewise, someone who embraced a nudist lifestyle could not insist on being on being laid out naked at their funeral ceremony, since this would breach the old common law rule that a corpse should be decently covered (see Ch 2, p 30). Illinois naturist Robert Norton's last request to be buried in the nude was ignored by his family, despite the fact that the 82-year-old had asserted a constitutional right to nakedness and been arrested numerous times over a 40-year period – see "Nudist to be Buried Clothed", *Chicago Tribune* (Chicago, 3 August 2005) http://articles. chicagotribune.com/2005-08-03/news/0508030440_1_arrests-buried-robert-norton (accessed 30 September 2015).

193   The deceased had collected garden gnomes, and her family had placed 30 of them on a roundabout which the funeral cortege would pass. The gnomes were subsequently removed by council workers – "Grandmother's Last Wish to Have Gnomes Lining Funeral Route Scuppered – By 'ELF and Safety Killjoys", *Daily Mail* (London, 4 May 2011), www.dailymail.co.uk/news/article-1383075/Grandmothers-wish-gnomes-lining-funeral-route-scuppered-health-safety.html#ixzz1RK8wwS4y (accessed 30 September 2015).

194   See *In re Seymour's Estate* 15 Cal App 287 (1911).

195   Funeral costs are recovered from the deceased's estate – see Ch 3, Pt II.

196   According to Crockett J in *Estate of Moyer v Moyer* 577 P2d 108, 100 (1978), the right of an individual to direct the posthumous fate of their remains is "so involved in the public interest … that it is not subject entirely to the desires, or the whim or caprice of individuals".

197   Cantor (n 91), p 56 cites the example of a Brazilian legislator who, in 2005, had requested that he be buried standing up because he had "bowed to no one" in life.

deep,[198] while requesting a novelty coffin or urn would be allowed[199] as long as it was not adorned with abusive slogans or symbols.[200] In contrast, funeral practices which are deemed indecent, offensive or contrary to public mores would almost certainly be prohibited. Take the example of someone living in England who wants their remains to decompose naturally on the surface of their own land or requests a Tibet-style sky burial whereby the body is cut into pieces and devoured by birds of prey. Both methods are more environmentally friendly than burial or cremation and need not endanger public health, yet would probably be impermissible in this country – considerations of public decency premised on societal notions of respect for the dead and a general abhorrence towards desecration of the body would almost certainly prevail,[201] even if what was proposed was taking place on private property hidden from public view.[202]

Another controversial type of funeral instruction is where the deceased insists on specific items of property being buried with them, or otherwise

---

198 Such directions would probably be void, especially if the proposed burial was on public land or on private land with an open view – see the discussion in A Bove and M Langa "Ted Williams: Is He Headed for the Dugout or the Deep Freeze? Property Rights in a Dead Body Resurrected", *Massachusetts Lawyers Weekly*, 19 August 2002.

199 For example Swindon mother-of-two Karen Lloyd loved Costa coffee and was laid to rest in a bespoke coffin in the coffee chain's burgundy colour, with the Costa logo on one side and Karen's standard coffee order written on the other – J White and D Wilkes, "Latte me to rest! Coffee-Lover Who Died After Battle with Cancer is Buried in Costa-Themed Coffin... With Her Order Written on the Side", *Daily Mail* (London, 17 February 2014) www.dailymail.co.uk/news/article-2561169/Making-mocha-ry-tradition-Coffee-lover-buried-Costa-themed-coffin.html (accessed 30 September 2015).

200 For example, a Nazi sympathiser insisting on a coffin adorned with swastikas.

201 The common law offence of outraging public decency (see Ch 1, pp 23–24) might also come into play.

202 Similar concerns were apparent in the first instance decision in *Ghai v Newcastle City Council* [2009] EWCH 978 (Admin), the spectre of human remains being burned on an outdoor funeral pyre taking priority to the applicant's own religious and cultural beliefs.

destroyed.[203] Although commonplace in ancient times,[204] modern courts (especially those in America where the issue has arisen more frequently) tend to invalidate them.[205] One of the most well-known is *In re Meksras' Estate*[206] in which the deceased's direction that she be buried along with valuable jewellery was deemed contrary to public policy in encouraging the violation of cemeteries,[207] though other ostentatious and seemingly wasteful directions have occasionally been upheld – notably those of Sandra West, a wealthy oil heiress, who died in 1977 having requested that she be buried in a lace nightgown at the wheel of her beloved blue vintage Ferrari.[208] In contrast, directions to destroy the deceased's pet and bury it with them – even if the motive is to protect the animal from subsequent mistreatment or abuse – are generally regarded as contrary to public policy.[209]

Any new legal framework for implementing funeral instructions should stipulate what is permissible – or at least set basic parameters. For example, the British Columbia statute only mandates compliance where this "would not be unreasonable or impracticable or cause hardship",[210] while the draft Queensland legislation is much more comprehensive in refusing to recognise

---

203   While an individual can request that specific items (e.g. diaries, personal correspondence) are destroyed after their demise, the emphasis here is on destruction of property as an adjunct to bodily disposal. An extreme example is that of a Japanese businessman who insisted that a Van Gogh painting, which he had purchased almost a decade earlier for $82.5 million, be cremated with him so that his children would avoid colossal death duties – D Usburne, "Missing Van Gogh Feared Cremated With Its Owner", *The Independent* (London, 27 July 1999) www.independent.co.uk/news/world/missing-van-gogh-feared-cremated-with-its-owner-1108973.html (accessed 30 September 2015). For an overview of the legal and policy issues posed by such requests, see A Sykas, "Waste Not, Want Not: Can the Public Policy Doctrine Prohibit the Destruction of Property by Testamentary Direction?" (2001) 25 *Vermont Law Review* 911. See also T Wear, "Wills: Direction in Will to Destroy Estate Property Violates Public Policy" (1976) 41 *Missouri Law Review* 309 and LJ Strahilevitz, "The Right to Destroy" (2005) 114 *Yale Law Journal* 781.

204   When property owners were often buried with items of personal property, and sometimes their favourite slaves or pets, fulfilling people's expectations that "the comforts of their previous lives [would] … accompany them into death" – Sykas (n 203), p 911.

205   See Sperling (n 85), pp 177–178.

206   63 Pa D & C 2d 371 (1974).

207   See also *Estate of Moyer v Moyer* 577 P2d 108 (1978) (courts reluctant to uphold directions that would "require extravagant waste of useful property or resources, or be offensive to the normal sensibilities of society").

208   To accommodate this, the deceased and her car were loaded into a concrete container and lowered into the grave by a crane – see L Evers, *I Told You I Was Ill: Dying for a Laugh* (Michael O'Mara Books, 2012), p 46.

209   See Sykas (n 203), pp 939–943. In 2013, the family of a blind American woman who had her beloved guide dog put down so that it could be buried with her (in accordance with the deceased's wishes) were subjected to a hate campaign and death threats – see O Williams, "How Could They Kill a Healthy Guide Dog?", *Daily Mail* (London, 4 April 2013) www.dailymail.co.uk/news/article-2303841 (accessed 30 September 2015).

210   Cremation, Interment and Funeral Services Act 2004, s 6(c).

funeral instructions that are unlawful, impossible to carry out or impracticable, offensive or indecent, contrary to public health or safety, or unreasonable when viewed against the net value of the deceased's estate.[211]

4.   Should funeral instructions be susceptible to challenge?

Should otherwise valid and permissible funeral instructions be susceptible to any form of legal challenge? If such directions are to be paramount, then the question has already been answered: the autonomous choices of the dead should not be disregarded simply because the living would have preferred something different. This is not to suggest that the designated decision-maker should be compelled to carry out funeral directions that they are fundamentally opposed to; they should simply stand aside and allow someone else to take charge of the funeral arrangements, or face being removed by the court.[212]

However, the absence of an automatic right of veto does not mean that funeral instructions can never be questioned. A material change in circumstances is one possibility – for example, where the deceased stipulated that he wanted to be buried with his wife, but the couple have since divorced. As with most legal transactions, funeral instructions may also be scrutinised where there is an allegation of fraud, which is relatively straightforward in itself. Questions will also be asked where duress or undue influence may have tainted the deceased's choices, or there are doubts about the deceased's mental capacity when these were set out. For example, in *Rosenblum v New Mount Sinai Cemetery Association*[213] the fact that the deceased's personality and attitude towards his family had changed significantly in the last two months of his life due to the combined effects of arteriosclerosis and terminal cancer prompted the court to reject his last burial request, because the deceased was not deemed competent during this time.[214] Meanwhile, in *Cottingham v McKee*[215] the court indicated that it

---

211   Draft Cremations and Other Legislation Amendment Bill 2011, s 4A(2).
212   For example, in *Kasmer v Guardianship of Limner* 697 So2d 220 (1997), an executor who opposed the deceased's expressed wish to be cremated (because he disagreed with this particular method of disposal) was instructed to comply with the deceased's request or stand aside.
213   481 SW 2D 593 (1972).
214   See also *Hunter v Hunter* (1930) 65 OLR 586 (deceased's final wishes disregarded, partly due to concerns over his mental state), as well as *Curlin v Curlin* 228 SW 602 (Tex Civ App 1921) mentioned at n 182. In *University Hospital Lewisham Trust v Hamuth* [2006] EWHC 1609 (Ch), a perceived lack of testamentary capacity resulted in the court refusing to grant custody of the deceased's remains to his executor. The alleged will had instructed the executor (a nurse in the nursing home where the deceased had been residing) to cremate the deceased's remains; this was opposed by members of the deceased's family who wanted to inter his body in the family burial plot.
215   821 So2d 169 (2001).

would have been prepared to disregard the deceased's direction that his remains be cremated if his mother had been able to substantiate an allegation of undue influence.

The use of such doctrines to invalidate funeral instructions is unobjectionable in itself, since courts should always be alert to the dangers of duress, undue influence or lack of capacity. However, judges should refrain from simply using them to strike out individual preferences, which clash with family expectations. The decision in *Williams v Williams*[216] is a classic example of this, favouring the surviving widow and children despite the deceased wanting to be cremated and entrusting a friend with this task. As one commentator has observed:

> The court's concern with the tender sentiments of the deceased's wife ... set [it] in search of a legal wrack on which to hang [its] emotions.[217]

Of course, the outcome in *Williams* was not disguised under claims of duress, undue influence or lack of capacity, yet there is a risk that courts might use these doctrines to disregard funeral instructions where family members object and judicial sympathies lie with the bereaved. Similar trends have been noted in wills contests, especially in America where courts have encroached on the principle of testamentary freedom by being more amenable to undue influence and incapacity arguments where an estate distribution excludes the deceased's close family.[218] The same approach appears to have been adopted in some funeral disputes,[219] and English courts might be tempted to follow suit if someone's funeral directions do not meet family expectations. However, the urge to manipulate these doctrines must be resisted if the deceased's preferences are to be paramount under any new legislative regime.

Conclusion

Despite an instinctive aversion to considering their own mortality, many people have definite views on whether they want to be buried or cremated,

---

216   (1882) 20 Ch D 659 discussed at p 127.
217   Barish (n 47), p 41.
218   See for example, R Madoff, "Unmasking Undue Influence" (1997) 81 *Minnesota Law Review* 571, p 576 ("the undue influence doctrine denies freedom of testation for people who deviate from judicially imposed testamentary norms") and FH Foster, "The Family Paradigm of Inheritance Law" (2001) 80 *North Carolina Law Review* 199, p 211 (wills that exclude the deceased's next-of-kin raise "judicial red flags" and are more susceptible to undue influence or incapacity arguments).
219   Hernández (n 63), pp 986–988.

and what form their funeral should take.[220] Occasionally mentioned in conversations with close relatives and friends, individual preferences are more often written into wills and pre-paid funeral plans. Yet, as English law currently stands, the deceased's advance choices (whether written or oral) do not have binding force; in most instances, the personal representative has decision-making authority, and can implement something different.

When it comes to disposal of a corpse or ashes, there are strong arguments for allowing an individual to dictate their own post-mortem fate – beyond providing an obvious reference point for courts faced with funeral disputes, when surviving relatives cannot agree.[221] Upholding funeral instructions is not just about respect for the dead; it embodies fundamental notions of self-determination and autonomy, by protecting the rights of the living to determine what happens to their bodies after they die through clearly articulated (and otherwise legal) ante-mortem preferences. This is something that is also reflected in the law's treatment of organ donation and posthumous reproduction, as the following chapter illustrates.

220   Of course, communicating these preferences is another matter, given widespread social taboos around death. Despite suggestions in a recent survey that 65 per cent of people questioned had thought about their own funeral plans, only 1 per cent of those questioned who had organised a funeral recently knew *all* of their loved one's funeral preferences, with 31 per cent admitting they did not know whether the deceased had wanted to be buried or cremated – Sun Life, *Cost of Dying: 2015* (13 October 2015) www.sunlifedirect.co.uk/press-office/cost-of-dying-2015/ (accessed 15 October 2015).
221   Family disagreements aside, an expression of wishes would help surviving relatives to organise the funeral at an emotionally difficult time.

6 Utilising parts of the dead

Organ donation and posthumous reproduction

Today our thoughts regarding the disposition of our bodies [are] not limited to the wake, the shroud, the casket and the funeral.[1]

Introduction

Corpses are not simply inanimate beings destined for speedy disposal; they play a vital role in medical research and teaching, as well as educational exhibits and museum displays,[2] and in scientific experiments to give a few examples.[3] While the subject-matter is dead material (either the intact corpse or specific bodily components), living material supplied by dead donors can also be put to various uses – setting surviving relatives on a collision course before thoughts have even turned to the funeral. Two of the most common scenarios involve organs used for transplant purposes, and reproductive material to facilitate post-mortem genetic parenting.

Both technologies entered the medical scene in the latter half of the twentieth century, and raise a host of legal and ethical issues. In analysing the law's response to cadaveric organ transplants and posthumous reproduction, this chapter focuses on the respective legal frameworks, and who has decision-making authority under each. It also examines potential conflicts between the wishes of the dead and those of the living or within the deceased's family itself (recurring themes in any discourse involving the dead), and questions the legal status of excised human material.

---

1   RC Groll and DJ Kerwin, "The Uniform Anatomical Gift Act: Is the Right to a Decent Burial Obsolete?" (1971) 2 *Loyola University Chicago Law Journal* 275, p 277.

2   For example, in the *Body Worlds* exhibit – see Ch 2, pp 52–53.

3   Both mainstream and less conventional – for example, decomposition tests in body farms, practice sites for plastic surgery, or crash test dummies for aircraft and car safety standards. For a humorous account, see M Roach, *Stiff: The Curious Life of Human Cadavers* (Penguin, 2003).

I.   Organ donation

Major medical strides have been made since the first successful kidney transplant in the 1950s, with several thousand organ transplants carried out annually in the UK.[4] While these include living and dead donors, the latter account for the majority of transplant procedures, offering the hope of life-saving (or life-prolonging) treatment to growing numbers of people.

Pattison describes the regulation and practice of cadaveric transplantation as an "ethical minefield"[5] – hardly surprising given the complex issues at play.[6] Individual legal systems must strike a delicate balance between the rights of the donor, the interests of their family and the public need for organs,[7] while recognising deep-rooted social, religious and cultural views on the inviolability of the human corpse that shape contemporary attitudes to donation.[8] Meanwhile, the success of cadaveric organ transplants has also forced certain changes in the law. Since most major organs have to be removed within a short time of death and only remain transplantable for a few hours, one of the most important has been the introduction of the brain-stem death standard alongside the traditional concept of cardiopulmonary death.[9] And while the preoccupation with being able to tell whether or not someone is 'really dead' is not a novel one,[10] it has added resonance here. As Mims explains:

> At one time the main worry was that people might be buried before they had actually died. … For organ transplantation, however, time is of the essence, and fear of premature burial has been replaced by fear of premature organ removal. The concept of 'brain death' was introduced to assuage this anxiety.[11]

---

4   Updated statistics are available at www.organdonation.nhs.uk/supporting-my-decision/statistics-about-organ-donation/ (accessed 30 September 2015). While tissues can also be donated, the following discussion only refers to organs (though the legal position is exactly the same).

5   SD Pattinson, *Medical Law and Ethics* (Sweet & Maxwell, 4th edn, 2014), p 419.

6   See D Price, *Legal and Ethical Aspects of Organ Transplantation* (CUP, 2000).

7   See RT Munoz and MD Fox, "Legal Aspects of Brain Death and Organ Donorship" in D Novitzky and DKC Cooper (eds), *The Brain-Dead Organ Donor* (Springer, 2013).

8   The allocation of donated organs also raises complex moral, ethical and legal issues – though these are not addressed here.

9   See Ch 1, Pt I.

10 For example, Quay points to the "long history of legal and moral disquisitions concerning wakes, embalming, the possibility of burying alive, [and] piercing the heart 'just to make sure'" – PM Quay, "Utilizing the Bodies of Dead" (1984) 28 *St Louis University Law Journal* 889, p 890.

11 C Mims, *When We Die: The Science, Culture and Rituals of Death* (Robinson, 2000), p 260.

By defining and applying strict diagnostic standards for brain-stem death,[12] the law reassures private citizens that they cannot simply be pronounced dead by overly-zealous transplant surgeons, loitering by the deathbed to harvest organs for needy recipients. However, altering the definition of death is only one aspect of the law's response to transplantation; legislators also had to introduce a specific legal mechanism for donating organs at the point of death. English law favours an 'opt-in' system, whereby organs can only be removed if the deceased consented to this while alive, or (failing that) surviving relatives sanction their retrieval.[13] This particular regulatory framework has been criticised, since the need to consciously opt in does little to address the acute shortfall in organs when thousands remain on waiting lists for transplants every year.[14] The perennial issue of demand outstripping supply is a global one, with other jurisdictions adopting an 'opt-out' or 'presumed consent' system (here, everyone is deemed to be a donor, unless they have registered their objection[15]) as a deliberate organ procurement strategy.[16] Wales is currently implementing this scheme,[17] though debates about similar legislative changes are ongoing throughout the rest of the UK.[18]

1.   The legal framework: The Human Tissue Act 2004

The current legislation is the Human Tissue Act 2004, passed in response to high-profile organ retention scandals at Alder Hey Hospital and

---

12   The complete and irreversible cessation of brain function – see Ch 1, p 10.

13   The Human Tissue Act 2004 (see below) retains this system.

14   See the collection of essays in AM Farrell, D Price and M Quigley (eds), *Organ Shortage: Ethics, Law and Pragmatism* (CUP, 2011).

15   Though Pattinson points out that 'presumed consent' is a misnomer, since consent is "fictionalised" in the absence of a registered objection – Pattinson (n 5), p 431.

16   See for example, MA Jacob, "On Silencing and Slicing: Presumed Consent to Post-Mortem Organ 'Donation' in Diversified Societies" (2003) 11 *Tulsa Journal of Comparative and International Law* 239 (noting the "complex mix of ambition, hope, and ruthless calculation", which make it "ethically and legally difficult to control the development of organ donation" – p 239) and H Welbourn, "A Principlist Approach to Presumed Consent for Organ Donation" (2014) 9 *Clinical Ethics* 10.

17   Human Transplantation (Wales) Act 2013. The change in the law came into effect on 1 December 2015.

18   Welbourn (n 16). Other alternatives include conscription whereby organs are removed automatically when a viable donor dies (regardless of consent) – see JJ Wisnewski, "When the Dead Do Not Consent: A Defense of Non-Consensual Organ Use" (2008) 22 *Public Affairs Quarterly* 289 and TM Wilkinson, *Ethics and the Acquisition of Organs* (OUP, 2011), ch 7. Financial incentives for organ donation are another option, but attract the inevitable 'slippery slope' arguments around commodifying the human body (as well as removing the altruistic intent underpinning donation) – see P Bruzzone, "Financial Incentives for Organ Donation: A Slippery Slope Toward Organ Commercialism?" (2010) 42 *Transplantation Proceedings* 1048 though Quigley, for example, suggests a number of other ways to incentivise organ donation – see M Quigley, "Incentivising Organ Donation" in Farrell, Price and Quigley (n 14).

others[19] throughout the UK in the late 1990s.[20] Its remit is an extensive one; as well as covering the storage and use of human tissue from living donors, the 2004 Act regulates the removal, storage and use of "relevant material" from the dead[21] – including human organs for transplant.[22]

The Act is underpinned by the principle of "appropriate consent".[23] Organ donation, however, contemplates three distinct levels of consent starting with the deceased's prior approval (or refusal) as a living, competent adult in accordance with s 3.[24] Where the deceased has not made their wishes known, decision-making authority passes to a nominated representative, or (failing that) to a person who stood in a "qualifying relationship" with the deceased.[25] An innovative feature of the 2004 Act, nominated representatives effectively act as surrogate decision-makers – important, for example, where

19   See Ch 1, n 31 and M Redfearn, *The Royal Liverpool Children's Hospital Inquiry Report* (London: The Stationary Office, 2003).

20   The 2004 Act applies throughout England, Wales and Northern Ireland (though Welsh law on consent has changed (see n 17) and replaces the 2004 Act in that particular respect). For an early overview of the 2004 Act, see D Price, "The Human Tissue Act 2004" (2005) 68 *Modern Law Review* 798 and J Mchale, "The Human Tissue Act 2004: Innovative Legislation – Fundamentally Flawed or Missed Opportunity" (2005) 26 *Liverpool Law Review* 169.

21   Section 53 defines relevant material as that "which consists of or includes human cells" but excluding gametes, embryos created outside the human body, and hair and nails from the body of a living person. Other human material created outside the body is also excluded – s 54(7).

22   2004 Act, s 1 and Sch 1 (and replacing the Human Tissue Act 1961). For an analysis see Pattinson (n 5), ch 12 and J Herring, *Medical Law and Ethics* (OUP, 5th edn, 2014), ch 8. Organ donation cannot interfere with the coroner's jurisdiction to investigate suspicious or untimely deaths (see Ch 1, Pt IV). Where circumstances mandate a referral to the coroner, s 11 of the 2004 Act requires confirmation that the coroner does not object to organs being removed from the potential donor. Dorries notes that this raises an "interesting if semantic point: the coroner only has jurisdiction once a death has occurred but the necessary speed with which transplant donation must be completed sometimes means that the coroner is asked whether he or she would have an objection before the patient 'dies'" (C Dorries, *Coroners' Courts: A Guide to Law and Practice* (OUP, 3rd edn, 2014), p 87.

23   2004 Act, s 1 and Sch 1. See also *Human Tissue Authority, Code of Practice 1: Consent* (Updated July 2014, www.hta.gov.uk/code-practice-1-consent) (accessed 30 September 2015).

24   See also *Human Tissue Act, Code of Practice 2: Donation of Solid Organs for Transplantation* (Updated July 2014, www.hta.gov.uk/code-practice-2-donation-solid-organs-transplantation) (accessed 31 July 2015). Consent can be written or oral under s 3, so long as it was "in force immediately before [the deceased] died" – 2004 Act, s 3(6).

25   2004 Act, s 3(6). In the case of a deceased child (defined as someone under the age of 18 – s 54(1)), the legislation again looks for prior consent (or refusal) to be an organ donor; if unavailable, substituted consent can be given by a parent or (absent that) a qualifying relative – 2004 Act, ss 2(1)–(2) and (7). Note that where a corpse is to be used for anatomical examination or display, *only* the deceased can give prior consent, and this *must* be in writing – 2004 Act, ss 2(4)–(6) and ss 3(3)–(5) (dealing with children and adults respectively).

the deceased does not want to burden (or, perhaps, does not trust) specific relatives with organ donation choices.[26] However, where family members are approached, the 2004 Act is more reflective of modern kinship structures than other laws concerning the fate of the dead.[27] Section 27(4) of the 2004 Act ranks qualifying relationships, in descending order, as follows: (a) spouse or partner, (b) parent or child, (c) sibling, (d) grandparent or grandchild, (e) niece or nephew, (f) stepmother or stepfather, (g) half-sibling and (h) a friend of long standing.[28] The fact that partners have the same rights as spouses is a significant development,[29] as is the inclusion of step-parents and a long-term friend where all other relational categories have been exhausted.[30] Such an expansive listing allows the consent-seeking net to be cast widely if this tertiary layer is invoked. However, the 2004 Act allows an individual's relationship to be omitted if he/she declines to deal with the issue, is unable to give consent,[31] or cannot be located in time to consider a request for organ donation.[32] Here, decision-making powers pass to the next person in the hierarchy (or within the same class).

Securing appropriate consent makes the relevant activity lawful under the 2004 Act.[33] However, the legislation also imposes criminal law sanctions for breach of this requirement, and for the unauthorised use of donated corpses and bodily material.[34] For example, under s 5(1) a person commits an offence if they fail to obtain appropriate consent when needed, unless that person "reasonably believes" that they are carrying out the relevant activity (such as obtaining organs for transplantation) with the requisite consent. Section 32 also prohibits commercial dealings in human material for transplantation.[35]

---

26   Nominated representatives (who must be adults) are dealt with under s 4 of the 2004 Act, which contemplates both written and oral appointments (subject to various witnessing requirements, etc). Where two or more individuals are appointed, they are jointly and severally responsible for making decisions (s 4(6)); a nominated representative can also renounce their appointment at any time (s 4(9)).

27   In particular, the common law duty of disposal ranking outlined in Ch 3, Pt I.

28   See also s 54(9). Under s 3(6), the qualifying relationship must have existed immediately before the deceased's death. Note than an executor is not mentioned here – probably because time is of the essence, and it is much easier to identify family members than an executor under the deceased's will.

29   The legislation characterises someone as the deceased's partner if "the two of them (whether of different sexes or the same sex) live as partners in an enduring family relationship" – 2004 Act, s 54(8).

30   The latter is not defined in the legislation.

31   Where that particular person is a child (qualifying relationships are not age-limited) or lacks capacity.

32   2004 Act, s 27(8).

33   See the opening words of s 1(1), read in conjunction with Sch 1.

34   For an overview, see J Herring, "Crimes Against the Dead" in B Brooks-Gordon, F Ebtehaj, J Herring, MH Johnson and M Richards, *Death Rites and Rights* (Hart, 2007), pp 225–230.

35   Previously dealt with in the Human Organ Transplantation Act 1989.

2.   The wishes of the dead versus the preferences of the living

Harvesting organs is often the first flash-point over the fate of the deceased's remains, as surviving relatives register their objections. Commonly cited reasons include bodily integrity, religious or cultural sensitivities, fears of 'harming' the dead, and an innate aversion towards organ donation; in brain-stem death scenarios, the patient's life-like appearance and concerns that death is being hastened for organ retrieval are also powerful influences.[36]

Any organ donation law "requires balancing the interests of multiple stakeholders".[37] Under the 2004 Act, appropriate consent can only be given by someone in a qualifying relationship with the deceased, if the latter's wishes are unknown (and there is no nominated representative). When confronted with a difference of opinion within the deceased's family, the law reverts to type and looks to the highest ranking individual in the statutory hierarchy;[38] their decision is final, and medical staff cannot traverse the pecking order to solicit permission from someone lower down the list.[39] Specific provisions also apply where there is a conflict between individuals who are ranked together – for example, if a partner's willingness to donate the deceased's organs is contested by an estranged spouse, or only one of the deceased's children is prepared to consent.[40] In keeping with the broad utilitarian objectives of transplantation (and the time-sensitive nature of organ retrieval), one person's consent will suffice here;[41] the procedure is lawful and can go ahead.

More complex issues arise, however, where the deceased has explicitly stated that they want to be an organ donor, but family members object.[42] Under the 2004 Act, the legal position is clear: the deceased's ante-mortem

---

36   These and other affective factors are discussed in G Haddow, "Death, Embodiment and Organ Transplantation" (2005) 27 *Sociology of Health & Illness* 92 and L Shepherd and R O'Carroll, "When do Next-of-Kin Opt-In? Anticipated Regret, Affective Attitudes and Donating Deceased Family Members' Organs" (2013) 19 *Journal of Health Psychology* 1508.

37   H Young, "The Right to Posthumous Bodily Integrity and Implications of Whose Right It Is" (2013) 14 *Marquette Elder's Advisor* 197, pp 239–240.

38   2004 Act, s 27(6).

39   *Code of Practice 1* (n 23), para [93] and *Code of Practice 2* (n 24), para [112].

40   Section 27(5) of the 2004 Act gives equal ranking to relationships within the same individual category.

41   2004 Act, s 27(7). The respective Codes of Practice also encourage the matter to be discussed sensitively with all relevant parties – *Code of Practice 1* (n 23), paras [95]–[96] and *Code of Practice 2* (n 24), paras [114]–[115].

42   The question of whose wishes should carry the most weight has been debated extensively. See for example, P Boddington, "Organ Donation After Death – Should I Decide, or Should my Family?" (1998) 15 *Journal of Applied Philosophy* 69; TM Wilkinson, "Individual and Family Consent to Organ and Tissue Donation: Is the Current Position Coherent?" (2005) 31 *Journal of Medical Ethics* 587; Wilkinson (n 18), chs 4–6; and G Den Hartogh, "The Role of the Relatives in Opt-In Systems of Postmortal Organ Procurement" (2012) 15 *Medicine, Health Care and Philosophy* 195.

preferences are the first indicator of appropriate consent and should be prioritised accordingly. However, the operational realities are very different. In both this country and elsewhere, standard medical practice is to negotiate organ retrieval with the next-of-kin regardless of the deceased's prior consent; if they object, organs will not be harvested, giving family members an effective right of veto over the deceased's opt-in.[43] While accepting that the emotional well-being of grieving relatives is a legitimate concern, Cantor has criticised this culture of deference because:

> ... nonadherence to decedents' wishes ... deprives decedents of their legal right to donate their organs or tissue for transplant. The decedent's prospective autonomy prerogative then remains a theoretical legal right without a remedy if neither the health facility nor another family member chooses to contest survivors' objections to implementing the decedents' wishes.[44]

Respect for autonomy, as a core value of most legal systems, dictates that an individual has the final say on the fate of their remains.[45] If the substantive requirements of the 2004 Act are met, the deceased's conscious choice to be an organ donor should not be discounted just because of family opposition. Besides, the right to self-determination is only part of the picture here; increasing the supply of organs has clear utilitarian objectives, which strengthen the case for the deceased's donative preferences being paramount.[46]

These basic ideas are reflected in the Codes of Practice accompanying the 2004 Act, which suggest that family members should be "encouraged to recognise the wishes of the deceased" and reminded, "if necessary, that they do not have the legal right to veto or overrule their [loved one's] wishes".[47]

---

43   See K O'Donovan and R Gilbar, "The Loved Ones: Families, Intimates and Patient Autonomy" (2003) 23 *Legal Studies* 332, pp 339–344 and the sources cited therein (discussing the equivalent situation under the then Human Tissue Act 1961). Wilkinson has suggested that most countries have, in effect, "a double veto for retrieving organs from the dead: people can veto the post-mortem retrieval of their organs and families can veto the retrieval of their dead relatives' organs" – Wilkinson (n 18), p 2.

44   NL Cantor, *After We Die: The Life and Times of the Human Cadaver* (Georgetown University Press, 2010), p 147.

45   Discussed at length in relation to funeral instructions, where similar tensions exist – see Ch 5, pp 147–149.

46   Family refusal prevents an average of three potential organ donations a day, while the number of people dying each day while waiting for a transplant could be halved if next-of-kin were persuaded to consent – D Shaw and B Elger, "Persuading Bereaved Families to Permit Organ Donation" (2014) 40 *Intensive Care Medicine* 96. Most families regret this decision (and within a short time) – see D Shaw, "We Should Not Let Families Stop Organ Donation From Their Dead Relatives" (2012) *BMJ: British Medical Journal* 345.

47   See *Code of Practice 2* (n 24), para [102].

There is – unsurprisingly – no English case law on whether a family can override valid donor consent based on either conscientious objections, or other grounds of disapproval. Unlike funeral instructions where the deceased's wishes are not legally enforceable, the 2004 Act provides a statutory mechanism for post-mortem donation which theoretically supersedes any entitlements that the deceased's next-of-kin (or executor) have by virtue of their common law right to possession of the remains.[48] While medical facilities could simply ignore family objections and harvest organs based on the deceased's explicit consent, this is unlikely to happen; ethical issues aside, inflicting emotional distress on grieving relatives would attract negative publicity, and doctors might fear the prospect of legal action if they proceed.[49] Until the relevant legislation (and procurement protocols) make it clear that valid donor consent is legally binding and should be followed where retrieval is medically viable,[50] organ donation is an area where the wishes of the dead will often yield to the sensitivities of the living.

3. The legal status of transplant material

Removal of bodily material from a human source raises inevitable questions around its legal status. What if an excised organ is damaged before it can be placed in a potential recipient; or lost or misappropriated while being transported between medical facilities? The 2004 Act imposes criminal law sanctions for breach of its consent provisions,[51] but is silent on the question of civil redress in the scenarios identified here.[52] Private law remedies are an obvious solution, but raise complex questions[53] – not just around the applicable cause of action, but what rights living and dead donors (or family

---

48 See generally Ch 3, Pt I.
49 Such fears are largely misplaced. Clear evidence of appropriate consent (or a reasonable belief that it was given) negates any potential criminal law sanction under the 2004 Act as already outlined, and would presumably prevent doctors from being exposed to civil liability (on the basis that their actions are lawful under the 2004 Act, despite family opposition).
50 See J Downie, A Shea, and C Rajotte, "Family Override of Valid Donor Consent to Postmortem Donation: Issues in Law and Practice" (2008) 40 *Transplantation Proceedings* 1255, pp 1260–1261 discussing the equivalent position in Canada. In several US states legislation allows medical institutions to harvest organs despite psychological pressure from grieving relatives where the deceased has declared his/her intent to be an organ donor – see JL Mesich-Brant and LJ Grossback, "Assisting Altruism: Evaluating Legally Binding Consent in Organ Donation Policy" (2005) 30 *Journal of Health Politics, Policy and Law* 687.
51 See p 164.
52 See RN Nwabueze, "Donated Organs, Property Rights and the Remedial Quagmire" (2008) 18 *Medical Law Review* 201, p 209.
53 For an overview, see B Morris, "You've Got to be Kidneying Me: The Fatal Problem of Severing Rights and Remedies from the Body of Organ Donation Law" (2008) 74 *Brooklyn Law Review* 543.

members representing the deceased) have in removed organs, and what legal entitlement (if any) the intended recipient has where that person has been identified when the damage or loss occurred.[54]

The fact that organ donations are handled exclusively by the NHS in England and Wales (and the rest of the UK) precludes recovery in contract because there is no such arrangement with the relevant hospital.[55] A cause of action in negligence is also unlikely. For example, negligent infliction of psychiatric illness (what used to be termed 'nervous shock') would be extremely difficult to establish,[56] while damage to property as a result of negligence would depend on courts recognising a sufficient proprietary interest in the damaged or lost organ to found such a claim – something which would also be required in an action in bailment.[57] The decision in *Yearworth v North Bristol NHS Trust*[58] at least opens up the possibility of an individual having proprietary interests in his/her donated material, and vociferous arguments have been made elsewhere that donated organs should be treated as a species of property to facilitate certain causes of action and recovery for damage or loss sustained in the period between removal and transplantation.[59] Whether surviving relatives could seek redress on behalf of a dead donor is questionable;[60] if asserting their own distinct, legal entitlement, would the right to possession of the body for disposal purposes extend to excised organs, given that these have been removed with appropriate consent and proprietary claims almost certainly relinquished at this

---

54   Most US states allow 'directed donation', whereby a potential donor can designate a specific recipient (e.g. a family member, or friend). This is not permissible in the UK, though current regulatory policy supports the idea of 'requested allocation' whereby cadaveric organ donations to a relative or long-term friend with a clinical need for the organ will be considered – see SD Pattinson, "Directed Donation and Ownership of Human Organs" (2011) 31 *Legal Studies* 392, pp 392–393.

55   Pattinson, *ibid*, p 403.

56   *Ibid*, and see also Nwabueze (n 52), pp 202, 204. This particular cause of action (considered briefly in Ch 3, p 82), would almost certainly fail because of the need for a recognisable psychiatric illness alongside the basic tort requirements of causation, foreseeability, etc.

57   Again, both causes of action were analysed briefly in Ch 3, Pt III.

58   [2009] 3 WLR 118 and discussed at p 178.

59   See for example, E Colleran, "My Body, His Property: Prescribing a Framework to Determine Ownership Interests in Directly Donated Human Organs" (2007) 80 *Temple Law Review* 1203 (courts should explicitly treat donated organs and bodily material as market-inalienable property, transferable by gift but not by sale). See also Nwabueze (n 52) and D Price, *Human Tissue in Transplantation and Research: A Model Legal and Ethical Donation Framework* (CUP, 2010). The 2004 Act does not address the issue of whether excised organs can constitute property – see Mchale (n 20), p 172.

60   This depends on courts accepting that the donor has some sort of proprietary interest in excised organs, which others can then enforce.

stage?[61] A property-based analysis is more apposite with living donors, and in directed donation scenarios generally when the rights of the intended recipient also come into play.[62]

II. Posthumous reproduction

Children coming into existence after the death of a parent is not a new phenomenon, though the ways in which this can occur have altered dramatically. Historically, the only example was where a child already in its mother's womb (*'en ventre sa mere'*, to use the common law's descriptive term) was born after the death of its father.[63] However, modern science facilitates postmortem reproduction (and even conception) where one of the biological parents is dead.[64] For example, frozen embryos[65] created as part of an earlier IVF cycle can be defrosted and implanted in the biological mother[66] (or a willing surrogate, if the former has died and her husband or partner wants to continue the couple's quest for parenthood). Frozen sperm samples – often deposited as precautionary measure where a man's fertility might be

61 There is also the question of whether this would constitute a sufficient proprietary or possessory interest to found such a cause of action. English case law on this point is scant; there is only a passing comment by Rose LJ in *R v Kelly* [1998] 3 All ER 741, p 750 that human body parts "are capable of being property [for theft purposes] ... if they have a use or significance beyond their mere existence" and citing the use in organ transplantation as an example. American courts seem to have rejected the idea of surviving relatives asserting proprietary interests in the organs removed from dead donors – see *Colavito v New York Organ Donor Network* 486 F 3d 78 (2nd Cir 2007) (claim rejected on the facts), though see KE Peterson, "My Father's Eyes and My Mother's Heart: The Due Process Rights of the Next of Kin in Organ Donation" (2006) 40 *Valparaiso University Law Review* 169 (arguing for a constitutionally protected property right in organs, in favour of the deceased's next-of-kin).

62 See the general discussion in Pattinson (n 54). Though whether an intended recipient can assert proprietary interests in the donor's organ has been rejected by US courts in cases such as *Colavito v New York Organ Donor Network* 486 F 3d 78 (2nd Cir 2007).

63 If subsequently born alive, English common law treated the child as having been born during the father's life for inheritance purposes – but only where the child was born to a married woman following the death of her husband. Today, a child *en ventre sa mere* has full inheritance rights (see for example, ss 46(1) and 55(2) of the Administration of Estates Act 1925), regardless of whether or not it was born to married parents (see for example, Pt II of the Family Law Reform Act 1969).

64 Although this chapter is focusing on posthumous reproduction, posthumous gestation is also possible, where a female is pronounced brain-dead while carrying a viable foetus and is kept 'alive' artificially until the child can be born (usually at the insistence of the woman's partner or parents). This raises complex issues around end-of-life decisions – see Cantor (n 44), pp 225–230.

65 Although the technical term for a pre-implantation embryo in the first fourteen days after fertilisation is a 'pre-embryo', the term embryo will be used here for convenience.

66 Or, where the earlier cycle was based on egg donation, the woman intending to carry the pregnancy and give birth to the child as its legal mother.

compromised[67] – can be used to inseminate his wife or partner in the event of the donor's death.[68] Moving a step further, sperm can be retrieved from a recently deceased male, or one who is in a coma, or a persistent vegetative state ('PVS') or has been pronounced brain-dead; this is then used to create the donor's biological child.[69]

Enabling the dead to procreate using artificial methods raises complex legal, medical, social and ethical issues, which all shade into each other at various points.[70] From the law's perspective, however, the most significant are entitlements to access and use reproductive material, and the legal status of frozen embryos and sperm.

1.   The legal framework: Contracts, consent and statute

Both IVF and assisted reproduction are regulated by statute throughout the UK, under the combined provisions of Human Fertilisation and Embryology Act 1990 and the Human Fertilisation and Embryology Act 2008. The 1990 Act also established the Human Fertilisation and Embryology Authority

---

67   For example, by men undergoing cancer treatment, or for serving members of the armed forces.

68   The reverse scenario could arise where a woman has frozen her eggs (e.g. when undergoing cancer treatment, or simply to defer motherhood), and her partner wants to use the eggs after the donor's demise (aided by a surrogate). However, egg freezing is not standard practice and has much lower success rates for subsequent fertility treatment – I Gafson, "The Facts Don't Lie: We Haven't Cracked Egg Freezing. Not Even Close", *The Telegraph* (London, 17 October 2014) www.telegraph.co.uk/women/womens-life/11169420/Facebook-egg-freezing-The-facts-dont-lie-We-havent-cracked-egg-freez ing.-Not-even-close.html (accessed 30 September 2015). If techniques improve, the same legal issues will arise.

69   While harvesting eggs from an irretrievably unconscious female is technically possible, this is much less commonplace. The medical procedure is more involved and time-consuming (given the need for intensive stimulation of the ovaries over a number of days in a given menstrual cycle) and low success rate for live births from frozen eggs (as well as requiring the use of a surrogate to gestate any embryo – and possibly also a sperm donor to achieve initial fertilisation). If this was to occur, however, the same basic legal issues would arise – though, for brevity, the discussion below focuses mostly on retrieved sperm.

70   There is a wealth of academic literature on this topic. For a flavour, see the following (as well as the sources referenced throughout): JA Robertson, "Posthumous Reproduction" (1993) 69 *Indiana Law Journal* 1027; LA Dwyer, "Dead Daddies: Issues in Postmortem Reproduction" (2000) 52 *Rutgers Law Review* 881; CP Kindregan and M McBrien, "Posthumous Reproduction" (2005) *Family Law Quarterly* 579; KD Katz, "Parenthood from the Grave: Protocols for Retrieving and Utilizing Gametes from the Dead or Dying" (2006) *University of Chicago Legal Forum* 289; R Deech and A Smajdor, *From IVF to Immortality: Controversy in the Era of Reproductive Technology* (OUP, 2008); J Pobjoy, "Medically Mediated Reproduction: Posthumous Conception and the Best Interests of the Child" (2008) 15 *Journal of Law and Medicine* 450; and K Blake, and HL Kushnick, "Ethical Implications of Posthumous Reproduction" in JM Goldfarb (ed) *Third-Party Reproduction* (Springer, 2014).

('HFEA');[71] as an independent regulatory body,[72] the HFEA licenses fertility clinics and centres carrying out IVF and other assisted conception procedures in this country, and issues ethical guidelines in the form of a Code of Practice, which is revised every few years.[73] HFEA-issued licences are required for the creation of embryos outside the body, and for their subsequent freezing or use; the same applies to the freezing or use of donated gametes such as sperm.[74] From a treatment perspective, however, the key requirement is "effective consent" as detailed in Sch 3 of the 1990 Act.[75] In short, written and signed consent is required from each of the parties being treated for infertility by a licensed clinic or centre;[76] this consent includes the use of any embryo in providing treatment services,[77] as well as the storage of gametes or embryos.[78]

While some other countries determine the use of frozen embryos or gametes through private law litigation,[79] the regulatory framework outlined above plays a major role in this jurisdiction. If a bereaved spouse or partner wants to access surplus embryos from an earlier IVF cycle, his/her ability to do so will depend on the deceased's express consent to their use in another treatment cycle – derived from the standard consent forms that the parties originally signed with the clinic, and mirroring the legal requirements of the

71   1990 Act, s 5. The HFEA's functions and procedures are dealt with in ss 6–10.
72   Further information is available at www.hfea.gov.uk/ (accessed 30 September 2015).
73   The Code is currently in its eighth edition, and (at the time of writing) was last updated on 1 April 2015 – see CH(15) 01, located at www.hfea.gov.uk/9559.html (accessed 30 September 2015). Although not legally binding, the Code of Practice carries considerable weight and non-compliance with its recommendations may result in a clinic having its licence revoked – 1990 Act, s 25(6).
74   1990 Act, ss 3 and 4. The scope of licences issued for treatment and storage (and research) is governed by s 11; ss 12–15A deal with licence conditions; and ss 16–22 with the grant, revocation and suspension of licences.
75   Defined in Sch 3, para 1 as consent that has not been withdrawn by the person giving it. Withdrawal of consent by one of the parties (see Sch 3, para 4) is decisive – *Evans v Amicus Healthcare Ltd* [2004] 1 WLR 681 (and refusal to allow embryos to be implanted where this represents the female's only chance of having a genetic child does not violate Article 8 of the ECHR – *Evans v UK* (2006) 46 EHRR 34).
76   Sch 3, para 1.
77   Sch 3, para 2(1).
78   Sch 3, para 2(2), which also states that the consent must specify the maximum period of storage (if less than the statutory period for embryos and gametes). Under s 14 of the 1990 Act the designated statutory storage period was 10 years, but is now a maximum of 55 years under the Human Fertilisation and Embryology (Statutory Storage Period for Embryos and Gametes) Regulation 2009, SI 2009/1582 (as amended). However, this time limit may not always be interpreted restrictively – see *Warren v Care Fertility (Northampton) Ltd* [2014] 3 WLR 1310 (sperm taken from a woman's husband prior to his death could be stored by a fertility clinic beyond the statutory period; refusal would constitute a disproportionate interference with her rights under Article 8 of the ECHR).
79   In the absence of comprehensive federal or state legislation, US courts have resolved the issue on the basis of torts and contract law principles – R Madoff, *Immortality and the Law: The Rising Power of the American Dead* (Yale University Press, 2010), p 42.

1990 Act.[80] Accessing preserved sperm in an attempt to conceive the deceased's biological child also depends on the deceased having consented to their use in this specific manner, and naming the surviving spouse or partner as the person to use his gametes in any treatment.[81] Once again, the regulatory framework upholds the core ideal of reproductive autonomy, as reflected in the deceased's ante-mortem choices.[82]

Much more complex legal (and ethical) issues arise where conceiving a biological child involves retrieving sperm from a sudden death or a comatose, PVS or brain-stem-dead male – usually at the request of the man's spouse,

---

80   Where couples undergo IVF treatment, the standard consent forms (which both must sign) invariably contain clauses detailing each individual's wishes around the posthumous use of any embryos remaining after a treatment cycle. Sch 3, para 2(2)(b) of the 1990 Act demands that consent to storage stipulates what is to be done with the embryo (or gametes) if the person who gave the consent dies (or subsequently lacks capacity to vary or revoke their consent). See also Sch 3, paras 8(2)–(3).

81   1990 Act, Sch 3, para 2(2)(b) and see Sch 3, para 8(1). The deceased may also consent to being named as the father of any child created by such treatment – see below for assigning parentage under the 2008 Act.

82   Where, however, the deceased did not designate a specific recipient or course of treatment (either through oversight, or failure to vary an original consent where sperm was frozen prior to forming a relationship with the surviving spouse or partner) a request to use the sperm may be denied. Courts in both the US and elsewhere have occasionally allowed this where there is clear evidence of the deceased's reproductive intent – see for example, *Hecht v Superior Court* 20 Cal Rptr 2d 275 (Ct App 1993) (noted briefly below) and *Paraplaix v CECOS*, Tribunal de Grande Instance de Creteil, (1 Ch Cir), 1 August 1984, though see *Estate of Kievernagel* 166 Cal App 4th 1024 (Cal Ct App 2008) where the sperm sample was destroyed because the deceased had requested this on the storage agreement. Whether English courts would reach similar conclusions to *Hecht* and *Paraplaix* is open to question, given the comprehensive statutory regime that applies here, and the clear wording of the consent provisions in the 1990 Act. The mere fact that the deceased froze his sperm at some stage in the past should not simply be regarded as implied consent for his spouse or partner to use it after his death, unless there is other compelling evidence.

girlfriend or partner.[83] While posthumous reproduction should be predicated on the deceased's express consent, the sudden and unexpected course of events means that the deceased will not have explicitly agreed to the retrieval of his sperm or its subsequent use. In these circumstances, the law looks for evidence of implied consent, based on things like the couple having mentioned having (more) children to family and friends, or having sought advice on fertility treatment at some stage in the past. The inherent dangers are obvious: it is easy to lose sight of the legal requirements when faced with the desperate pleas of a grieving spouse or partner,[84] and decision-makers may struggle with how much weight to give self-serving statements about the deceased's apparent willingness to conceive a child posthumously where he died without any explicit indication.[85] What constitutes implied consent is open to debate;[86] merely being in a committed relationship does not, in itself,

---

83   See for example, C Strong, "Ethical and Legal Aspects of Sperm Retrieval After Death or Persistent Vegetative State" (1999) 27 *Journal of Law, Medicine & Ethics* 347 and C Strong, JR Gingrich and WH Kutteh, "Ethics of Postmortem Sperm Retrieval: Ethics of Sperm Retrieval After Death or Persistent Vegetative State" (2000) 15 *Human Reproduction* 739. See also JD Hans and B Dooley, "Attitudes Toward Making Babies… With a Deceased Partner's Cryopreserved Gametes" (2014) 38 *Death Studies* 571. Even more controversial are requests from parents to extract and preserve sperm from dead sons to give the parents a grandchild using donor eggs and a surrogate (or in dead daughter scenarios, retrieving eggs to be fertilised by donor sperm and gestated by a surrogate). There appears to be no legal precedent for this in the UK; the only scenario that comes close is the legal battle currently being fought by a mother who wants to access her dead daughter's eggs (frozen by a London clinic when the latter was about to undergo cancer treatment) and take these to the US to allow the mother to become pregnant with her own genetic grandchild. Export permission has been refused by the HFEA – see "Mother in Legal Battle to Bear Dead Daughter's Child", *The Guardian* (London 8 May 2015) www.theguardian.com/uk-news/2015/may/08/mother-in-legal-battle-to-bear-dead-daughters-child (accessed 30 September 2015). There are isolated examples of posthumous retrieval of sperm and eggs by prospective grandparents occurring in other countries. In 2011, an Israeli court allowed eggs to be harvested from a 17-year-old girl who died in a car crash, following a request by her parents (A Bloomfield, "Family Given Permission to Extract Eggs from Ovaries of Dead Daughter in World First", *The Telegraph* (London, 8 August 2011) www.telegraph.co.uk/news/health/news/8689479/Family-given-permission-to-extract-eggs-from-ovaries-of-dead-daughter-in-world-first.html (accessed 30 September 2015); two years earlier a Texan court had allowed a mother to retrieve her dead son's sperm after he died following a street assault ("Texas Judge Tells Medical Examiner to Help Mom Get Dead Son's Sperm", *American Bar Association Journal*, 8 April 2009).

84   See Hans and Dooley (n 83).

85   See DD Williams, "Over My Dead Body: The Legal Nightmare and Medical Phenomenon of Posthumous Conception Through Postmortem Sperm Retrieval" (2011) 34 *Campbell Law Review* 181, pp 190–194.

86   See for example, C Strong, "Gamete Retrieval After Death or Irreversible Unconsciousness: What Counts as Informed Consent?" (2006) 15 *Cambridge Quarterly of Healthcare Ethics* 161, JD Hans, "Posthumous Gamete Retrieval and Reproduction: Would the Deceased Spouse Consent?" (2014) 119 *Social Science & Medicine* 10 and H Young, "Presuming Consent to Posthumous Reproduction" (2014) 27 *Journal of Law & Health* 68.

prove that the deceased and their surviving partner were agreed on having children – let alone that the same partner should conceive, carry and raise the individual's child after his death. Cantor, for example, emphasises that the deceased's prospective autonomy interest can be damaged if there is no compelling evidence that he wanted to become a post-mortem parent.[87]

The legal position is also unclear, and was first explored in this country by *R v Human Fertilisation and Embryology Authority, ex parte Blood*[88] where Diane Blood wanted to use sperm extracted from her comatose husband shortly before he died, in order to conceive the couple's biological child.[89] Since the storage of gametes and their use in any subsequent fertility treatment in the UK is dependent on effective consent in accordance with the 1990 Act,[90] and no such consent had been obtained from the deceased here, the Court of Appeal decided that the sperm was being stored and preserved unlawfully. The HFEA was asked to reconsider its decision to refuse an export licence for the sperm,[91] and the gametes were eventually transferred to a clinic in Belgium.[92] A decade later, similar issues arose in *L v Human Fertilisation and Embryology Authority ('HFEA')*[93] where the deceased's wife wanted to use her husband's sperm (obtained after his sudden death following an appendectomy),[94] with questions once again being posed over the lawfulness of storing gametes in this country without the written consent of the gamete provider.[95] However, the court also addressed the more fundamental issue of authorising sperm retrieval from a dead or irretrievably unconscious male. In *Blood*, the Court of Appeal sidestepped the issue, pointing out that the lawfulness (or otherwise) of the initial extraction was not regulated by the 1990 Act[96] but governed by "common law principles relating to the patient's consent ... which have not

---

87   Cantor (n 44), p 222 and making the point that the "mere fact that the decedent wished to someday be a parent does not, by itself, support an inference that he wanted to have sperm extracted from his fresh corpse in order to become a posthumous parent".

88   [1997] 2 All ER 687 and see D Morgan and R Lee, "'In the Name of the Father?' *Ex parte Blood:* Dealing with Novelty and Anomaly" (1997) 60 *Modern Law Review* 840.

89   Stephen Blood had contracted meningitis and fallen into a coma.

90   Written and informed consent, as detailed above.

91   Exercising its powers under s 24(4) of the 1990 Act.

92   The Court of Appeal acknowledged that the HFEA was right to refuse treatment in the UK because the deceased had not given written consent, but ruled that the clinic had failed to take account of Mrs Blood's rights under EU law (and in particular her directly enforceable right to receive cross-border medical treatment elsewhere). Diane Blood went on to have two children, using her dead husband's sperm.

93   [2008] EWHC 2149 (Fam).

94   The deceased's wife wanted to provide a full genetic sibling for the couple's young daughter.

95   Sch 3, para 9 of the 1990 Act (introduced by Sch 3, para 12 of the 2008 Act), and prescribing specific situations in which consent was not required for storage, did not apply here.

96   Something which Charles J reiterated in *L v HFEA* [2008] EWHC 2149 (Fam), [11].

been argued before us".[97] However, Charles J in *L v HFEA* was not convinced that gametes could be lawfully removed from a dead person who had not given effective advance consent. The wife's written evidence in this case provided "powerful support for the view that if [the husband] had been asked about the retrieval and use after his death of his sperm in an attempt to enable [his wife] to have another child he would have agreed".[98] While this might point towards inferred consent, there were "considerable doubts and uncertainties" as to whether a court or anyone else (for example, the deceased's personal representative or next-of-kin) "could authorise the removal of gametes from a dead person".[99] According to Charles J, the common law ability to deal with the corpse for disposal purposes was just that; it did not translate into the ability to authorise the removal of gametes.[100]

Finally, if any of these techniques are successful and result in a live birth, can the resultant child be recognised as the deceased's child? The question has "both emotional and legal significance",[101] given the implications for the

97   [1997] 2 All ER 687, 695.
98   [2008] EWHC 2149 (Fam), [33]. Key points in the wife's written evidence included the couple having wanted more than one child and discussing this with friends; the husband raising the issue of IVF a few days before his death, when the couple had sought professional advice; and the husband's family being supportive of the wife's desire to have another child – though it is difficult to see how the final factor is in any way indicative of the husband's inferred consent.
99   *Ibid*, [155]. In the present case, the attending hospital had sought a court declaration to this effect – granted in emergency proceedings as a form of interim relief.
100   *Ibid*, [158]–[159]. English courts will almost certainly be faced with this issue in the future, and could look to Australia where legal authority to remove sperm from dead or irretrievably unconscious males has been litigated extensively – even if the outcomes and underlying rationales are inconsistent. In both *Re Gray* [2001] Qd R 35 and *Baker v Queensland* [2003] QSC 2 the respective courts refused to sanction retrieval despite the applicant having the support of the deceased's wider family, with Chesterman J in the former case questioning whether those entitled to possession of a body have rights beyond ensuring prompt and decent disposal. In *Maw v Western Sydney Area Health Service* [2000] NSWSC 358 permission was also refused because there was nothing to suggest the deceased would have consented to sperm extraction for posthumous insemination of his wife (the couple seemed to have made a conscious decision not to have children in the past), while the court also doubted whether it had jurisdiction to authorise retrieval. In contrast, the respective courts sanctioned this in *Re Denman* [2004] QSC 70, *AB v Attorney-General (Vic)* (Supreme Court of Victoria, 23 July 1998) and *Fields v Attorney-General of Victoria* (Supreme Court of Victoria, 1 June 2004). For an analysis of the various decisions, see M Leiboff, "Post-Mortem Sperm Harvesting, Conception and the Law: Rationality or Religiosity?" (2006) 6 *Queensland University of Technology Law and Justice Journal* 193. Australian courts have also allowed sperm retrieval on the basis of human tissue legislation, which authorises this – *Y v Austin Health* [2005] VSC 427 and *Re Edwards* [2011] NSWSC 478. However, the same legislative solution is not available under English law, since the Human Tissue Act 2004 does not extend to gametes, as confirmed in *L v HFEA* [2008] EWHC 2149 (Fam).
101   Madoff (n 79), p 44.

child's sense of identity, relational networks, birth registration details and inheritance rights (both from the dead parent's estate and from other relatives).[102] Part 2 of the 2008 Act deals with assigning parentage in assisted reproduction,[103] and includes specific provisions around posthumous reproduction.[104]

2.   Family objections to posthumous reproduction?

Do the deceased's relatives have any say in whether a surviving spouse or partner should access frozen embryos or sperm, or be able to retrieve gametes posthumously? Given that the deceased's parents or existing children, in particular, may not always be supportive of the survivor's request?

The issue has arisen in several American cases, where close kin have tried to prevent the deceased's spouse or partner from accessing frozen embryos or (more usually) preserved gametes.[105] Courts have ruled that the deceased's reproductive intent should prevail – an outcome that Cantor sees as consistent with the law's tendency to give "broad autonomy" to citizens wishing to shape the post-mortem fate of their bodies and bodily parts,[106] regardless of opposition from surviving relatives. While family attitudes towards the deceased's reproductive material will be fundamentally different

---

102   The inheritance implications of children being born to dead parents have been debated extensively. See for example, R Chester, "Freezing the Heir Apparent: A Dialogue on Postmortem Conception, Parental Responsibility, and Inheritance" (1996) 33 *Houston Law Review* 967; R Atherton, "En Ventre sa Frigidaire: Posthumous Children in the Succession Context" (1999) 19 *Legal Studies* 139; KS Knaplund, "Postmortem Conception and a Father's Last Will" (2004) 46 *Arizona Law Review* 91; L Tritt, "Sperms and Estates: An Unadulterated Functionally Based Approach to Parent-Child Property Succession" (2009) 62 *Southern Methodist University Law Review* 367; and BC Carpenter, "A Chip off the Old Iceblock: How Cryopreservation Has Changed Estate Law, Why Attempts to Address the Issue Have Fallen Short, and How to Fix It" (2011) 21 *Cornell Journal of Law & Public Policy* 347.

103   See generally C Jones, "The Identification of 'Parents' and 'Siblings': New Possibilities under the Reformed Human Fertilisation and Embryology Act" in J Wallbank, S Choudhry and J Herring (eds) *Rights, Gender and Family Law* (Routledge-Cavendish, 2009). See also J McCandless and Sally Sheldon, "The Human Fertilisation and Embryology Act (2008) and the Tenacity of the Sexual Family Form" (2010) 73 *Modern Law Review* 175.

104   See for example, s 39 (use of sperm, or transfer of embryo, after the death of the man providing the sperm); s 40 (embryo transferred after the death of a husband or unmarried partner who did not provide the sperm); and s 46 (embryo transfers after the death of a civil partner, or intended female parent).

105   See for example, *Hecht v Superior Court* 20 Cal Rptr 2d 275 (Ct App 1993) (deceased's children tried to prevent his partner from accessing the deceased's cryopreserved sperm); *Hall v Fertility Institute of New Orleans* 647 So2d 1348 (La App 1994) (mother attempted to block dead son's donation of frozen sperm to his girlfriend); and *Dhanoolal v United States Department of the Army* No 4:08-CV-42(CDL) (Middle District of Georgia, 2008) (mother attempted to block donation to son's wife).

106   Cantor (n 44), p 216.

to attitudes towards the deceased's organs (understandable, given the former's potential for creating life), the interests of the deceased and their surviving spouse or partner take precedence where the deceased consented to post-mortem use – just as family members have no say in a living couple's lifetime choice to be parents. However, retrieving gametes from a dead donor is a different proposition, and objections from the deceased's parents or existing children might carry more weight here.[107]

3. The legal status of reproductive material

Legal contests around posthumous reproduction and access to preserved embryos and gametes have raised the question of what (if any) property rights exists in reproductive material.

Although English courts have yet to consider whether embryos might be property in any strict legal sense, several US cases seem to have rejected the idea, the Tennessee Supreme Court in the seminal case of *Davis v Davis*[108] preferring to describe them as an "interim category [between 'persons' and 'property'] because of their potential for human life".[109] However, the same courts have been more willing to adopt a property law narrative when

---

107 Acknowledging that parents have an "understandable interest in whether their children produce offspring", Hans and Yelland note that opposition from the deceased's parents may be a contraindication to posthumous sperm retrieval and that some American hospitals have protocols to this effect – JD Hans and EL Yelland, "American Attitudes in Context: Posthumous Uses of Cryopreserved Gametes" (2013) *Journal of Clinical Research & Bioethics* 1, p 3. Leiboff makes a similar point, arguing that removal of sperm should not be sanctioned if the deceased's family disagree, since relational interests must be taken into account – Leiboff (n 100), pp 194–195.

108 842 SW2d 588 (Tenn Sup Ct 1992) and analysed in MA Pieper, "Frozen Embryos – Persons or Property? *Davis v Davis*" (1990) 23 *Creighton Law Review* 807.

109 *Ibid*, 596–597. *Davis* involved a dispute between, divorcing couples, over frozen embryos created during the couple's marriage (and see *Rener v Reiss* 42 A 3d 1131 (Pa Superior Ct 2012) for a similar factual scenario). However, the wider property issue has also arisen in legal contests involving the transfer of reproductive material to another medical facility (*York v Jones* 717 F Supp 421 (ED Va 1989) and *Jeter v Mayo Clinic Arizona* 121 P3d 1272-74 (Ariz Ct App 2005)) and damage to or destruction of embryos (*Del Zio v Columbia Presbyterian Center* 74 Civ 3588 (SDNY Apr 12, 1978)). In *Del Zio*, a member of staff at the medical centre destroyed the couple's embryos because he believed IVF was unethical and morally wrong; Mr and Mrs Del Zio were awarded damages for emotional distress but not for unlawful destruction of their property. In *Jeter*, where the couple alleged the negligent destruction or loss of five of their frozen embryos while being transferred between reproductive centres, the court allowed the claim but refused to categorise the embryos as property; the court also dismissed the Jeters' wrongful death suit on the basis that embryos are not persons under Arizona's wrongful death statutes.

considering requests to use the deceased's preserved sperm after his death.[110] Although English courts have yet to address the same issue, the Court of Appeal in *Yearworth v North Bristol NHS Trust*[111] was prepared to classify frozen sperm samples as property for the purposes of an action in bailment when samples (deposited by male cancer patients before undergoing treatment for chemotherapy) were destroyed following an equipment failure at the storage centre.[112] In recognising ownership interests, the court acknowledged the men's source interests in their sperm and absolute control (albeit in a negative sense) over its use and continued storage.[113] Whether this operates as a pragmatic and fact-specific ruling, or paves the way for a wider recognition of property rights in gametes (and excised human material more

110   For example, in *Hecht v Superior Court* 20 Cal Rptr 2d 275 (Ct App 1993) where the deceased's will purported to leave several vials of frozen sperm, deposited in a California sperm bank, to his girlfriend, the California Court of Appeal allowed the bequest. While the sperm was not inheritable property as such, the deceased "at the time of his death … had an interest, in the nature of ownership, to the extent that he had decision-making authority as to the use of his sperm for reproduction", which was "sufficient to constitute 'property'" – *ibid*, 283. See also *Hall v Fertility Institute of New Orleans* 647 So2d 1348 (La App 1994) (again, dealing with the issue of frozen sperm samples as estate property). Discussing these and other US cases, Cooper and Harper suggest that questions about the deceased's reproductive material are "subject to a quasi-property right, the exercise of which is colored by a decedent's intentions" – IS Cooper and RM Harper, "Life After Death: The Authority of Estate Fiduciaries to Dispose of Decedents' Reproductive Matter" (2010) 26 *Touro Law Review* 649, p 649 However, Rao is less convinced, arguing that US courts (when discussing stored sperm and embryos) appear "utterly confused as to how to classify these objects" – R Rao, "Property, Privacy and the Human Body" (2000) 80 *Boston University Law Review* 359, p 414.

111   [2009] 3 WLR 118. For an analysis, see C Hawes, "Property Interests in Body Parts: *Yearworth v North Bristol NHS Trust* (2010) 73 *Modern Law Review* 130 and GT Laurie, "*Yearworth v North Bristol NHS Trust*: Property, Principles, Precedents and Paradigms" (2010) 69 *Cambridge Law Journal* 476.

112   While an action in contract was not an option here because storage was facilitated by a NHS hospital, the position would presumably have been different if the sperm had been deposited in a private clinic as part of a non-NHS, funded storage agreement.

113   However, Lee is critical of what he sees as an instrumentalist approach in this case, with the Court of Appeal constructing property rights in sperm to facilitate a claim under the law of obligations – J Lee, "*Yearworth v North Bristol NHS Trust* [2009]: Instrumentalism and Fictions in Property Law" in S Douglas, R Hickey and E Waring (eds), *Landmark Cases in Property Law* (Hart, 2015).

generally), remains to be seen.[114] Looking further afield, Australian courts have dealt with a comparatively high volume of legal contests involving preserved sperm samples, with judges categorising the stored material as property and granting possession to a surviving spouse or partner as the intended recipient.[115]

Conclusion

A single corpse has multiple uses, not only in its post-mortem state but as it makes the transition from life to death. Both organ donation and post-humous reproduction, with their potential to save and to create life respectively, raise all sorts of legal issues, which existing regulatory frameworks cannot always address. Neither technology impacts significantly on final disposal of the body, though questions will continue to be asked about the legal status of excised organs and stored gametes or embryos. When it comes to decision-making authority, the position is much clearer: the law imposes strict consent requirements if the dead are to become organ donors or genetic parents. Family objections can be noted, but must usually yield to ante-mortem expressions of the deceased's donative or reproductive intent.

---

114   The idea of a property framework has support – see for example, EBS Ball, "Property and the Human Body: A Proposal for Posthumous Conception" (2008) 15 *Journal of Law and Medicine* 556, BM Fuselier, "Trouble with Putting All of Your Eggs in One Basket: Using a Property Rights Model to Resolve Disputes over Cryopreserved Pre-Embryos" (2008) 14 *Texas Journal on Civil Liberties & Civil Rights* 143, and LB Moses, "The Problem with Alternatives: The Importance of Property Law in Regulating Excised Human Tissue and In Vitro Human Embryos" in I Goold, K Greasley, J Herring and L Skene, *Persons, Parts and Property: How Should We Regulate Human Tissue in the 21st Century?* (Hart, 2014), pp 197–214. However, others are less convinced – see B Bennett "Posthumous Reproduction and the Meaning of Autonomy" (1999) 23 *Melbourne University Law Review* 286 (arguing that the reproductive potential of such material distinguishes it from other human tissue, preferring an analysis based on the deceased's reproductive (and relational) autonomy interests in the use of his/her gametes) and J Nedelsky, "Property in Potential Life? A Relational Approach to Choosing Legal Categories" (1993) 6 *Canadian Journal of Law and Jurisprudence* 343 (the law should focus on decision-making rights and control over reproductive material, instead of property).

115   See for example, *Re Denman* [2004] QSC 70, *AB v Attorney-General (Vic)* (Supreme Court of Victoria, 23 July 1998) and *Fields v Attorney-General of Victoria* (Supreme Court of Victoria, 1 June 2004), as well as *Bazley v Wesley Monash IVF Party Ltd* [2010] QSC 118, *Re Edwards* [2011] NSWSC 478, RE H, AE (No 2) [2012] SASC 177, and *Re Section 22 of the Human Tissue and Transplant Act 1982 (WA), ex parte C* [2013] WASC 3. The more recent tranche of cases is discussed in Lee (n 113), pp 33–37 and see also C Kr_løkke and S Adrian, "Sperm on Ice: Fatherhood and Life after Death" (2013) 28 *Australian Feminist Studies* 263. More generally, Skene notes that the basis on which preserved sperm samples actually constitute property is unclear in these cases – see L Skene, "Proprietary Interests in Human Bodily Material: *Yearworth*, Recent Australian Cases on Stored Sperm and Their Implications" (2012) 20 *Medical Law Review* 277.

7 Exhuming the dead

Exhumation ... needs strict controls ... in the interests of public welfare, the wishes of the deceased and the feelings of the families of those involved.[1]

Introduction

Exhumation involves opening up a grave (or occasionally a vault) and removing human remains already buried there.[2] Also known as 'disinterment', exhumation is controversial – even if the intent is usually to rebury displaced remains elsewhere.[3] Most societies and cultures that embrace burial as a means of bodily disposal exhibit an entrenched reluctance to disturb the dead's earthly repose for two reasons. The first is public health concerns

---

1 C Lovatt, "Equality Issues from Beyond the Grave" (2000) 19 *Equal Opportunities International* 29, p 29.

2 This is invariably a 'one-off' movement; successive displacements of the same individual are rarely sought and would probably be refused. Although most of this chapter focuses on interred corpses, buried or envaulted ashes can also be exhumed and relocated; this raises equally sensitive issues, and the same basic legal requirements apply – at least in England and Wales. Finally, most exhumations will involve single graves, though mass exhumations from multiple graves are possible (see some of the illustrations in Pt I).

3 Though this does not always occur – for example, when the motive for exhumation is to cremate the deceased's body, to scatter ashes (whether on their own, or along with the ashes of another family member), or simply to keep them at home. Human remains can also be exhumed to lift and deepen a grave, or to confirm the deceased's identity; here, the dead will be reinterred in the same grave.

around the potential transmission of disease from decaying corpses.[4] Secondly, and more fundamentally, exhumation offends the basic moral premise of allowing the dead to 'rest in peace' and is generally regarded as a forbidden or sacrilegious act.[5] As a result, exhumation in England and Wales is subject to strict legal controls.[6]

This chapter analyses the complex laws, policy debates and conflicting values around disinterment. In many ways, the issues which arise here mimic those generated by disposal of the dead. For example, parts of the applicable legal framework are lacking in clarity, while disputes within families can occur where surviving relatives are divided over whether to exhume the deceased's remains. Religious and cultural imperatives can also play a significant role in disinterment decisions, just as human rights influences are increasingly coming to the fore. However, exhumation also raises its own distinct issues, beyond the basic question of what constitutes sufficient reason for disturbing the dead. There are two different legal processes in England and Wales with potentially diverse outcomes, depending on whether the original interment was in consecrated or unconsecrated ground. The level of consensus required where some members of the deceased's family oppose exhumation is also unclear, while disputes between surviving relatives and the relevant burial authorities create numerous opportunities for conflict.

I.   Why exhume?

Requests for exhumation fall into two broad categories: public interest, and personal reasons. Examples of the former include investigation of potential

---

4   While the risk to public health subsides over time, bodily decomposition is most rapid and noxious during the initial period after burial (even allowing for embalming). However, Green and Green argue that there is little cause for concern; while an Environmental Health Officer must be present at exhumation and those involved in digging up the remains must wear protective clothing, the attendant risks are minimal and exhumation cannot proceed if there is any perceived danger to public health – see J Green and M Green, *Dealing with Death: A Handbook of Practices, Procedures and Law* (Jessica Kingsley, 2nd edn, 2008), p 136. Of course, disinterment of ashes does not raise the same public health issues, and may not even be physically possible where ashes buried in a biodegradable container have subsequently dispersed into the soil.

5   "While the corpse may not physically sense disturbance to its sepulcher, 'rest in peace' has always been considered an appropriate approach to disposal of human remains" – NL Cantor, *After We Die: The Life and Times of the Human Cadaver* (Georgetown University Press, 2010), p 239.

6   A basic presumption against disinterment exists elsewhere – see for example, PR Kehoe, "Cemetery Abandonment and Disinterment of Human Remains" (1970) 35 *Albany Law Review* 320, p 321 discussing the equivalent position in the United States ("the situations in which the eyes of the law will look favourably upon a disinterment are few and the standards to be met are stringent"). However, this chapter will focus primarily on exhumation laws in England and Wales given the influence of ecclesiastical law on this particular area and the divergences which exist between secular and non-secular disinterments in this jurisdiction.

criminal offences where new evidence has come to light about an individual's death[7] or bodies have been discovered in suspicious circumstances,[8] as well as mass exhumations following wartime atrocities.[9] Exhumation can also be ordered, in the interests of justice, to ensure that an earlier jury verdict in a murder trial was correct in light of advances in DNA profiling.[10] Moving beyond the realm of the criminal law, public interest concerns also come to the fore where planned development projects (e.g. new infrastructure or buildings) require multiple graves in older and frequently disused burial grounds to be cleared, as the demands of the living take precedence over the longer-term dead.[11] Other isolated examples include disinterment for sanitary

7   One of the most high-profile examples being patients who died while under the care of Dr Harold Shipman, the UK's worst serial killer who was jailed in 2000 for killing 15 people (though the real figure is suspected to be much higher). Exhumed bodies were found to contain lethal doses of diamorphine, administered by Shipman – see DJ Poulter, "The Case of Dr Shipman" (2003) 24 *American Journal of Forensic Medicine and Pathology* 219.

8   For example, the recent discovery of infants' remains in a septic tank beside a home for unmarried mothers in Galway – "Mass Grave of 796 Babies Found in Septic Tank at Catholic Orphanage in Tuam, Galway", *The Belfast Telegraph* (Belfast, 4 June 2014) www.belfasttelegraph.co.uk/news/local-national/republic-of-ireland/mass-grave-of-796-babies-found-in-septic-tank-at-catholic-orphanage-in-tuam-galway-30327483.html (accessed 30 September 2015).

9   For example, the mass killings carried out in Bosnia and Herzegovina in the 1990s. Here, exhumation goes beyond familial interests in recovering the dead and laying loved ones to rest; it acts as a means of piecing together the past, allowing remains to be forensically examined and investigating war crimes. Although dealing with a different type of scenario, the Protection of Military Remains Act 1986 is a specific UK statute that prevents unauthorised interference with the wreckage of military aircraft and designated military vessels; its purpose is to safeguard war graves (though the loss of the aircraft or vessel need not have been caused by a hostile act – accidental and peacetime losses are also covered).

10   *R v Hanratty* (*The Times*, 26 October 2000) (Court of Appeal (Criminal Division)).

11   As in *Re St Mary the Virgin, Woodkirk* [1969] 1 WLR 1867 (road widening scheme in Morley which would disturb almost 400 graves) and, more recently, the proposed route for the new high-speed rail link between London and Birmingham which could affect the remains of some 50,000 people (see L Townsend, "Digging Up The Dead", *BBC News Magazine* (26 June 2012) www.bbc.co.uk/news/magazine-18505222 (accessed 30 September 2015)). Legislation can compel the removal of human remains to facilitate development projects in certain circumstances (for example, the Channel Tunnel Rail Link Act 1996). The dead are more susceptible to being disturbed here because they have no living relatives, and subsequent generations may not be interested in preserving their ancestors' graves. For example, Mims makes the point that "the thread of family love and intimacy" is broken with more distant descendants, and that a corpse "should be finally rendered 'anonymous' after two or three generations" – C Mims, *When We Die: The Science, Culture and Rituals of Death* (Robinson Publishing, 2000), p 128.

reasons,[12] for archaeological investigation,[13] a desire to honour (or even protect) the remains of important historical or public figures,[14] and exhumation to facilitate important medical research.[15]

Disturbing the dead for any of these reasons is a relatively infrequent occurrence, though the public interest dimension can tip the scales in favour of allowing exhumation.[16] In contrast, exhumations driven by personal reasons – the low-profile, single grave disinterments involving comparatively recent burials and occurring within the private, familial sphere – are much more commonly sought. The intent may be to move the deceased from a single grave into a family plot or to repatriate a corpse or ashes when the deceased's loved ones are moving elsewhere (in both instances, ensuring family unity after death),[17] or to move the deceased's remains to a place which has more significance for their family (for example, where another

---

12   See *Rector etc of St Helen's Bishopgate, with St Mary, Outwich v Parishioners of St Helen's Bishopgate, with St Mary, Outwich* [1892] P 259 (sub-surface remains discovered in a church vault, and linked with parishioners complaining of unpleasant smells and harmful fumes).

13   As in the recent high-profile dispute surrounding the exhumation of King Richard III from a burial site in Leicester – see *R (on the application of Plantagenet Alliance Ltd) v Secretary of State for Justice* [2014] EWHC 1662 (QB).

14   For example, the remains of Marie and Pierre Curie were exhumed from a family grave in Sceaux and re-interred in France's national mausoleum, the Panthéon, in honour of their work, while Elvis Presley's remains were moved from his original grave in Memphis to the grounds of his Graceland mansion following an attempt to steal his body. In contrast, exhumation may also be necessary to prevent a grave becoming a shrine to atrocities committed by the deceased and any associated ideologies. For example, the remains of Adolf Hitler's deputy, Rudolf Hess, were exhumed and cremated (the ashes were then scattered at sea) to stop the grave being used as a pilgrimage site for neo-Nazis – "Top Nazi Rudolf Hess Exhumed From 'Pilgrimage' Grave", *BBC News Online* (21 July 2011) www.bbc.co.uk/news/world-europe-14232768 (accessed 30 September 2015).

15   For example, permission was granted in 2007 to exhume the body of Yorkshire Baronet, Sir Mark Wright, who was killed by the Spanish flu virus in the 1919 pandemic. Scientists hoped that tests on the remains (which had been buried in a lead-lined coffin) might yield valuable information which could be used in the fight against the modern 'bird flu' (H5N1) virus – *Re St Mary, Sledmore* [2007] 3 All ER 75 and see M Beckford, "Aristocrat to be Exhumed in Bird Flu Fight", *The Telegraph* (London, 28 February 2007) www.telegraph.co.uk/news/uknews/1544094/Aristocrat-to-be-exhumed-in-bird-flu-fight.html (accessed 30 September 2015).

16   However, each case is fact-specific, and public interest arguments are no guarantee that an application will succeed.

17   As in *Re St Edward, King and Confessor, New Addington* (Cons Ct (Southwark), 23 October 2013) where the deceased's daughter wanted to exhume her father's ashes to bury them with her mother's body in Victoria, Australia; the daughter (the only family member still living in the UK) was also emigrating to Australia to join six of her siblings already living there. The family unity motive is also apparent where the deceased's exhumed remains are to be cremated and the ashes placed with those of a close relative who died after the deceased – see *Re Matheson* [1958] 1 WLR 246 (son, acting on his mother's wishes, wanted to exhume his father's remains so that they could be cremated and his ashes placed with those of the mother).

close relative has subsequently died and been buried elsewhere).[18] Spiritual beliefs may also be a factor, with requests to move the deceased from unconsecrated ground to a burial place affiliated with a particular faith, or *vice versa*.[19] Finally, disinterments may be intended to confirm the deceased's identity and even the fact of death;[20] to retrieve important papers accidentally placed in the deceased's coffin;[21] and because the deceased was initially buried in the wrong burial plot, often because of a mistake by church officials or cemetery staff.[22]

Irrespective of motive, exhumations for personal reasons are requested with greater frequency and raise more complex issues; from a policy perspective, the absence of any discernible public benefit makes disturbing the dead more difficult to justify in these cases, which form the focal point of the following discussion.

II.   Legal authority to exhume

Exhuming the dead requires legal authorisation,[23] secured by one of three methods. The first is the relatively narrow category of cases where a coroner orders exhumation, either to carry out a post-mortem or where a body needs to be examined as part of any criminal proceedings.[24] However, the vast majority of disinterment applications fall into the other two categories: a Ministry of Justice licence or an ecclesiastical faculty, depending on where the remains were originally interred.

---

18   See for example, *Re St Mark's Churchyard, Fairfield* [2013] PTSR 953.
19   As in *Re Crawley Green Road Cemetery* [2001] 2 WLR 1175 and *Re Putney Vale Cemetery* (Cons Ct (Southwark), 6 August 2014) (both noted later in this chapter). In most instances, the remains will have been mistakenly interred first time around, though not necessarily – see *Re Durrington Cemetery* [2000] 3 WLR 1322 (also discussed below).
20   Cantor (n 5), p 65 cites the bizarre example of a woman who wanted to make sure that her husband was actually dead, following reports that he had been 'seen alive' months after his burial – a situation mirrored on this side of the Atlantic in *Druce v Young* [1899] P 84.
21   *Re Edward Hall (Deceased)* (1893) unreported and cited in *R v Tristam* [1898] 2 QB 371.
22   As in *Re St Andrew, Thringstone* (Cons Ct (Leicester), 3 September 2013) though mistaken burial, in itself, does not generate an automatic right of removal (*Beard v Baulkham Hills Shire Council* (1986) NSWLR 273). Sometimes the mistake occurs at a much earlier stage – for example, a 2009 incident involving Hammersmith Hospital in west London when mortuary workers released the bodies of two men with identical surnames to the wrong funeral directors (G Lynn, "Exhumation After Wrong Bodies Buried in Hospital Mix-Up", *BBC News Online* (4 July 2011) www.bbc.co.uk/news/uk-england-london-13993666 (accessed 30 September 2015).
23   Otherwise it constitutes a criminal offence – see Pt VI.
24   Coroners and Justice Act 2009, Sch 5, para 6 and see Ch 1, n 40. Dorries notes that coroners' exhumations are "extremely rare", with an average of four carried out nationally each year – C Dorries, *Coroners' Courts: A Guide to Law and Practice* (OUP, 3rd edn, 2014), p 88.

1.   The ecclesiastical route: Faculty petition

Where human remains were originally buried in ground consecrated by the Church of England, any subsequent displacement requires the grant of a faculty in accordance with ecclesiastical law.[25] Thus exhumation of a corpse or ashes[26] from a Church of England churchyard or the consecrated part of a municipal cemetery is dependent on permission from the Chancellor of the consistory court for the diocese in which the grave is located.[27] Removal and reinterment of the remains in exactly the same place (for example, to deepen the existing grave to allow further burials) requires a faculty.[28] More often, the underlying purpose will be reinterment in a different plot in the same or other consecrated ground, or moving the remains to unconsecrated ground.[29] Other possibilities include exhuming a body to cremate it, or previously interred ashes to rebury, scatter or keep them. However, where the remains are going to be exhumed, to and for what underlying purpose will determine whether or not a faculty will be granted.[30]

The formal process requires completion of a special Faculty Petition Form, detailing the reasons for exhumation, where the remains are currently located and where they are being moved to, as well as the petitioner's relationship to the deceased.[31] However, unlike the Ministry of Justice procedure, there is no standard application form. Different diocesan versions apply depending on where the remains are interred; and even if the questions posed and consents required (for example, from the holder of the exclusive right of burial, and close members of the deceased's family) are broadly the same, there is no set definition of who qualifies as 'petitioner'. While a surviving spouse, followed by the deceased's closest living relative or next-of-kin would be well-placed to apply for a faculty, the position may be less clear for surviving unmarried cohabitants who fall outside traditional kinship

---

25   The faculty jurisdiction was noted in Ch 2, p 31 and see also M Hill, *Ecclesiastical Law* (OUP, 3rd edn, 2007), ch 7. Note however, that the Church in Wales cannot grant a faculty for exhumation and must follow the secular route outlined below.

26   Both must be treated with the same degree of reverence and dignity under ecclesiastical law – *Re Atkins* [1989] 1 All ER 14.

27   Consistory courts are first instance courts of the Church of England, presided over by a Chancellor who exercises this particular jurisdiction on behalf of the Ordinary (usually the bishop of the diocese).

28   Even though there is no geographical displacement, disturbing the remains makes this – technically – an exhumation.

29   Reburial in unconsecrated ground (along with reinterment in the same place) also used to require secular authority, though not anymore – see pp 190–191.

30   See pp 191–199.

31   A faculty can be revoked by the relevant ecclesiastical court where it was obtained by misrepresentation or its terms are subsequently exceeded – *St Pancras Vestry & St Martin-in-the-Fields (Vicar and Churchwardens)* (1860) 6 Jur NS 540.

categories (at least for estate administration purposes).[32] Slight regional variations are a possibility here.

Where the grant of a faculty is declined, there is a subsequent right of appeal to either the Chancery Court of York in the Northern Province of England or the Court of Arches in the Southern Province, depending on the geographical location of the first instance consistory court.[33] The existence of twin appellate courts for different regions raises questions around the consistency of the decision-making process, and whether consistory courts should be influenced by decisions of the appellate court in another province. However, both are guided by the same fundamental principle (namely, the permanence of Christian burial)[34] and, according to *Re St Nicholas's, Sevenoaks*[35] should be seen as two divisions of a single court for precedent purposes.

2   The secular route: Application for a Section 25 Licence

Faculties only apply where ground has been consecrated by the Church of England. The resultant lacuna where human remains were not protected under ecclesiastical law was exposed in the nineteenth century, as increasing secularisation coincided with public concerns (fuelled by the growth in body snatching) around the safe and permanent interment of the dead.[36] As a means of protecting burials outside the remit of the established church,[37] s 25 of the Burial Act 1857 introduced a secular authority to exhume human remains and was originally enacted in the following terms:

> Except in the cases where a body is removed from one consecrated place of burial to another by faculty granted by the ordinary for that purpose, it shall not be lawful to remove any body, or the remains of any body, which may have been interred in any place of burial, without licence under the hand of one of Her Majesty's Principal Secretaries of State, and with such precautions as such Secretary of State may prescribe as the

32   See Ch 3, Pt I. There may also be question marks over whether the same individual would have to signify their consent to the proposed exhumation and reinterment where the faculty was sought by someone else.

33   For an overview, see S Gallagher, "Raising the Dead: Exhumation and the Faculty Jurisdiction: Should We Presume To Exhume?" [2010] 1 *Web Journal of Current Legal Issues* (accessed 30 September 2015). A final appeal lies to the Judicial Committee of the Privy Council.

34   *Re St John the Divine, Pemberton* (Cons Court (Wakefield), November 2003), the court also stressing that each case has to be decided on its own facts.

35   [2005] 1WLR 1011.

36   See the lengthy discussion in S Gallagher, "Protecting the Dead: Exhumation and the Ministry of Justice" [2008] 5 *Web Journal of Current Legal Issues* (accessed 30 September 2015).

37   Which, until then, were only protected by common law offences around unauthorised or unlawful disinterment – see Pt VI.

condition of such licence; and any person who shall remove any such body or remains, contrary to this enactment, or who shall neglect to observe the precautions prescribed as the condition of the licence for removal [commits an offence].[38]

As a result of this provision, "remains in unconsecrated ground became protected just as remains in consecrated ground had been".[39]

Section 25 remains in force today, and while its wording has recently been altered[40] the basic remit is still the same. The provision incorporates both bodies and ashes,[41] and governs exhumations that fall outside the faculty jurisdiction as well as those that are not governed by other specific statutory provisions.[42] It applies to the removal of remains from unconsecrated ground, such as municipal cemeteries and burials on privately or publicly owned land, as well as from other denominational burial grounds – essentially affiliated churchyards and specific burial places for faiths beyond the Church of England.[43] Exhumation may be followed by reburial in other unconsecrated or consecrated ground,[44] or in the exact same grave;[45] alternatively, it may be to cremate a body, or to scatter or keep ashes.[46] However, the initial disinterment requires a s 25 licence, regardless of the subsequent fate of the remains.

The formal process involves submitting a detailed application form to the Ministry of Justice,[47] under the remit of the Secretary of State for Justice.[48] Permission is usually sought by one of the deceased's surviving relatives; however, anyone can apply, as long as they explain why they are making the

---

38   The offences aspect is considered in Pt VI.

39   Cameron QC, Dean of the Arches in *Re Blagdon Cemetery* [2002] 4 All ER 482, 485.

40   See p 190.

41   Both the original and amended versions of s 25 refer to a "body, or the remains of any body". There is no limit on the age of the remains covered by the legislation, though different consent requirements tend to be adopted for those over 100 years old (see below) while archaeological exhumations require a different application process.

42   Such as those that govern specific development works and authorise exhumation to release old and disused burial grounds – see n 11.

43   For example, Gallagher notes that a s 25 licence was required in 2008 to exhume what were thought to be the remains of Cardinal Newman, originally interred in 1890 in a Worcestershire cemetery consecrated by the Catholic Church – Gallagher (n 33). See also *R (on the application of Rudewicz) v Ministry of Justice* [2011] EWHC 3078 (Admin) and discussed below.

44   Where relocation is to a consecrated burial site, there is no need to secure a faculty because the latter has never been required for burial in consecrated ground.

45   Again, as part of an identification process or simply to deepen the grave for another burial.

46   Either where the deceased was originally cremated and the ashes interred, or where the reason for exhumation was cremation.

47   The jurisdiction was formerly exercised by the Home Office.

48   The form and accompanying guidance notes are located at www.gov.uk/government/uploads/system/uploads/attachment_data/file/326818/application-exhumation-licence.pdf (accessed 30 September 2015).

request.[49] Ensuring that the relevant consents are in place is a vital part of the process. The initial emphasis is on the deceased's next-of-kin as "identified and prioritised by standard probate principles",[50] so that the basic ranking of spouse or civil partner, children, parents, siblings, etc, of the deceased reasserts itself on exhumation.[51] Although the consents of *all* of the next-of-kin are usually required,[52] a s 25 licence can be issued where the application is made by the highest ranking relative or all of those ranked equally highest, despite objections from someone of lesser kinship; and where a spouse or civil partner applies, that is usually the end of the inquiry.[53] Any known objections must be detailed on the application form, yet the Ministry of Justice will not become embroiled in family disputes and can defer the grant of a licence until these have been resolved.[54] Looking beyond the next-of-kin, the applicant must also secure the consent of the owner of any exclusive rights of burial on the grave as well as the relevant burial authority (or landowner, in the case of burial on private land). A s 25 licence only grants permission to exhume the deceased's remains; it "does not authorise the

49   For example, the owner of an adjoining grave as in a recent petition brought by a Muslim family who wanted to exhume the remains of a Catholic man ('Mr Smith') interred beside their dead relative in a multi-denominational cemetery because Mr Smith was a 'non-believer'. The Ministry of Justice confirmed that it would not allow exhumation without the full written consent (see below) of Mr Smith's family, which had been refused here – G Rayner, "Grave of Catholic Shadrack Smith Will Not Be Exhumed, Family Told in Row Over Muslim Burial Plot Next Door", *The Telegraph* (London, 11 February 2015) www.telegraph.co.uk/news/religion/11406710/Grave-of-Catholic-Shadrack-Smith-will-not-be-exhumed-family-told-in-row-over-Muslim-burial-plot-next-door.html (accessed 30 September 2015).

50   *R (on the application of Plantagenet Alliance Ltd) v Secretary of State for Justice* [2014] EWHC 1662 (QB), [107].

51   In other words, the intestacy based ranking discussed in Ch 3, Pt I. Once again, this excludes the deceased's unmarried partner; and while he/she could still apply for exhumation, the application is vulnerable to objections from those who qualify as the deceased's next-of-kin (unless an unmarried partner could invoke Article 8 of the Convention here (see generally Pt V), or is the grave owner, in which case their consent would also be required (see below)). Unlike disposal of the deceased's remains, the executor does not play an integral role on exhumation, unless the executor makes the application (which still requires the consent of the next-of-kin) or falls within one of the designated kinship categories.

52   The application process contemplates situations in which individual next-of-kin are unable to give consent or missing. Substituted consent may also be possible in certain circumstances – *R (on the application of Rudewicz) v Ministry of Justice* [2011] EWHC 3078 (Admin) (when seeking to exhume the remains of members of religious orders, the consent of the head of the order is accepted in place of the next-of-kin).

53   Advice provided by Ministry of Justice Coroners, Burials, Cremation and Inquiries Team (email on file with the author). When more than one person is equal highest next-of-kin (e.g. parents, children or siblings who fall within the same kinship tier and have joint status), a s 25 licence will only usually be issued when all of them consent. If there is an objection from someone of equal or higher kinship to the applicant, a s 25 licence is very unlikely.

54   *R (on the application of Plantagenet Alliance Ltd) v Secretary of State for Justice* [2014] EWHC 1662 (QB), [107].

licence holder to go onto the land on which the licence holder has no interest".[55] Since this would constitute a trespass, a s 25 licence is very unlikely where the holder of the exclusive right of burial or the burial authority/landowner objects.

There is no statutory right of appeal against the grant or refusal of a s 25 licence; judicial review is the only option, if the Secretary of State acted unlawfully, unreasonably or irrationally in reaching the decision.[56] Yet, the broad statutory power conferred by s 25 makes this difficult to establish, as *R (on the application of Rudewicz) v Ministry of Justice*[57] illustrates. Permission to exhume the body of a Polish priest from the churchyard where it had been buried 50 years earlier was granted by the Secretary of State, following a request by the priest's religious order and endorsed by senior members of the Catholic Church. The order, in possession of what was now defunct property and bound by the terms of a charitable endowment, was selling the property and contractually obliged to remove the deceased's remains under the conditions of sale; Father Jarzebowski would be reinterred with other members of the same religious order in a cemetery two miles away, and where people could still visit his grave – an important factor, given his saint-like status amongst Polish Catholics both here and abroad. However, the disinterment was strongly opposed by members of the Polish community, including the claimant who challenged the Secretary of State's decision in her capacity as a distant but closest living relative of the dead priest.

Looking at the exhumation controls imposed by s 25, the High Court accepted that there should usually be "some proper reason"[58] for disinterment and that a Secretary of State who granted permission "for a frivolous reason or for no reason at all would be acting unreasonably and irrationally".[59] However, the decision-making process could not be impugned in the present case; the Secretary of State had been entitled to decide that relocating the deceased's remains (a specific course of action favoured by the Catholic Church) to a burial site where he would be reunited with his colleagues and where the public would have unrestricted access to his grave favoured the grant of a s 25 licence, regardless of opposition from the claimant and others.[60]

---

55   *R (on the application of HM Coroner for East London) v Secretary of State For Justice* [2009] EWHC 1974 (Admin), [26].
56   The first category (i.e. illegality) would include acting in breach of Convention rights. The influence of human rights on disinterment decisions is considered more generally in Pt V.
57   [2011] EWHC 3078 (Admin).
58   *Ibid*, [20].
59   *Ibid*.
60   The fact that the deceased had wanted to be interred in his original resting place was deemed irrelevant, since the constitution of the religious order, which had bound the deceased in life, clearly stated that the head of the order had the final say on a priest's burial place. Human rights arguments were also rejected by the High Court – these are noted in Pt V.

3.   Dual permission requirements abolished

Although the faculty and licensing jurisdictions have clearly defined boundaries, the original wording of s 25 of the 1857 Act meant that both types of permission were sometimes required. A literal interpretation of the opening proviso – "[e]xcept in the cases where a body is removed from one consecrated place of burial to another by faculty granted ... for that purpose" – resulted in the disinterment of remains from a Church of England grave or plot and their subsequent reburial in the same place, or in unconsecrated ground, needing both a faculty and a licence.[61] From 1 January 2015,[62] however, s 25 reads as follows:

(1)   It is an offence for a body or any human remains which have been interred in a place of burial to be removed unless one of the conditions listed in subsection (2) is complied with.

(2)   The conditions referred to in subsection (1) are:
(a)   the body or remains is or are removed in accordance with a faculty granted by the court;[63]
[...]
(c)   unless the body or remains is or are interred in land which is subject to the jurisdiction of the court..., the body or remains is or are removed under a licence from the Secretary of State and in accordance with any conditions attached to the licence.

The restatement retains the original substance of the provision, while simplifying exhumation procedures in the two consecrated land scenarios outlined here. Both are now dependent solely on a faculty, thus removing another layer of bureaucracy and legal complexity from the disinterment process – given that faculties and s 25 licences have distinct formal requirements. Dual applications involved two sets of paperwork and supporting evidence, with different time periods for reaching a decision.[64] The two authorisations were also independent (one could not be made conditional on the grant of the other), and the person seeking exhumation needed both for

---

61   Since the initial disturbance to consecrated ground required a faculty, and the fact that the remains were not being removed from *one consecrated burial site to another* invoked the licensing requirement under s 25.

62   As a result of s 2 of the Church of England (Miscellaneous Provisions) Measure 2014, which substitutes a new s 25.

63   The relevant 'court' (as defined in s 25(1)(d)) will be the one having jurisdiction when seeking the initial grant of a faculty – usually the consistory court of the relevant diocese.

64   The grant of a faculty usually takes longer, for reasons discussed below.

it to be legal.[65] Even more problematic was the prospect of the respective decision-makers reaching different verdicts on the same factual scenario; and while the abolition of dual permission requirements prevents this, ecclesiastical authorities and the Ministry of Justice can still adjudicate factually similar requests for exhumation in dissimilar ways.

III. Faculty versus licence: Contrasting approaches

Ecclesiastical law presumes that the initial act of interment is intended to be permanent; and while consistory courts (as first instance decision-makers) can sanction the removal of human remains from a consecrated grave, faculties are not granted as a matter of course. As Cameron QC, Dean of the Arches explained in *Re Blagdon Cemetery*:[66]

> Lawful permission can be given for exhumation from consecrated ground. ... However, that permission is not, and never has been, given on demand by the consistory court. The disturbance of remains which have been placed at rest in consecrated land has only been allowed as an exception to the general presumption of permanence arising from the initial act of interment.[67]

The onus is on the petitioner to put forward "special circumstances" which, on the balance of probabilities, "justify the making of an exception from the norm that [interment in consecrated ground] ... is final".[68] However, the discretion to exhume will be exercised sparingly. While disturbing the dead may sometimes be necessary or expedient, consistory courts do not regard human remains as "portable [items], to be taken from place to place",[69] even

---

65   See DA Smale, *Davies' Law of Burial, Cremation and Exhumation* (Shaw & Sons, 7th edn, 2002), pp 242–245, and *R v Tristam* [1898] 2 QB 371 and *Re Talbot* [1901] P 1. For more recent confirmation see *Re Marley Lane Cemetery* (Cons Court (Battle), 9 December 2013) (the secular discretion vested in a government minister is entirely separate from and additional to the requirements of the church authorities where consecrated land is concerned – citing *R (HM Coroner for the Eastern District of London) v Secretary of State for Justice* [2009] EWHC 1974 (Admin)).

66   [2002] 4 All ER 482, 486.

67   See also *Re Atkins* [1989] 1 All ER 14, 19 ("although the Court's jurisdiction to grant or refuse such a Faculty is quite unfettered, it is to be exercised reasonably, according to the circumstances of each case, taking into account changes in human affairs and ways of thought but also mindful that consecrated ground and human remains committed to it should, in principle, remain undisturbed"). In deciding whether to grant a faculty, consistory courts must exercise their jurisdiction in a judicial manner – *Re Christ Church, Alsager* [1999] 1 All ER 117.

68   *Re Blagdon Cemetery* [2002] 4 All ER 482, 489 and see also *Re St Nicholas's, Sevenoaks* [2005] 1 WLR 1011. Whether the presumption against disinterment can only be rebutted in 'special' circumstances or, as recent cases suggest, 'exceptional' circumstances (see below) is probably a matter of semantics.

69   *Re Atkins* [1989] 1 All ER 14, 19.

if the potential displacement distance is a comparatively short one.[70] And while the physical difference between interred ashes and an interred corpse might make exhuming the former a simpler task,[71] this does not make the grant of a faculty more likely; both types of request are treated in the same way.[72]

In rebutting the basic presumption against disinterment, ecclesiastical courts regard certain factors as persuasive.[73] An error or oversight surrounding the original burial will usually suffice – for example, failure to appreciate the spiritual significance of interment in consecrated ground (especially where the deceased adhered to different faith values)[74] or, more likely, mistaken burial in a grave space already reserved for someone else.[75] In the latter scenario[76] where two sets of grieving families are the

70   Relocating the deceased's remains a short distance within the same churchyard or cemetery does not automatically warrant the grant of a faculty – see for example, *Re St Peter, Gunton* (Cons Ct (Norwich), 26 October 2013) (proposed exhumations would have resulted in remains being reinterred less than 6 feet from their original burial site) and *Re Christ Church, Alsager* [1999] 1 All ER 117 (petitioner's mother buried some 90 feet from her husband's ashes). However, the short displacement distance was a material factor in *Re St Bartholomew, Horley* (Cons Ct (Southwark), February 2010) where the deceased's ashes were buried in a garden of remembrance, around 90 yards from the grave of his wife who had died 13 months later (faculty granted to exhume the ashes because the two sets of remains were so close together, and had only been interred separately because of the wife's desire to be buried; exhuming the ashes would also free up space in the garden of remembrance).

71   Disinterment of ashes involves less physical disturbance to the grave and its surroundings, and does not generate a public health risk.

72   *Re Atkins* [1989] 1 All ER 14.

73   Two of the leading authorities are still *Re Christ Church, Alsager* [1999] 1 All ER 117 (analysed in P Petchey, "Exhumation Reconsidered" (2001) Ecclesiastical Law Journal 122) and *Re Blagdon Cemetery* [2002] 4 All ER 482.

74   As in *Re Putney Vale Cemetery* (Cons Ct (Southwark), 6 August 2014) where the petitioner and his family were Buddhists, and wanted to exhume three relatives whose remains had been interred in consecrated ground (the petitioners having subsequently been advised that such interment was not in accordance with Buddhist tradition). Permission was granted. See also *Re Crawley Green Road Cemetery* [2001] 2 WLR 1175 (discussed at pp 206–207) and *Re Miresse (Deceased)* (Cons Ct (Southwark), July 2003) (deceased's family, who were Italian Catholics, had not appreciated the significance of burying their daughter in the consecrated part of Lambeth cemetery).

75   Usually where the petitioner has already reserved the plot beside a loved one's grave, but someone else is mistakenly interred there as in *Re St Luke's, Holbeach Hurn* [1990] 2 All ER 749 and *Re Jean Gardiner (Deceased)* (Cons Ct (Carlisle), 28 May 2003). The encroaching corpse or ashes will usually be removed; while it is a "serious matter" to order disinterment in the face of familial opposition, it is an "equally serious matter" to lose the right to be buried beside a "beloved family member" because of a mistake – *Re St Andrew, Thringstone* (Cons Ct (Leicester), 3 September 2013), [22]. See also *Re Blagdon Cemetery* [2002] 4 All ER 482, 490–491 (a faculty will readily be granted in these circumstances because it is correcting an administrative error rather than displacing the presumption of permanent burial).

76   Which seemingly runs contrary to the philosophical norm that 'two wrongs do not make a right'.

unfortunate victims of circumstance, exhumation will be desperately sought by one side and fiercely opposed by the other.[77] Alternatively, the physical or (more likely) mental wellbeing of the living can justify exhumation. While advancing years or medical conditions that make it difficult for relatives to visit the grave of a loved one may not suffice,[78] severe psychiatric or psychological problems related to the location of the deceased's remains can constitute exceptional circumstances.[79] For example, in *Re St Mary the Virgin, Stansted*[80] the court allowed exhumation of the deceased's remains from a churchyard close to where he had been killed by a dangerous driver. The deceased's mother and brother had been unable to visit the grave because neither could bear to pass the place where they had last seen the deceased alive; medical evidence indicated profound emotional difficulties, which would improve if the remains were moved elsewhere.[81] Other potential grounds for allowing exhumation include the fact that the original interment was not in accordance with the deceased's express wishes;[82] to ensure that the coffin contains the proper remains

77   In *Re St John's Church, Walsall Wood* (Cons Ct (Lichfield), 6 April 2010) the remains of a young man who had been killed in Spain were mistakenly interred in the grave that the petitioner had reserved so that she could be buried beside her husband of over 50 years who had recently died of cancer. The court authorised exhumation – even though the young man's remains had been exhumed once before (he had initially been buried in Spain, by order of a Spanish court and his mother had fought a legal battle to repatriate her son's body). Occasionally, the person whose reserved plot has been mistakenly appropriated may petition for their loved one's remains to be removed elsewhere, to avoid any distress to the other set of grieving relatives. Again, a faculty is likely to be granted – see *Re St Peter, Dunchurch* (Cons Ct (Coventry), 21 July 2013).

78   See *Re St Helen, Edlington v Wispington* (Cons Ct (Lincoln), 16 July 2014) (faculty refused, despite the petitioner's osteoarthritis restricting visits to her father's grave). The petitioners' respective mobility issues were not sufficient reasons, in themselves, to justify a faculty in *Re Ronald Carr (Deceased)* (Const Ct (Lincoln), 27 August 2014) and in *Re St Andrew's Churchyard, Alwalton* [2012] PTSR 479.

79   As explained in *Re Blagdon Cemetery* [2002] 4 All ER 482, 490.

80   (Cons Ct (Rochester), 17 October 2013).

81   See also *Re South London Crematorium* (Cons Ct (Southwark), 27 September 1999) (mentioned briefly at n 102). However, contrast these cases with *Re Crawley Green Road Cemetery* [2001] 2 WLR 1175 (petitioner's long-term depressive illness and grief since her husband's death were not sufficient medical reasons).

82   See for example, *Re Knight (Deceased)* (Cons Ct (Chester), 23 November 1993), though contrast this with *Re Mary Wood (Deceased)* (Cons Ct (Sheffield), June 2002).

where this fact is later disputed;[83] and, in more extreme situations, to ensure that the deceased is not buried with a family member who sexually abused them in life.[84]

Turning to more generic factors, consensus amongst the deceased's closest relatives that the remains should be disinterred can also be persuasive,[85] assuming the existence of a valid reason for exhumation.[86] However, material objections are not confined to the deceased's next-of-kin. Consistory courts will be sensitive to the views of those whose relatives are buried in adjacent graves,[87] though the amount of 'local support' for a petition (for example, that of the church council or congregation) will normally be irrelevant.[88] Another influential factor is lapse of time; a request for exhumation within weeks or months of the original burial is more likely to succeed than one where remains have been undisturbed for years[89] – though the longer-term dead may be displaced if there is a public interest in moving remains buried over a century ago.[90] Late applications place a much

83   See *Re Walker (Deceased)* (Cons Ct (Liverpool), January 2002) (remains of stillborn twins sent to Alder Hey Hospital in Liverpool for a post-mortem; evidence suggested that only one of the twins' bodies had been returned for burial). However, doubts as to whether or not a coffin contained all of the deceased's bodily organs following a post-mortem may not result in a faculty – *Re Makin (Deceased)* (Cons Ct (Liverpool), January 2002) (deceased infant had died at Alder Hey Hospital).

84   See *Re X (Deceased)* (Cons Ct (Liverpool), October 2001) (body of father who had been buried with his late wife and daughter exhumed where it was subsequently disclosed that he had been abusing this daughter and another sister). This resonates with the idea of protecting the dead from further harm, something which is also apparent in funeral disputes – see Ch 4, p 91.

85   *Re Christ Church, Alsager* [1999] 1 All ER 117, 123.

86   *Re Marks (Deceased)* (Cons Ct (Chester), 31 August 1994) (next-of-kin's reasons for wanting a faculty must be well-founded and sufficient).

87   Either because they fear that exhumation might physically disturb their loved one's remains (something that contributed to the petition being refused in *Re St Helen, Edlington v Wispington* (Cons Ct (Lincoln), 16 July 2014), or because they find the proposed disinterment distressing and offensive (see the comments in *Re Robin Hood Cemetery, Solihull* (Cons Ct (Birmingham), 1 November 2006) – though this would be less convincing than the first reason).

88   *Re Blagdon Cemetery* [2002] 4 All ER 482, 491, rejecting suggestions to the contrary in *Re Christ Church, Alsager* [1999] 1 All ER 117.

89   As noted in *Re Atkins* [1989] 1 All ER 14, 19 and see also *Re Christ Church, Alsager* [1999] 1 All ER 117, 123 ("[t]he passage of time, especially when this runs into a number of years, may make it less likely that a faculty will be granted") – though Petchey (n 73), pp 125–126 argues that exhumation may be more distressing shortly after burial.

90   See for example, *Re St Mary the Virgin, Woodkirk* [1969] 1 WLR 1867 (road improvement scheme); *Re St Mary the Virgin, Hurley* [2001] 3 WLR 831 (repatriation of the remains of a Brazilian national hero – though Gallagher (n 33) stresses that not all such repatriation requests will succeed); *Re St Mary, Sledmore* [2007] 3 All ER 75 (important medical research – see n 15); and *Re Radcliffe Infirmary Burial Ground* [2011] PTSR 1508 (construction of a new university department). Faculties were granted in each of these cases.

stronger onus on the petitioner, who must present a credible explanation for the delay alongside supporting reasons.[91] Another important consideration in faculty petitions is where the remains will be placed after exhumation. Reinterment in the same grave (usually at a greater depth to facilitate further burials) does not tend to be problematic,[92] and consistory courts may be more inclined to grant a faculty where exceptional circumstances are established and the deceased is to be reinterred in consecrated ground.[93] However, other displacement locations and outcomes are less clear-cut. For example, while there is no outright prohibition on reinterment in unconsecrated ground, the grant of a faculty may not be as forthcoming,[94] and questions have been raised about disinterment to allow the deceased's remains to be cremated.[95] Case law also suggests that a faculty will not be granted where the petitioner's intent is to scatter ashes currently interred in

91   As in *Re Atkins* [1989] 1 All ER 14 (petitioner's decision to bury her husband's ashes in a particular churchyard – which was now likely to fall into disuse – was made under the stress of bereavement; remains to be reinterred in the consecrated part of a cemetery where other relatives were buried) and *Re Durrington Cemetery* [2000] 3 WLR 1322 (deceased's relatives delayed applying for a faculty out of respect for his widow – discussed further at p 206). Contrast this with *Re Smith* [1994] 1 All ER 90 (passage of over 8 years since the burial placed a particularly heavy onus on the petitioner, which he failed to discharge), *Re Robin Hood Cemetery, Solihull* (Cons Ct (Birmingham), 1 November 2006) (no credible explanation for wanting to exhume the deceased's ashes 20 years after they were interred) and *Re Loughborough Cemetery* (Cons Ct (Leicester), 9 June 2014) (exhumation of infant refused, almost 46 years after her death and 40 years after her parent's had moved to Australia).

92   See *Re Jenkin's Petition* [2013] PTSR 297 (given the general shortage of burial space, placing remains deeper in the ground was to be encouraged) and *Re Strood Cemetery* (Cons Ct (Rochester), January 2005) (faculty granted where several sets of family ashes were to be exhumed from an existing grave and placed in a new coffin with the petitioner's remains after she died). A retrospective faculty was granted in *Re All Saints, Beckley* (Cons Ct (Chichester), November 2006) after the gravedigger had moved the ashes of a father and son to bury the mother in the same grave.

93   As in a significant number of cases discussed throughout this section.

94   Such a faculty can still be granted (as in *Re Blagdon Cemetery* [2002] 4 All ER 482), though the court must be persuaded that the proposed reinterment site is suitable and that the grave will be properly maintained (*Re Loughborough Cemetery* (Cons Ct (Leicester), 9 June 2014).

95   Smale (n 65, p 239 suggests that, since cremation is now an accepted bodily disposal method within the Church of England, there is no reason why a faculty to disinter remains for this purpose would now be refused (arguing that the situation has evolved since *Re Dixon* [1892] P 386 where one was denied). Permission to this effect was granted in *Re St Mark, Worsley* (Cons Ct (Manchester), 31 July 2006) (discussed at n 120), though may not be forthcoming where the petitioner intends to cremate remains that have been interred for a long time (see *Re Streatham Park Cemetery* (Cons Court (Southwark), 16 February 2011) and *Re Loughborough Cemetery* (Cons Ct (Leicester), 9 June 2014).

consecrated land (as opposed to reburying them elsewhere),[96] or to keep ashes at home indefinitely.[97]

Other scenarios militate against the grant of a faculty, though a factor that might not suffice on its own can rebut the presumption of permanent burial when taken in conjunction with others.[98] As a general rule, a mere change of mind by the deceased's family after the initial interment will not constitute exceptional circumstances,[99] nor will the fact that the family has moved to a new area and wants to relocate the deceased's remains as well.[100] Living in a more mobile society does not mean that families can expect to move their dead as part of every geographical transition, unless there are other mitigating factors.[101] These might include medical reasons,[102] or the absence of any link between the deceased and their place of burial – for example, where close family had not established a permanent home at the time of the

96 *Re Stocks* (*The Times*, 5 September 1995) and *Re Tixall Road Cemetery, Stafford* (Cons Ct (Lichfield), 2 November 2014).

97 As in *Re St Andrew's Churchyard, Alwalton* [2012] PTSR 479 (discussed at p 198) where the petitioner wanted to retain her husband's ashes indefinitely, so that they could eventually be scattered with hers (the deputy Chancellor noting that "burial within consecrated land is permanent so that any subsequent reinterment is a necessary corollary of the grant of any faculty to exhume" – *ibid*, [4]).

98 *Re St Andrew's Churchyard, Alwalton* [2012] PTSR 479.

99 *Re Blagdon Cemetery* [2002] 4 All ER 482, 490.

100 *Ibid*, 489–490 citing *Re South London Crematorium* (Cons Ct (Southwark), 27 September 1999) where George Ch stressed that human remains should not be the subject of multiple moves, just because their families move home. The Church of England has sounded alarm over the growing number of petitions based on this – "Church Warns Against Treating Bereaved Relatives as 'Portable Remains'", *The Telegraph* (London, 2 August 2015) www.telegraph.co.uk/news/religion/11779209/Church-warns-against-treating-bereaved-relatives-as-portable-remains.html (accessed 30 September 2015) and C Wheeler, "RIP… Until You Move Home: Brits Taking Dead Relatives With Them", *Daily Express* (London, 1 August 2015) www.express.co.uk/news/uk/595436/Relatives-moving-home-deceased (accessed 30 September 2015).

101 See *Re Marley Lane Cemetery* (Cons Court (Battle), 9 December 2013) (this "engages the portability concept and runs contrary to established Christian doctrine" – *ibid*, [10]). However, Gallagher (n 33) suggests that some judges are sympathetic to the fact that geographically mobile families are not always attached to one particular area – citing *Re Lambeth Cemetery, Re Streatham Park Cemetery* (Cons Ct (Southwark), October 2002) as an example.

102 For example, in *Re South London Crematorium* (Cons Ct (Southwark), 27 September 1999) a faculty was granted where the petitioner's brother had a serious history of depression and would benefit from having his father's ashes exhumed and reburied in the same grave as his mother.

deceased's death,[103] and despite a lengthy time since the original interment.[104] Reuniting family members in the same grave can sometimes rebut the presumption of permanent burial,[105] though other exceptional factors can be required.[106] The fact that the living find separate resting places distressful is irrelevant (especially where the deceased's family originally opted for this arrangement[107]), as is a testamentary request that a dead relative's remains are exhumed and placed with those of the more recently deceased.[108] Family members wanting to relocate a loved one's remains to a newly opened cemetery,[109] or one which would have been their original choice had space been available, will not automatically justify exhumation – even if, in the latter scenario, the proposed reinterment site is where the deceased would also have preferred to have been laid to rest.[110]

Concerns around the upkeep and safety of a particular churchyard or consecrated area of a cemetery will probably fail, even if the physical state

103   See *Re Wandsworth Cemetery* (Cons Ct (Southwark), 22 November 2013).

104   As in *Re Blagdon Cemetery* [2002] 4 All ER 482 where the petitioners' son had died in an industrial accident in 1978; his remains were buried in the local churchyard where the petitioners were living at that time. The petitioners moved every few years (the deceased's father was a publican and managed different pubs around the country) before retiring to Suffolk where they wanted to reinter their son's remains in a triple plot where they (the petitioners) could also be buried. A faculty was granted, based on the fact that the deceased had no ties to the community in which he was buried, the peripatetic nature of his parents' life at that time and the fact that the petitioners had (shortly after their son's death) inquired about the possibility of moving his remains once they had acquired a permanent home.

105   See *Re St Mark's Churchyard, Fairfield* [2013] PTSR 953.

106   As in *Re Strood Cemetery* (Cons Ct (Rochester), January 2005) (mentioned at n 92), where enabling family remains to be kept together was sufficient reason to rebut the presumption against disinterment on the facts. See also the repatriation case of *Re St Edward, King and Confessor, New Addington* (Cons Ct (Southwark), 23 October 2013) (highlighted at n 17) where exceptional circumstances were established because the deceased had died while visiting his daughter on holiday in the UK, the same daughter was now moving to Australia, and exhumation would reunite the deceased and his wife in the family grave.

107   As in *Re St Helen, Edlington v Wispington* (Cons Ct (Lincoln), 16 July 2014) where the cremated remains of the petitioner's parents (who had been married for 42 years) were buried in different graves. The court refused to allow the father's ashes to be exhumed and reinterred with the mother's, given that he had died in 1988 and a conscious decision to bury the remains separately had been taken when the mother died in 2007.

108   As in *Re Robin Hood Cemetery, Solihull* (Cons Ct (Birmingham), 1 November 2006) (wife's will requested that her husband's ashes – interred in the consecrated part of a cemetery some 20 years earlier – be exhumed and scattered with hers).

109   See *Re Field Road Cemetery* (Cons Ct (Lichfield), 18 May 2014) (father's remains already buried in a perfectly suitable plot where the petitioner's mother could also be buried).

110   See *Re Smith* [1994] 1 All ER 90 (petitioner and his wife wanted to be buried in the churchyard of the Bolton church that they had both attended from an early age and where they had been married, but no spaces were available when the wife died; court refused the petitioner's request for a faculty to exhume his wife's remains and reinter them in Bolton when vacant graves became available eight years later).

makes it difficult for older petitioners (or those with limited mobility) to tend to a grave[111] or a family member has been mugged on a previous visit.[112] Likewise, a faculty will not be granted because a petitioner objects to the enforcement of regulations about what can be placed on a family grave.[113] A good illustration is *Re St Andrew's Churchyard, Alwalton*[114] where a wife wanted to exhume her husband's ashes because a vase and flowers had been removed from his grave by church authorities, who had implemented rules on floral tributes and other memorials. Refusing the faculty on this and other grounds,[115] the deputy Chancellor stressed that consistory courts should be "very slow" to grant petitions that would undermine the enforcement of churchyard regulations and set a potentially dangerous precedent.[116] Other factors against the grant of a faculty include public health concerns, the fact that exhumation would be contrary to the deceased's wishes, "reasonable opposition" from members of the deceased's family or the fact that the petition itself is based on "improper motives".[117] Petitions designed to remove the dead from family plots based on "family differences" and ongoing grievances will "usually fail",[118] as in *Re Mangotsfield Cemetery*[119] where an ex-husband wanted to have his ex-wife's ashes exhumed from the plot where they had been placed with the couple's dead son, despite the husband only discovering this seven years later during one of his infrequent

---

111   See *Re St Peter, Gunton* (Cons Ct (Norwich), 26 October 2013) (poor cemetery maintenance that made it tricky – and potentially dangerous – for the two octogenarian petitioners to tend to their husbands' graves was not a sufficient reason to grant a faculty).

112   *Re St Andrew (Old Church), Hove* (Cons Ct (Chichester), 14 July 2005).

113   Commemoration disputes between families and church or cemetery authorities are fairly common – see Ch 8, Pt I.

114   [2012] PTSR 479.

115   As noted earlier, the wife's medical circumstances alongside the fact that she did not intend to reinter or otherwise dispose of the ashes immediately did not favour exhumation.

116   [2012] PTSR 479, [44]. The general issue of precedent has been noted by consistory courts, who must be mindful of the consequences of their decisions – see the respective observations in *Re West Norwood Cemetery* (Cons Ct (Southwark), 6 July 2000) and *Re Blagdon Cemetery* [2002] 4 All ER 482.

117   *Re Christ Church, Alsager* [1999] 1 All ER 117, 123.

118   Gallagher (n 33).

119   (Cons Ct (Bristol), February 2005).

visits to the grave.[120] Finally, case law also suggests that a faculty will be refused even though the petitioner has been mistakenly advised by a funeral director that the request would be likely to succeed,[121] and where exhumation is requested for historical or scientific reasons,[122] unless a public benefit is established.[123]

The jurisdiction to grant a faculty is discretionary. While theoretically unfettered, the high thresholds imposed by ecclesiastical courts and the aura of permanence surrounding burial in consecrated ground make faculties difficult to obtain. Each petition is assessed on its own merits, and the decision-making process can be a lengthy one as all the evidence is weighed up and the court makes a formal determination.[124] The result is a substantial body of case law, with individual outcomes that are highly fact-specific; and while some basic trends can be identified, broadly analogous scenarios can sometimes seem rife with inconsistencies as the discussion has highlighted.

Turning to s 25 of the 1857 Act, the jurisdiction to exhume is also a discretionary one though the assessment of individual applications differs significantly. Both the original and amended versions of this provision set out a basic regulatory framework; unlike the faculty jurisdiction however, the statutory requirement of a licence for non-secular exhumations does not subscribe to the core principle of leaving human remains undisturbed. As McCombe J explained in *R (on the application of Rudewicz) v Ministry of Justice*:

---

120   The fact that seven years had passed since the ex-wife's interment – and sixteen since the petitioner had last visited the grave – alongside familial objections, and the ex-husband's callous behaviour were collective grounds for refusing a faculty. In *Re St Mark, Worsley* (Cons Ct (Manchester), 31 July 2006), a faculty was granted to exhume the deceased's remains when his widow (shortly after burying her husband in a family grave) discovered that he had been having an affair throughout their marriage; the husband had also diverted some of his wife's inheritance into business funds, without telling her. The petition was supported by the couple's adult children. However, in permitting the remains to be exhumed for cremation and the ashes reinterred in a local authority cemetery, the Chancellor stressed that the decision should not be seen as a retrospective punishment for the deceased.

121   As in *Re Christ Church, Alsager* [1999] 1 All ER 117.

122   See for example, *Re St Nicholas's, Sevenoaks* [2005] 1 WLR 1011 (petition to exhume the deceased's remains for DNA testing in order to confirm long-standing family rumours that he was the illegitimate grandson of Queen Victoria; family curiosity and purely speculative genealogical interest were not enough to justify exhumation). See also *Re Holy Trinity, Bosham* [2004] Fam 125 (exhumation requested for DNA testing, to establish whether the interred remains were – according to local legend – those of King Harold; petition refused on the basis that, even if human remains were found in the coffin, DNA analysis was unlikely to give any meaningful result, and academic opinion pointed towards the King having been buried elsewhere).

123   As in *Re St Mary, Sledmore* [2007] 3 All ER 75 (see n 15) where a faculty was granted.

124   It is not simply a matter of completing certain formalities in order to persuade the court to grant a faculty – *Re Edward William Knight (Deceased)* (Cons Ct (Chester), 23 November 1993).

[Section 25] does not provide for any assumption or presumption that should be applied in the exercise of the statutory discretion. It ensures that bodies are not to be removed from their resting place without the sanction of a competent authority. ... It does not, in the hands of a secular authority ... require the rebuttal of some presumption against the removal of remains.[125]

In the same case, Hallet LJ suggested that there should be "some proper reason"[126] for disinterment, and while seemingly trivial or capricious requests might be rejected, the fact remains that a s 25 licence is entirely at the discretion of the Secretary of State. The provision imposes no limits on the circumstances in which human remains can be disturbed, leaving decisions to be made on a case-by-case basis (with the help of internal guidelines) and without having to devise exceptions to a basic presumption against disinterment. The upshot is what Gallagher describes as a "marked divergence"[127] between the factors that secular authorities and ecclesiastical courts regard as influential. For example, lapse of time since the original interment may not deter the grant of a s 25 licence,[128] and permission has also been more forthcoming where exhumation is requested for archaeological or scientific reasons.[129] The actual decision-making process is a secular one, although religious views will be taken into account where the proposed disinterment is from a burial place designated to specific denominations or faith groups.[130] When considering private, family applications the underlying reasons may not even be explored;[131] as long as the necessary consents are in place,[132] the application is for "personal, family reasons" and there are "no known legitimate objections", the Secretary of State will normally grant permission for disinterment.[133] So while the grant of a faculty is an evaluative judgment,

125   [2011] EWHC 3078 (Admin), [39].
126   *Ibid*, [20].
127   Gallagher (n 33).
128   Whereas a faculty is often refused on this basis – see p 194.
129   Gallagher (n 37). Again, consistory courts tend to regard this as insufficient reason to displace the presumption against disinterment – see p 199.
130   Advice provided by Ministry of Justice Coroners, Burials, Cremation and Inquiries Team (email on file with the author). Like Church of England burials, adherents of different faiths can have strong views on respecting the sanctity of the grave and ensuring that the dead are not exhumed without good reason.
131   Advice provided by Ministry of Justice Coroners, Burials, Cremation and Inquiries Team (email on file with the author).
132   Namely the next-of-kin, and burial authority or landowner.
133   Home Office Consultation Paper, *Burial Law and Policy in the 21st Century: The Need for a Sensitive and Sustainable Approach* (2004), p 12 and noting that, because a licence is at the Secretary of State's discretion, permission may still be issued where not all the consents are available. Although discussing Home Office practice (as the department that used to have jurisdiction over s 25 applications), the Ministry of Justice approach is the same – confirmed by Ministry of Justice Coroners, Burials, Cremation and Inquiries Team (email on file with the author).

the approval of a s 25 application is more procedural. The decision-making process is also much quicker, as applications are typically determined within 20 days.[134]

With the exception of mistakenly interred remains where s 25 licences have been refused in the past,[135] it is much easier to obtain a s 25 licence than a faculty.[136] Aside from the different procedural requirements and decision-making criteria,[137] those seeking to exhume the remains of a loved one can be subjected to very different outcomes on what are broadly similar sets of facts.[138] The issue was raised in *Re Christ Church, Alsager*[139] where the petitioner listed the number of disinterments from a nearby local authority crematorium as part of his appeal against the refusal of a faculty to exhume his father's ashes from a Church of England churchyard. In rejecting this evidence, the Chancery Court of York stressed that the churchyard was consecrated ground and therefore subject to different legal constraints; ecclesiastical law could not be influenced by equivalent practices in municipal cemeteries. Yet one of the reasons behind the petitioner's appeal was clearly a perception of unequal treatment, given the variance in practice between two different burial locations less than 6 miles from each other.[140] And, as Lovatt points out, an even more extreme example could occur "within a space of a few feet" in municipal cemeteries, "depending on the territorial boundaries" of the consecrated area.[141]

Despite the court's assurances that the petitioner had not been subjected to "unfair discrimination" in *Re Christ Church, Alsager*[142] the fact remains that requests to exhume the dead are treated very differently in English law, depending on whether original interment was in consecrated or unconsecrated ground. In many ways this is unavoidable, given the sharp contrast in the secular and ecclesiastical rationales for exhumation – though the

---

134 As specified in the Ministry of Justice application form – see n 48.
135 This contrasts sharply with the ecclesiastical jurisdiction where faculties are often granted for this purpose (see p 192), and is widely attributed to *Reed v Madon* [1989] 2 All ER 431, where a s 25 licence to exhume the remains of a wrongly interred corpse was refused as being contrary to the wishes of the deceased's next-of-kin (an integral part of the statutory regime) – see the discussion in Gallagher (n 33). According to the author, this is one of "two notable instances when it appears easier to obtain a faculty", the other being exhumation of old burials to deal with overcrowding in churchyards – *ibid*.
136 This disparity has been noted elsewhere – see the respective observations in Gallagher (n 33) and Lovatt (n 1).
137 Ecclesiastical petitions are more complex, both factually and procedurally, compared with the s 25 process (assuming the necessary consents are in place).
138 Though the same criticism can also be levied at decisions within the exclusive remit of ecclesiastical courts, as noted above.
139 [1999] 1 All ER 117.
140 See the discussion in Lovatt (n 1), pp 30–32.
141 *Ibid*, p 33.
142 [1999] 1 All ER 117, 124.

distinction is not always widely known or understood by bereaved families when choosing a particular gravesite.[143] Making informed choices is essential,[144] and those who opt for consecrated ground should be cautioned against "bury[ing] today thinking that they may exhume tomorrow"[145] if circumstances change. However, there may be an argument for bringing the secular and ecclesiastical jurisdictions to exhume into line, perhaps by assimilating the two within a single decision-making body and introducing some measure of "certainty to an emotive area of law".[146] Ecclesiastical authorities would almost certainly oppose this, given the basic theological values underpinning burial in consecrated ground. Yet a unified decision-making body could still be cognisant of these values and the underlying presumption of permanent burial, while simultaneously recalibrating the secular jurisdiction along similar lines. Introducing more stringent requirements for exhumation from unconsecrated ground accords with the basic premise of allowing the dead rest in peace. As George Ch stressed in *Re West Norwood Cemetery*,[147] "[m]ost people, and not merely Christian people, feel a sense of respect for the dead and a reluctance to interfere with their remains".

IV.   Disinterment disputes within families

Exhumation is an emotionally sensitive issue for the deceased's family, especially if it involves moving a buried corpse (as opposed to ashes, though this can also be upsetting). While plans to disinter will often be consensual, they can also generate bitter disputes between the living. For example, parents or siblings might object to the deceased's remains being moved elsewhere by someone further up the notional kinship hierarchy (usually a surviving spouse or civil partner), with discordant religious and cultural

---

143   Assuming that they have the practical option of burying the deceased in unconsecrated ground as well (though this may depend on the general location).
144   And points towards any consequential discrimination not being unfair.
145   Gallagher (n 33).
146   *Ibid*, though Gallagher subsequently rejects the idea that exhumation be removed from the ecclesiastical jurisdiction altogether.
147   (Cons Ct (Southwark), 3 July 2000) and cited in Petchey (n 73), p 133.

sensitivities providing a further source of conflict. Latent family tensions can also be reignited, despite the passage of time since the original interment.[148]

Both the secular and faculty routes proceed on the basis that the necessary consents are in place, at least from the deceased's immediate family. Individual objections should be documented, though Ministry of Justice guidelines emphasise that it will not adjudicate contentious applications,[149] while ecclesiastical courts regard family feuds as a factor against the grant of a faculty.[150] Permission to exhume may be more forthcoming where close relatives all agree to relocate the deceased's remains (though this, in itself, is no guarantee – especially when seeking a faculty). However, the secular route seems to adopt a normative kinship ranking, which prioritises the deceased's surviving spouse or civil partner.[151] As a result, their views on exhumation will be a major determinant; a s 25 licence is unlikely to be granted if the surviving spouse or partner objects, regardless of the views of other close family members.

V.   Human rights arguments

Arguments based on the European Convention of Human Rights are increasingly employed in funeral disputes, either in support of family members who favour certain arrangements or as a case for upholding the deceased's own bodily disposal preferences.[152] Similar trends can be seen in exhumation decisions, with a steady trickle of submissions since the Convention became part of domestic law in the UK in October 2000.[153] Most have involved faculty petitions, and whether refusal to allow

---

148   There are few reported English cases involving family disputes, though see *Re West Norwood Cemetery* [2005] 1 WLR 2176 (discussed in Ch 2, pp 33–34). This dearth of authority (explained immediately below) contrasts sharply with the number of American cases involving contested applications – see for example, *Cooney v Lawrence* 2 Pa D 22 (Pa Com Pl 1890) (husband's petition to exhume his wife's body and rebury it in a non-Catholic cemetery opposed by her Catholic siblings); *Smith v Shepherd* 64 NJ Eq 401 (NJ Ch 1903) (deceased's sister, who owned the burial plot, opposed the removal of her brother's remains at the request of his widow who wanted to be buried in the same grave as her husband; sister also opposed the exhumation of the remains of the deceased's first wife from the same plot to allow his widow to be buried there); *DiObdila v St Cecilia's Cemetery* 45 Pa D&C 3d 420 (PaComPl 1987) (wife's request to move her husband's body to a different plot in the same cemetery so that she could be buried alongside him opposed by the deceased's children); and *Estate of Jiminez* 65 CalRptr 2d 710 (CalApp 2 Dist, 1997) (deceased's cemetery plot chosen by three of his four children; fourth child sought an order that her father's remains be moved elsewhere, in accordance with his wishes).

149   See p 188.

150   *Re Christ Church, Alsager* [1999] 1 All ER 117, 123.

151   See pp 188.

152   See Ch 4, pp 107–114 and Ch 5, pp 139–144.

153   And in at least one case immediately before this date – namely *Re Durrington Cemetery* [2000] 3 WLR 1322.

exhumation breaches the petitioner's human rights, though a small number have also challenged the grant of a s 25 licence on this basis.[154] As with funeral disputes, claims invariably centre on the right to private and family life under Article 8 of the Convention and the guarantee of freedom of religion under Article 9.[155]

Article 8(1) rights can be engaged in exhumation decisions, presumably as part of a legitimate familial interest in what happens to the remains of a loved one.[156] However, some sort of discernible kinship link is essential (exhumation to establish a speculative relationship through DNA testing will not suffice[157]), and a petitioner will not be able to rely on Article 8 just because he/she has some sort of proven yet distant connection to the deceased.[158] More importantly, refusal of permission to disinter may not interfere with the petitioner's right to family life[159] and, even it does, may be acceptable under Article 8(2), which permits interference "in accordance with the law" and other specified grounds. One of the leading authorities is a decision of the European Court of Human Rights ('ECtHR') in *Dödsbo v Sweden*[160] where the applicant relied on Article 8 after Swedish authorities refused to sanction

---

154 As in *R (on the application of HM Coroner for East London) v Secretary of State For Justice* [2009] EWHC 1974 (Admin) and *R (on the application of Rudewicz) v Ministry of Justice* [2011] EWHC 3078 (Admin).

155 The protection of property provision in Article 1, Protocol 1 ('A1P1') has also been discussed in several exhumation cases. In *Re West Norwood Cemetery* [2005] 1 WLR 2176 (see Ch 2, pp 33–34), the son as legal owner of the exclusive right in the grave plot argued that burial of his father's ashes without the son's consent interfered with the son's A1P1 rights. While the court accepted that "possessions" in A1P1 included a legitimate expectation relating to property, and that refusal to allow exhumation of remains that had been buried without the consent of the owner of exclusive rights of burial constituted an impermissible interference with the owner's A1P1 rights, the petition was rejected because the son held the exclusive burial rights on trust for other members of his family. In the absence of an exclusive right of burial, the grave plot is not 'owned' by the petitioner and A1P1 will not apply (*Re St Andrew's Churchyard, Alwalton* [2012] PTSR 479). Opposing a proposed exhumation on the basis that this would infringe A1P1 because of alleged property rights in the deceased's remains looks likely to fail because of the basic rule that dead bodies are not property – see *Re Jean Gardiner (Deceased)* (Cons Ct (Carlisle), 28 May 2003) and discussed at p 206.

156 See Ch 4, pp 108–109.

157 See *Re St Nicholas's, Sevenoaks* [2005] 1WLR 1011 discussed at n 122 (Article 8 not engaged here).

158 In *R (on the application of Rudewicz) v Ministry of Justice* [2011] EWHC 3078 (Admin) (see p 189) a distant but closest living relative of the deceased could not claim an infringement of her Article 8 rights because she had never met the deceased and was never part of his family when the deceased was alive.

159 As in *Re Blagdon Cemetery* [2002] 4 All ER 482 (Chancellor's initial refusal of a faculty authorising parents to exhume their son's remains 20 years after his death did not constitute an interference with their Article 8 rights). See also *Re Walker (Deceased)* (Cons Ct (Liverpool), January 2002) (no breach of Article 8, though faculty granted on other grounds).

160 (2007) 45 EHRR 22.

the relocation of her husband's ashes from the burial plot in which they had been interred 34 years earlier (close to where the couple and their children had then resided) to the applicant's own family burial plot in Stockholm. The authorities had relied on domestic legislation[161] that restricted the removal of remains or ashes, and gave primary consideration to the expressed wishes of the deceased. In its submission to the court, the Government accepted that refusal of permission interfered with the applicant's Article 8(1) rights, but claimed that this was justified under Article 8(2). Counsel argued that the applicable statute was premised on the sanctity of the grave and "the right of the living to be assured that, after death, their remains would be treated with respect",[162] both of which were in the public interest. In any event, there was no indication that the husband had not been buried in accordance with his wishes. The court agreed, also suggesting that there was nothing to prevent the wife from having her own final resting place beside her husband in the town where the couple had lived and where the applicant had chosen to remain for 17 years after her husband's death.

Ecclesiastical courts have considered *Dödsbo* and the wide margin of appreciation under Article 8(2) when deciding that refusal of a faculty does not contravene the petitioner's right to family life. For example, in *Re St Dunstan's, Whiston*[163] a mother wanted to exhume her young son's ashes because her decision to inter them at a church that she attended was 'ill-considered'[164] and she now believed that he would have wanted his ashes to be scattered at sea. The petition was dismissed because the petitioner had failed to demonstrate special circumstances displacing the presumption against disinterment; however, the court also stressed that, even if the petitioner's Article 8 rights had been infringed, any interference was justified. A similar conclusion was reached in *Re St Andrew's Churchyard, Alwalton*,[165] the strong presumption of permanent burial in consecrated ground and cemetery regulations around the placing of floral tributes on graves dispelling any suggestion that there had been a disproportionate interference with the petitioner's right to family life. However, the deputy Chancellor also suggested that the faculty jurisdiction was Convention complaint, despite unfavourable comparisons with the secular licensing system. In his view, the stricter approach adopted by consistory courts was "within the margin of appreciation that would be granted by the [ECtHR]".[166]

---

161   Funeral Act 1990 (Prop. 1990/91:10).
162   (2007) 45 EHRR 22, [20].
163   (Cons Ct (Southwark), 22 January 2007).
164   The mother felt that her son had no connection with this particular church (except through her).
165   [2012] PTSR 479 (discussed at p 198).
166   *Ibid*, [56].

Finally, Article 8(2) also permits interference with the right to family life where this is necessary "for the protections of the rights and freedoms of others". This specific point was raised in *Re Jean Gardiner (Deceased)*,[167] several years before domestic courts had the benefit of *Dödsbo*. Because of an error in numbering the grave plots in *Gardiner*, the deceased had been buried in a grave already reserved for someone else. A faculty was granted to disinter the deceased's remains and bury them in another grave space, despite her husband and daughter claiming that this would infringe their Article 8(1) rights. The presiding Chancellor doubted whether granting a faculty would constitute such an infringement; in any event it would be justified on the basis that Convention disputes often involve balancing the competing rights of different parties, and the petitioner's right to have her mother buried in the grave specifically reserved for her would prevail here. Arguments that exhumation would infringe the protection of property provision in Article 1, Protocol 1 of the Convention were also rejected; despite the broad interpretation of "possessions", the deceased's remains could not be classed as such because they were not "an established interest with some economic value"[168] and, in any event, English law does not recognise property in a dead body.[169] And even if the provision was engaged here, exhumation of the deceased's remains would fall under the 'public interest' exception because of the need to correct the mistaken burial.

In contrast to Article 8(1), courts have been more receptive to Article 9(1) arguments, deciding on at least two occasions that refusal of a faculty would infringe the right to freedom of religion and conscience. For example, in *Re Durrington Cemetery*[170] the deceased, a practising Jew, had been buried in the consecrated part of a municipal cemetery by his non-Jewish wife; when she moved to Australia some 18 years later, the deceased's Jewish relatives wanted to exhume the body and reinter it in a Jewish cemetery. A faculty was granted despite the long delay since burial, given the absence of any improper motive on the part of the family[171] and lack of opposition from the deceased's wife. However, the court was also conscious of infringing the Article 9(1) rights of the deceased's relatives if permission was refused – and earlier in the judgment, Hill Ch had also suggested that the deceased "would have wished to be buried in a Jewish cemetery" given his "serious commitment ... to the practice of the Jewish faith".[172] In *Re Crawley Green Road Cemetery*,[173] the deceased's ashes were mistakenly interred in a

---

167   (Cons Ct (Carlisle), 28 May 2003).
168   *Ibid*, [22]
169   Possessory rights which vest in certain individuals as a result of the legal duty of disposal would terminate after the ashes' initial interment – see Ch 3, pp 71–73.
170   [2000] 3 WLR 1322.
171   The family had been acting out of respect for the deceased's wife.
172   [2000] 3 WLR 1322, 1325.
173   [2001] 2 WLR 1175.

consecrated part of the municipal cemetery, despite the deceased having had a humanist funeral since neither he nor his wife had any Christian allegiance. Granting the wife's request to have the ashes exhumed and reinterred at a local crematorium, the court indicated that refusal might have infringed the wife's Article 9(1) rights since the existing place of burial was, in her eyes, hypocritical and contrary to her humanist beliefs.

Reliance on Convention rights in these two cases was criticised in *Re Blagdon Cemetery*,[174] the court suggesting that the mistaken interment of both sets of remains in consecrated ground would have justified a faculty without recourse to Article 9.[175] However, both *Durrington* and *Crawley Green* raise the prospect of freedom of religion arguments influencing exhumation decisions, with the latter acknowledging the potentially wide scope of Article 9(1) in embracing non-religious beliefs such as pacifism.[176] Of course, the basic jurisprudential tests still apply. The request to exhume the deceased's remains (or a stated objection to this) must be based on a matter of conscience or some sort of religious belief,[177] though Article 9(1) is not engaged by the mere fact that there is a religious background to the dispute[178] and not every refusal of permission to exhume constitutes a denial of the petitioner's right to 'manifest' their beliefs.[179] In *R (on the application of Rudewicz) v Ministry of Justice*[180] – one of the few secular cases to address human rights issues around disinterment – the court held that, even if Article 9 was engaged here, there was no infringement. According to Hallet LJ, the permanence of burial was not a basic tenet of the Catholic faith, and the fact that a large number of Catholics opposed exhumation of the dead priest's remains was not determinative (especially when the heads of the religious order to which the deceased belonged, supported the exhumation). Lastly, even if there is a breach of Article 9(1), this can be justified under Article 9(2) where refusal of a faculty and the consequent freedom to manifest one's

174   [2002] 4 All ER 482.
175   A faculty is likely in cases of mistaken interment. While *Crawley Green* is an obvious example of this (albeit that the remains were not simply in the wrong grave, but should never have been placed in consecrated ground in the first place), it is difficult to see how *Durrington* falls under the 'mistaken location' category since the deceased's wife had deliberately chosen a consecrated part of the municipal cemetery so that she could easily visit her husband's grave (though it was suggested that she had not discussed this with the deceased's Jewish relatives).
176   Citing *Arrowsmith v UK* (1978) 3 EHRR 118 and *Kokkinakis v Greece* (1993) 17 EHRR 397, 418.
177   For example, Article 9 was not engaged in *Re St Andrew's Churchyard, Alwalton* [2012] PTSR 479 since the petitioner's request for a faculty was not founded on either.
178   *R (on the application of Rudewicz) v Ministry of Justice* [2011] EWHC 3078 (Admin).
179   Petchey (n 73), p 132 has questioned *Crawley Green* on this point.
180   [2011] EWHC 3078 (Admin) and discussed at p 189.

religion or beliefs stems from limitations "prescribed by law" or deemed necessary "for the protection of the rights and freedoms of others".[181]

Generally speaking, while human rights arguments can influence the grant or refusal of a faculty by ecclesiastical courts (and the grant of a s 25 licence, if this is subsequently challenged), the overall impact has been fairly muted to date. For example, exceptional circumstances can embrace a petitioner's Article 8 or Article 9 rights, but these will not automatically displace the presumption against disinterment from consecrated ground. In some instances, a potential human rights angle has been highlighted but not dealt with in any meaningful way;[182] other cases have engaged with the basic arguments, though consistory courts in particular tend to be more superficial in their analysis.[183] Meanwhile, other fundamental issues have yet to be explored. Familial conflict is an obvious one, where the deceased's surviving relatives are split between those requesting exhumation, and those opposing it. In these circumstances, both sides might be able to assert Article 8 or Article 9 rights, which courts would have to evaluate and prioritise – a scenario that could have arisen in *Durrington* if the deceased's wife had contested the relatives' request for a faculty. Another important issue is whether refusing permission to exhume could infringe Article 8 or Article 9 because the deceased had specifically wanted to be buried elsewhere. None of the exhumation cases have addressed this issue; although the comments in *Durrington*[184] suggest that the deceased's views are significant, this was not from an Article 9 perspective, while the deputy Chancellor in *Re St Andrew's Churchyard, Alwalton*[185] emphatically stated that "the Convention does not grant human rights to the dead".[186] Whether a convincing case can be made for recognising and upholding the deceased's human rights in the exhumation context remains to be seen – though it is easy to imagine a request for a faculty being bolstered by claims that the deceased's express wish to be interred with their family (for example, if they are relocating and establishing a new family grave elsewhere), or in a place of spiritual significance (for

181   See R Sandberg, "Human Rights and Human Remains: The Impact of *Dödsbo v Sweden*" (2006) 8 *Ecclesiastical Law Journal* 453, suggesting that the *Dödsbo* precedent of permissible interference would apply to Article 9 as well.

182   For example, in *R (on the application of HM Coroner for East London) v Secretary of State for Justice* [2009] EWHC 1974 (Admin), where the deceased's parents and brother opposed the coroner's request to re-issue a s 25 licence (needed to conduct a new inquiry into the deceased's death and because the remains were not lying in that particular coroner's district), the court simply noted that the deceased's brother had claimed that exhumation would interfere with his Article 8 and Article 9 rights (the deceased's family were orthodox Serbs and objected to disinterment on religious grounds).

183   However, Gallagher explains the general attitude of Chancellors as having "more to do with social and political realism than sound interpretation of Convention rights and their application within the faculty jurisdiction" – Gallagher (n 33).

184   [2000] Fam 33, 36.

185   [2012] PTSR 479.

186   *Ibid*, [52].

example, where a grave has subsequently become available in the churchyard where the deceased wanted to be buried) attracts Articles 8 and 9 respectively. Case law suggests an increasing emphasis on the wishes of the dead in funeral disputes,[187] and if allowing an individual to stipulate the posthumous fate of their remains arguably invokes Convention values, this could also influence requests for exhumation.[188]

VI. Criminal liability for unauthorised exhumation

The fact that corpses and ashes are not property in the legal sense[189] means that neither can be the subject of theft if unlawfully removed from their final resting place. However, the unauthorised exhumation of human remains attracts specific criminal law sanctions, reflecting customary notions of social morality in allowing the dead to rest in peace.

At common law, removing a corpse from a burial ground without lawful authority was an indictable misdemeanour as early as 1788.[190] This extended to both consecrated and unconsecrated graves,[191] introducing vital legal safeguards for the latter, which were outside ecclesiastical cognisance and, at the time, not protected elsewhere (and in stark contrast to consecrated burials, where disinterment without a faculty was always an offence against ecclesiastical law[192]). In committing the common law offence, the underlying motives were irrelevant; an individual could be convicted of unlawfully disinterring a corpse where the intent was to sell it for dissection purposes,[193]

187 See Ch 5, pp 129–130.
188 Similar trends can be seen elsewhere. For example, the deceased's religious beliefs have been an important factor when American courts are deciding whether to grant a request for exhumation as highlighted in P Zablotsky, "'Curst Be He That Moves My Bones': The Surprisingly Controlling Role of Religion in Equitable Disinterment Decisions" (2007) 83 *North Dakota Law Review* 361. Zablotsky (at p 366) mentions *Ingraffia v Doughtery* 29 North Co R 294, 300 (Pa 1944) where the court stated that "to deny to the dead the faith which sustained them in life is to deny to the living the faith that sustains them in death. In this sense the fundamental American ideal of religious freedom transcends the confines of the grave".
189 Introduction, pp 2–3.
190 See *R v Lynn* (1788) 2 Term Rep 733. This particular offence reached its zenith in the late eighteenth and early nineteenth centuries when the spectre of grave robbing haunted the poor and working classes in Britain as so-called 'resurrectionists' plundered graves within days (or sometimes hours) of burial to supply corpses for medical schools – see R Richardson, *Death, Dissection and the Destitute: The Politics of the Corpse in Pre-Victorian Britain* (University of Chicago Press, 2001). See also ET Hurren, *Dying for Victorian Medicine: English Anatomy and Its Trade in the Dead Poor, c.1834-1929* (Palgrave MacMillan, 2014).
191 "[W]hether in ground consecrated or unconsecrated, indignities offered to human remains in improperly and indecently disinterring them, are the ground of an indictment" – *Foster v Dodd* (1867) LR 3 QB 67, 77.
192 *Adlam v Colthurst* (1867) LR 2 A & E 30.
193 See *R v Lynn* (1788) 2 Term Rep 733, *R v Cundick* (1822) Dow & Ry NP 13 and *Foster v Dodd* (1867) LR 3 QB 67.

or simply to relocate a family member's remains to a more suitable resting place for pious or laudable reasons. However, the absence of any improper motive could influence the penalty imposed on conviction. For example, in *R v Sharpe*[194] where a son dug up his mother's body in order to bury it beside that of his father in a different churchyard,[195] the court stressed that wrongful removal of a corpse was still a misdemeanour even if the "the motive for the act deserved approbation",[196] but imposed a nominal fine of one shilling because the defendant had been driven by "filial affection and religious duty".[197]

Turning to statute, s 25 of the Burial Act 1857 in both its original and re-enacted forms makes it a criminal law offence to exhume or disturb human remains in contravention of the prescribed licensing requirements. Although there are few recorded convictions under s 25,[198] the provision creates an alternative criminal law sanction and would apply to the removal of both corpses and ashes.[199] However, this does not mean that the common law offence is redundant; as Gallagher has pointed out, the "basic premise of the common law is that human remains are sacred whenever they are interred ... [and] if s 25 or [ecclesiastical law] ... does not protect them, the common law will apply".[200] A good illustration is *R v Stephen Pearson*[201] where the appellant was sentenced to 12 months' imprisonment, after having pleaded guilty to removing a body from a disused cemetery as part of a drunken dare. In upholding the sentence, the Court of Appeal described the offence as "disgraceful and deplorable" and one which was "bound to cause revulsion to the public".[202] More generally, the common law offence should extend to ashes if removed from a grave without lawful authority; although the offence emerged at a time when cremation was not a popular disposal method and case law consequently focused on corpses, ashes are entitled to the same basic legal protection. To decide otherwise would also create a gap in the criminal law, since the s 25 and ecclesiastical law offences apply equally to both corpses and ashes.

194   (1857) Dears and Bell 160.
195   The mother's coffin was badly decomposed, and the son had also disturbed other bodies buried in the same grave.
196   (1857) Dears and Bell 160, 163.
197   *Ibid*, 162.
198   In *R v Lichfield Justices, ex parte Coyle* (Queen's Bench Division (Crown Office List), 19 July 1988) the appellant had been convicted under s 25 of aiding and abetting five other men in removing bones from a churchyard, in order to sell them to a firm in London; the conviction was subsequently quashed on grounds of potential bias.
199   See p 190.
200   Gallagher (n 37).
201   (1981) 3 Cr App R (S) 5.
202   *Ibid*, 7.

Conclusion

Exhumations are comparatively rare,[203] and cannot take place without the necessary legal authorisation.[204] Consequently, the initial interment of corpses and ashes should ordinarily be seen as final acts in the bodily disposal process.

There are many reasons why families might want to remove the remains of a loved one from an existing grave or burial plot. Any exhumation framework must strike a delicate balance between allowing disinterment where there is some justification for this, while safeguarding public health and upholding the basic moral imperative of allowing the dead to rest in peace. The spiritual significance of burial for different faith groups and cultures adds another dimension, and is reflected in the strong presumption against disinterment from consecrated ground.[205] Within the last decade, the government has also rejected the idea of a more liberal disinterment policy, concluding that there was no need to extend the existing grounds for secular exhumation in England and Wales.[206] Of course, one might argue that the s 25 route is already fairly liberal compared to its ecclesiastical counterpart; in August 2015, Church of England authorities also cautioned against the increasing societal belief in 'exhumation on demand' in this country, if the Ministry of Justice were simply to allow families to relocate their dead when they move home – something which ecclesiastical courts typically refuse.[207] This is just one of many divergences, as the discussion here has highlighted. However, the fact that English law has two distinct legal mechanisms for exhuming the dead ensures that both will continue to influence this area for years to come.

---

203   When compared to annual mortality rates. For example, there were 1,238 exhumations of corpses or ashes in 2014 (figure cited by Rayner, (n 49)) compared to 501,424 deaths in England and Wales in the same year.

204   Exhumation can also be expensive; typical costs include securing the licence, the actual disinterment, the funeral director's fees (including purchasing a new coffin) and reburial.

205   At least in England; as noted earlier, the Church in Wales follows the secular route for exhumation – see n 25.

206   See Ministry of Justice, *Burial Law and Policy in the 21st Century: The Need for a Sensitive and Sustainable Approach* (2004): *Government Response to the Consultation Carried Out by the Home Office DCA* (2007), p 12.

207   Rayner, (n 49) and see p 196.

8   Memorialising the dead

Memorials offer a form of immortality for those who have died as well as the possibility of a continuing link between those who have gone and those who remain.[1]

Introduction

While previous chapters have focused on the fate of the deceased's mortal remains, memorialising the dead is often viewed as the final stage in the bodily disposal process. Marking the physical location of an interred corpse or ashes creates a focal point for mourning and remembrance.[2] More importantly, memorials denote the deceased's life and passing in a particular way and, as such, symbolise ongoing (and very public) narratives of association between the living and the dead.[3] We have already seen that an individual's life is embedded in a network of relationships that their survivors are keen to nurture (and, in some instances control) after death.[4] Memorials, like certain

---

1   M Holloway, *Negotiating Death in Contemporary Health and Social Care* (Policy Press, 2007), p 160.
2   See for example, A Petersson, "The Production of a Memorial Place: Materialising Expressions of Grief" in A Maddrell and JD Sidaway (eds), *Deathscapes: Spaces for Death, Dying, Mourning and Remembrance* (Ashgate, 2010). With ashes, of course, creating a focal point is not always possible because interment is only one option. They can be scattered at an emotionally significant site, or kept in a container in a family member's home; and as McKechnie J pointed out in *Milenkovic v McConnell* [2013] WASC 421, [39], "ashes may be reverently disposed of in a way that precludes any ability thereafter to express affection at a fixed point such as a grave". In *Re St Peter's Church, Limpsfield* [2004] 3 All ER 978 a faculty was refused, following a request to erect a memorial to a former churchwarden whose ashes had been scattered elsewhere; the court was conscious of setting a dangerous precedent, as well as dissipating precious burial space in consecrated ground.
3   See generally K Woodthorpe, "Private Grief in Public Spaces: Interpreting Memorialisation in the Contemporary Cemetery" in J Hockey, C Komaromy and K Woodthorpe (eds), *The Matter of Death: Space, Place and Materiality* (Palgrave Macmillan, 2010).
4   Funeral disputes are a classic example – see Ch 4.

funeral rites, create a permanent and indelible memory picture of the deceased; who determines what form they take is an important question, and frequently a source of contention.

This chapter analyses the legal issues around memorialising the dead, focusing on graveside memorials[5] to private citizens in churchyards and municipal cemeteries.[6] Much of the emphasis is on permanent structures such as headstones, monuments and commemorative plaques; however, temporary adornments and more transient commemorative acts (for example, placing keepsakes and laying flowers on the deceased's grave) are also popular. Like bodily disposal, contemporary discourse highlights a number of recurring themes: a lack of clarity over the applicable legal framework (compounded by the fact that there are even fewer substantive legal rules around memorialisation); disputes between families and cemetery or church authorities over the form of memorial, as well as conflicts within families themselves; and the potential impact of human rights on this area. Memorialising the dead also raises its own distinct issues. The inherent public–private dichotomy adds another layer of complexity; memorials are, at the same time, "both intensely personal and manifestly public"[7] and must straddle the boundary between suitable expressions of private grief, which are also appropriate acts of public remembrance.[8] In addition, the emergence of Facebook memorials and other forms of 'virtual commemoration' have re-shaped how the living remember their dead, while posing new legal challenges that have yet to be addressed.

---

5   As opposed to what are usually more ephemeral memorials (such as those found on roadsides, or at the site of a major disasters) denoting the place in which individuals died, often in tragic and sudden circumstances – see for example, J Clark and M Franzmann, "Authority from Grief, Presence and Place in the Making of Roadside Memorials" (2006) 30 *Death Studies* 579 and J Santino (ed), *Spontaneous Shrines and the Public Memorialization of Death* (Palgrave Macmillan, 2006).

6   Most memorials are found here, since these are where the bulk of interments occur. Public memorials (for example, war monuments) are another means of preserving the memory of the dead, and can raise politically sensitive issues – see the discussion in K McEvoy and H Conway, "The Dead, The Law, and the Politics of the Past" (2004) 31 *Journal of Law and Society* 539, pp 547–554 and the various examples listed. These will not be analysed here, given the focus throughout this chapter (and the rest of the book) on legal contests surrounding the dead within the private, familial sphere.

7   T Walter and C Gittings, "What Will the Neighbours Say? Reactions to Field and Garden Burial" in Hockey, Komaromy and Woodthorpe (n 3), p 172.

8   Woodthorpe takes this further, noting that cemeteries (and presumably also churchyards) are spaces in which "private emotion (grief) and public behaviour (mourning) intersect in potentially problematic ways" – Woodthorpe (n 3), p 117. As an aside, the fact that individual gravestones are such public and highly visible statements of memorialisation mean that taking pictures of a headstone or including it in some sort of visual recording is unlikely to generate a cause of action in privacy – see *Bradley v Wingnut Films* [1994] EMLR 195 (partial shot of a family headstone in a horror film did not constitute a breach of privacy where the headstone was not identifiable and only appeared on screen for 14 seconds).

I.   Churchyard or cemetery memorials

Gravestones and commemorative plaques are permanent markers placed at the head of a grave, with an inscription identifying the interred remains.[9] Individual cemeteries and churchyards[10] invariably have their own rules on what constitutes a permissible memorial; and while these are internal governance systems (as opposed to binding rules of law[11]), they constitute part of the overall legal arrangement that binds each and every grave holder.[12] Exclusive rights of burial confer a right to inter one or more sets of remains in a particular plot;[13] however, the right-holder does not 'own' the grave in any legal sense (though this is major public misconception[14]), and must adhere to churchyard or cemetery regulations around the design and dimensions of tombstones or plaques, and what objects can be left on the deceased's grave.[15]

---

9   According to the nineteenth century case of *Menzies v Ridley* (1851) 2 Gr 544, not only is a gravestone "usual and considered a proper mark of respect … but it is useful as marking the place of burial and furnishing evidence of pedigree". Natural burial sites do not tend to have conventional gravestones, though individual graves can be marked with a tree or a wooden plaque. Some death scholars view this with concern, suggesting that "a sense of history of earlier communities, of the collective nature of death … and of personal biography … [are] jeopardised by digressing from conventional records on headstones" – A Clayden, J Hockey and M Powell, "Natural Burial: The De-Materialising of Death" in Hockey, Komaromy and Woodthorpe (n 3), p 162. Burials on private land can also be marked by a suitable memorial, as long as the proposed structure does not fall foul of planning laws.

10   Whether Church of England graveyards (and, as such, consecrated ground), or private burial grounds set aside by other religions.

11   With the exception of the erection of memorials on consecrated ground, which are under the ecclesiastical law jurisdiction – see n 15.

12   The type of headstone or commemorative plaque will also be limited by costs, since these are payable from the deceased's estate – see Ch 3, Pt II.

13   See Ch 2, pp 32–34.

14   "[T]he popular belief that a grave is something which belongs to the closest family member left behind by the deceased is wrong" – *Re Christ Church Cemetery, Harwood* [2002] 1 WLR 2055, 2062.

15   For local authority cemeteries (which are overseen by cemetery managers) the content of these regulations is a matter for individual councils, though the Institute for Cemetery and Cremation Management issues generic guidance on the management of memorials (see www.iccm-uk.com/iccm/index.php?pagename=guidance (accessed 30 September 2015)). Likewise, it appears that none of the major churches or religious denominations in England and Wales (or throughout the rest of the UK) have centralised regulations for their particular burial grounds. Outline plans for permanent memorials may have to be submitted for approval in advance; and in the case of consecrated burial grounds, erecting a monument is dependent on the grant of a faculty – see for example, *Re Woldingham Churchyard* [1957] 1 WLR 811.

1.   **Disputes between the deceased's family and cemetery or churchyard authorities**

Memorialisation involves balancing the rights of the deceased's family against those of members of the public and other families with graves inside the same physical space, something that individual cemeteries and churchyards address through internal regulations. There are two key concerns here: health and safety, and what might be termed 'graveyard aesthetics'. The net result, however, is conflict and dissonance where families are prevented from memorialising their dead in a certain way.

Individual churches and local authorities are legally obliged to ensure that their burial grounds are safe places for both staff and visitors.[16] Practical regulations often prohibit glass and metal containers, which might shatter and cause injury; leaving the deceased's favourite food and drink on their grave is discouraged because it attracts vermin; and placing kerbing around a grave could constitute a tripping hazard.[17] However, more complex issues arise where proscriptions are based on aesthetic values and the overall 'appearance' of the cemetery or churchyard.[18] Just as the socio-cultural rhetoric of funerals has changed in favour of more personalised, life-centred narratives of the deceased,[19] so too has the way in which the living commemorate their dead. Families increasingly want personalised headstones or markers, which speak to the deceased's identity in some way – but

16   For example, the Local Authorities Cemeteries Order 1977 deals with the management and operation of local authority cemeteries, and specifically requires local authorities to keep their cemeteries in "good order and repair" (art 4).

17   And may also make maintenance and upkeep of the cemetery or churchyard (as a whole) more difficult – see *Re Holy Trinity, Eccleshall* (Cons Ct (Lichfield), 20 December 2014).

18   For example, standardisation of individual memorials creates an orderly appearance, and prevents gravestones from competing with one another and becoming "an exercise in one upmanship on the part of those remaining" – *Re Christ Church Cemetery, Harwood* [2002] 1 WLR 2055, 2066. Individual restrictions can also be dependent on siting and location; since the type of memorial must be appropriate to its surroundings, what might be permitted in an urban context will be very different from a rural setting or in graveyards adjoining churches that are listed buildings. There is also an element of how individual burial grounds will be perceived by future generations (see the comments in *Gilbert v Buzzard* (1820) 3 Phill Ecc 335, 357–358).

19   See Ch 1, pp 21–22.

potentially breaching the rules around design and permissible wording.[20] While any epitaph is both emotive and intensely personal, it is also a very public and a permanent representation of the deceased's identity. And while offensive wording would not be allowed,[21] the boundary between what is humorous or quirky, and what is perceived as distasteful is not always easy to discern.[22] Photographs on headstones (popular in a number of European countries) can also be contentious, with many cemeteries and churchyards

20   For example, where a family had a Sudoku puzzle and equation inscribed on their mathematician father's gravestone and were ordered to remove it by the parish council ("It Doesn't Add Up! Family's Anger After Council Orders Them to Remove 'Offensive' Sudoku Headstone Tribute to Late Father Mathematician", *Daily Mail* (London, 17 October 2013) www.dailymail.co.uk/news/article-2465290/Familys-anger-council-orders-removal-sudoku-headstone-mathematician-Allan-Robinson.html (accessed 30 September 2015)), or where the family of a former roadie to rocker Ozzy Osborne were refused the grant of faculty to write "it's only rock and roll" on his headstone after the consistory court in question ruled it was 'flippant' ("Church Bans Headstone of Ozzy Osbourne Roadie", *The Sentinel* (Stoke, 6 July 2013) www.stokesentinel.co.uk/Video-Church-bans-headstone-Ozzy-Osbourne-roadie/story-19485728-detail/story.html (accessed 30 September 2015)).

21   According to the Supreme Court of Rhode Island in *McGann v McGann* 28 RI 130, 134 (1907) an inscription on a tombstone will not be permitted where it would "shock the sense of the community or show disrespect or contempt for the dead, or in any real sense do injury to the feelings of the surviving relatives". Individual churchyards and cemeteries will have rules on what is permissible; and as a matter of general law, certain inscriptions will not be allowed (for example, a "posthumous expression of any racial or sexual prejudices of the deceased" that would have been illegal "if expressed in life" – *Re Christ Church Cemetery, Hardwood* [2002] 1 WLR 2005, 2061) and may also contravene the Public Order Act 1986. Certain types of insignia or emblems may also be prohibited – see for example, "Swastika Covered Up Austrian Tombstone", *The Times of Israel* (Israel, 6 May 2014) www.timesofisrael.com/swastika-covered-up-on-austrian-tombstone/ (accessed 30 September 2015).

22   For examples of humorous epitaphs, see L Evers, *I Told You I Was Ill: Dying for a Laugh* (Michael O'Mara Books, 2012), pp 58–78, which includes Spike Milligan's famous quotation replicated in the opening words of the book's title.

throughout England and Wales operating a blanket ban on any visual representation of the dead.[23]

Turning to more transient memorials, individual graves are often adorned with an array of keepsakes and paraphernalia such as teddy bears, football scarves, wind-chimes, solar lights, figurines and plastic flowers. Viewed by some as vibrant and deeply personal tributes to the dead, others claim that they detract from the calming and reflective nature of graveyards – and, in particular, that cluttering graves with tasteless items is intrusive and indicative of a consumer-oriented culture where the bereaved insist on whatever memorial they find appropriate.[24] Opting for the more traditionalist approach, many churchyards and cemeteries have strict rules around what can be placed on graves – regardless of the sensitivities of grieving relatives.

23 Church of England burial grounds often have stringent rules around photographic images. For an extensive discussion, see *Re Christ Church Cemetery, Hardwood* [2002] 1 WLR 2005 and the various authorities referenced throughout the judgment. Holden Ch in this case allowed an engraved black and white etching of the deceased incorporated into the stone and in a subdued hue, which did not clash with the gravestone's immediate environment. However, the court stressed that it was not setting a precedent and that a ceramic picture or plaque would not have been permitted; more generally, strong reasons would be required for the exceptional grant of a faculty to allow a memorial that did not comply with diocesan graveyard regulations – sentiments reiterated in *Re Christ Church, Timperley* (Cons Ct (Timperley), 29 January 2004) (permission for a photograph of the deceased refused). While some secular burial grounds (and those of other religious denominations) may be more willing to allow photographic images, many still impose restrictions – see *Jones v United Kingdom* (Application no. 42639/04, ECHR, 13 September 2005) (noted at pp 220–221). The Canadian case of *Ayache Estate v Muslim Association of Calgary* 2000 ABPC 101 suggests that unauthorised images can be removed, if necessary (defendant entitled to remove deceased's image from a gravestone where this contravened cemetery rules).

24 The rise of what has been termed the 'Poundland cemetery' (after the well-known UK retailer); for an illustration, see J Chapman, "It's Like Poundland! Council Bans Grave Trinkets and Chimes", *Daily Express*, (London, 2 February 2011) www.express.co.uk/ news/uk/226622/It-s-like-Poundland-Council-Bans-grave-trinkets-and-chimes (accessed 30 September 2015).

Numerous disagreements have been reported in the national media;[25] and, in at least one instance, a widow requested a faculty to exhume her husband's remains because of what she felt were overly stringent churchyard regulations about placing floral tributes on his grave.[26]

2.   Commemoration disputes within families

Memorials that comply with cemetery or churchyard regulations can still be the subject of a legal dispute where the deceased's relatives cannot agree on the type of structure or – more usually – the form of wording to be used on the headstone. Simmering tensions and ongoing family feuds frequently come to the fore on death;[27] at some level, commemoration disputes are simply another example of this – especially within complicated family structures or where the deceased's relatives are estranged from each other. However, the fact that the chosen inscription is a permanent and public record, not only of the deceased's lifespan, but of selected family ties and close personal relationships, makes it highly symbolic and creates another layer of emotional complexity. Problems arise where someone is excluded from this narrative, or the wording chosen does not acknowledge their role in the deceased's life. A good illustration is the Canadian case of *Wiebe v Bronstein*[28] where the deceased's same-sex partner challenged the inscription

---

25   In particular, around what families view as overly officious 'policing' of these regulations. See for example, "Cemetery Bans Artificial Flowers Over Fear of Being Sued", *The Telegraph* (London, 10 May 2010) www.telegraph.co.uk/news/uknews/7705476/ Cemetery-bans-artificial-flowers-over-fear-of-being-sued.html (accessed 30 September 2015); P Barkham, "Should Fake Flowers Be Banned From Cemeteries?", *The Guardian* (London, 12 January 2011) www.theguardian.com/lifeandstyle/2011/jan/12/ cemetery-ban-for-fake-flowers (accessed 30 September 2015); F Rohrer, "Should Graveyard Wind Chimes and Plastic Displays be Banned?", *BBC News Magazine* (9 February 2011) www.bbc.co.uk/news/magazine-12396991 (accessed 30 September 2015); A Smith Squire, "Couple's Grief as Vicar Bans Teddies From The Grave of Their Stillborn Baby In Case They Upset Other Churchgoers", *Daily Mail* (London, 27 September 2013) www.dailymail.co.uk/news/article-2436092/Couples-grief-vicar-bans-teddies-grave-stillborn-baby-case-upset-churchgoers.html (accessed 30 September 2015); and C Pleasance, "Bereaved Mother Ordered to Tear Down Stillborn Daughter's Shrine Decorated with Windmills, a Pink Fence and Pink Gravel: Because It's Not 'In Keeping' with Graveyard", *Daily Mail* (London, 29 January 2015) www.dailymail. co.uk/news/article-2931795/Bereaved-mother-ordered-tear-stillborn-daughter-s-shrine-decorated-windmills-pink-fence-pink-gravel-s-not-keeping-graveyard.html (accessed 30 September 2015).

26   *Re St Andrew's Churchyard, Alwalton* [2012] PTSR 479. Dealing with the petitioner's alleged breach of churchyard regulations, the presiding Chancellor remarked: "Disputes as to what may be permitted in churchyards are not new and it is most unlikely that there will ever be a consensus as to taste … I am not ruling whether the disputed vases placed on … [the deceased's] plot were in my subjective judgment tasteful or not. That is not the issue" – *ibid*, [39].

27   The underlying reasons were explored in Ch 4, Pt I.

28   2013 BCSC 1041.

chosen by the deceased's sister (who had selected the gravestone as executrix under her brother's will); the partner was omitted from the inscription, which referred to the deceased as a "beloved son, brother, uncle and friend".[29]

As a matter of law, any final decision lies with the person who has the exclusive right of burial in the particular plot. The leading statement can be found in *Smith v Tamworth City County Council*,[30] which involved a dispute between the deceased's biological and adoptive parents, the latter having erected a headstone describing the deceased as their "much loved son" and making no reference to the biological parents. According to Young J, the right to decide on the appearance of the grave and any headstone (subject to compliance with cemetery by-laws) belongs to the person with the exclusive right of burial, and that was the adoptive mother in this case.[31] However, the biological parents could not be denied access to the grave; according to the judge, the holder of the right of burial "cannot use his or her right in such a way as to exclude friends and relatives of the deceased expressing their affection for the deceased in a reasonable and appropriate manner such as by

29   The partner's claim was unsuccessful here. See also *Escott v Brikha* [2000] NSWSC 458 (deceased's headstone was erected by his mother and brothers, with the consent of the deceased's wife; however, the wife subsequently discovered that there was no mention of her on the memorial, which referred to the deceased as a "dear son and brother and father") and the similar factual scenario in *Re Campbell (Judicial Review)* [2013] NIQB 32 (gravestone erected by the deceased's father referred to him as a "much loved son and brother", but omitted the deceased's wife). In *Watene v Vercoe* [1996] NZFLR 193 estranged parents disagreed over the wording on their 6-year-old son's headstone. The father had custody of the child, who died while in the care of his mother; the child's maternal grandmother and her partner were facing murder charges as a result. Burial took place in a plot purchased by the father. However, the mother objected to the proposed inscription on the headstone, which referred to the son being murdered, and made no mention of her as the child's mother.

30   (1997) NSWLR 680.

31   Identical conclusions were reached in *Re Campbell (Judicial Review)* [2013] NIQB 32 and in *Boni v Larwood* [2014] SASC 185 (the court in the latter case taking the view that a father was entitled to spell his son's name as 'Davide' on the headstone to his wife (the son's mother), because the father held the sole right to the grave at this stage). Likewise, any subsequent change in otherwise permissible wording is dependent on the grave holder's permission. For example, the family of Claire Morris (who had been murdered by her husband, Malcolm Webster, in a staged car accident) were forced to ask her killer's permission to change the gravestone (since he bought and paid for the burial plot and the memorial). The family were intending to change the inscription from 'Claire Webster' to the deceased's maiden name, and to remove Webster's tribute to his 'dear wife' – "Gravestone Needs Killer's Consent", *The Telegraph* (London, 3 June 2011) www.telegraph.co.uk/news/uknews/law-and-order/8552935/Gravestone-needs-killers-consent.html (accessed 30 September 2015).

placing flowers on the grave".[32] Family and friends are entitled to visit a loved one's grave, and to honour their dead in an appropriate way.[33]

The holder of the exclusive right of burial may not be the deceased's next-of-kin;[34] that person may not even be the deceased's executor or presumptive administrator, unless they secured the interment rights in the plot in question or acquired the exclusive right of burial in a plot reserved by the deceased before his/her death.[35] The fact that the next-of-kin do not 'own' the grave plot can have unforeseen and unintended consequences, impacting not just on the form of memorial, but on their own ability to be interred alongside the deceased at a later date.[36]

3.   Human rights arguments

As one might expect human rights arguments have also been raised in commemoration disputes (particularly since the passage of the Human Rights Act 1998), though not to the same extent as in legal contests involving the deceased's mortal remains,[37] and with mixed results. In *Jones v United Kingdom*[38] the applicant argued that his rights under Articles 8 and 9 of the Convention had been breached when a local authority refused him permission to incorporate a photograph of his daughter on her memorial stone.[39] In declaring the application inadmissible, the European Court of Human Rights ruled that there had been no infringement of Article 8:

32   (1997) NSWLR 680, 694. See also *Hoskins Abrihall v Paignton UDC* [1929] 1 Ch 375 – the person who holds the exclusive right of burial (and also the by-laws of a particular cemetery) cannot prevent the public from placing flowers on a grave.

33   This may be something of a moot point where remains are in a churchyard or public cemetery, where access is controlled by the governing body as opposed to the person with the right of disposal or the person who owns the grave. Burial on private land is different; here the landowner may be able to exclude others in both a practical and a legal sense, even though *Smith* suggests this should not occur.

34   See *Re Campbell (Judicial Review)* [2013] NIQB 32, [16] ("[a] plot of burial and a right can be contracted lawfully with anyone").

35   The decision in *Watene v Vercoe* [1996] NZFLR 193 (where the court suggested that the executor has the right to determine the deceased's memorial – *ibid*, 196) should be treated with caution on this point. In *Smith v Tamworth City Council* (1997) NSWLR 680, 694 Young J suggested that, on the death of the executor or administrator, the right to control the grave would pass to the legal personal representative of the original deceased.

36   Another source of upset for the deceased's wife in *Re Campbell (Judicial Review)* [2013] NIQB 32. Who owns the exclusive right of burial will also be an important factor in any subsequent request for exhumation – see Ch 7.

37   See Ch 4, pp 107–114 (funeral disputes); Ch 5, pp 139–144 (funeral directions); and Ch 7, Pt V (exhumation).

38   (Application no. 42639/04, ECHR, 13 September 2005).

39   The applicable cemetery regulations specified that all memorials were subject to prior approval from the local authority.

The regulations applicable to the cemetery required prior approval of all headstones and memorials. Notwithstanding the applicant's personal preference for the addition of a photograph to the headstone and the fact that other burial authorities apparently gave permission for such features, the court does not find that the refusal of permission in this case can be regarded as impinging on the applicant's personal or relational sphere in such a manner or to such a degree as to disclose an interference with his right to respect for private or family life. [40]

Likewise, there had been no breach of Article 9; refusing permission for a photograph did not prevent any manifestation of the applicant's religious beliefs, since he could properly pursue his religion and worship without a photograph on his daughter's memorial. Although dealing with a different factual scenario (and with consecrated ground under the remit of ecclesiastical law), the reasoning in *Jones* was applied several years later in *Re St Mary the Virgin Churchyard, Burghfield*.[41] A petition for a faculty to remove unauthorised graveside ornaments and edgings used to fence off graves was granted despite opposition from two of the families affected, the presiding Chancellor taking the view that neither Article 8 nor Article 9 was engaged on the facts.[42]

Looking beyond Articles 8 and 9, insisting on a particular form of commemoration (for example, a gravestone inscription, which, by its very nature, is a public statement) could conceivably invoke the right to freedom of expression under Article 10 of the Convention.[43] While there is no direct authority on this point, substantively similar issues were raised in the Israeli case of '*Jerusalem Community' Funeral Society v Lionel Aryeh Kestenbaum*.[44] Here, the respondent wanted to have his wife's tombstone engraved with her name, Gregorian date of birth and Gregorian date of death in Latin characters as stipulated by his wife; the funeral society refused this request, relying on a term in its standard contract that no letters other than those of the Hebrew alphabet should be engraved on its tombstones. A majority of the Supreme Court of Israel held that this term was void because it violated the respondent's right to freedom of expression, conscience and human dignity. However, the fact that graveyards and cemeteries are public spaces with numerous interments brings other interests into play. Even if the wording on a particular headstone came within the scope of Article 10(1), any right to freedom of expression is inherently limited by Article 10(2),

40   (Application no. 42639/04, ECHR, 13 September 2005), p 5.
41   [2012] PTSR 593.
42   It was also held that Article 1, Protocol 1 (protection of property) did not apply as long as the families in question were given a reasonable opportunity to repossess their ornaments once they had been removed from the respective graves.
43   See generally Ch 5, pp 142–143.
44   (1991) 46(2) PD 464.

which permits such restrictions as are "prescribed by law and ... necessary in a democratic society". Offensive or inappropriate language would almost certainly be rejected on this basis.

Finally, the limited human rights arguments presented to date have involved disputes between the bereaved and cemetery authorities. It is more difficult to see how these could be raised in disputes within families – though one option might be an alleged breach of Article 8 where, for example, the deceased's relationship with his wife or long-term partner is omitted from the memorial by the person who commissions the wording. Whether or not this would succeed is another matter entirely.

II.   Virtual memorials: Commemorating the dead 'online'

With the advent of the digital age, we spend increasing amounts of our time in the virtual world[45] – interacting, transacting and generating assets online. The question of how those assets are disposed of or transferred when someone dies has become an important estate planning issue, because established succession laws do not deal effectively with digital assets.[46] However, the post-mortem management of an individual's digital persona raises equally complex issues, as social networking services ('SNS') fundamentally reshape how the living remember their dead.[47]

---

45   See for example, J Miller, "Britons Spend More Time on Tech Than Asleep, Study Suggests", *BBC News Online* (7 August 2014) www.bbc.co.uk/news/technology-28677674 (accessed 30 September 2015). Facebook is the world's largest social population, and is used by around one-fifth of the world's inhabitants – "Facebook: One Out Of Every Five People On Earth Have An Active Account", *International Business Times* (30 January 2015) www.ibtimes.com/facebook-one-out-every-five-people-earth-have-active-account-1801240 (accessed 30 September 2015).

46   Various articles have been written on this, though mostly from a US perspective – see for example, M Perrone, "What Happens When We Die: Estate Planning of Digital Assets" (2012) 21 *CommLaw Conspectus* 185; JP Hopkins, "Afterlife in the Cloud: Managing a Digital Estate" (2013) 5 *Hastings and Science Technology Law Journal* 210; GW Beyer and N Cahn, "Digital Planning: The Future of Elder Law" (2013) 9 *NAELA Journal* 9 135; and NM Banta, "Inherit the Cloud: The Role of Private Contracts in Distributing or Deleting Digital Assets at Death" (2014) 83 *Fordham Law Review* 799. Digital assets can include things like digital photographs, i-tunes accounts, email and social media accounts, and electronic banking and investment records.

47   The focus here is on memorialisation, and not on the way in which individual deaths are announced on SNS. This can have a severe emotional impact on the deceased's family, where traditional conventions are disregarded when it comes to informing them of the death – see M Gibson, "Automatic and Automated Mourning: Messengers of Death and Messages from the Dead" (2015) 29 *Journal of Media & Cultural Studies* 339. The author also highlights the proliferation of automated messages, which allow the deceased to announce their own biological death online, suggesting that this "represents a cultural shift as it displaces the idea that it is only the other who can speak for the dead, not the dead speaking for themselves" (*ibid*, p 349).

1   Virtual memorials

For most people, the interment of human remains still takes places in a cemetery or churchyard. Yet acts of memorialisation now extend beyond the traditional headstone or grave marker adorned with a smattering of grave goods. Virtual cemeteries, available since the 1990s, are interactive resting places for the dead, where the bereaved can visit, leave tributes and erect headstones at a loved one's virtual grave.[48] Much more common is the conversion of the deceased's own online social media profile into a memorial site, the most common examples occurring on Facebook.[49] Instead of simply removing the deceased's profile, the individual's family (or certain non-relatives) can apply to Facebook to have the page converted into a memorial;[50] this can only be accessed by the deceased's existing Facebook 'friends', and can potentially remain in place indefinitely. The same 'friends' can visit the memorial page as often as they wish, posting their own personal tributes and messages of support for each other, as well as leaving messages for the dead.

In some ways, virtual memorials resemble physical ones. For example, they "relocate the deceased to a place which is accessible but separate from the spaces usually occupied by the living",[51] allow people to share their feelings and memories in a public forum,[52] and to post formulaic messages similar to those found in sympathy cards and death notices.[53] Yet, while things like Facebook posts allow the deceased's family and friend to engage in ritualised behaviours "akin to [those] … performed at … cemetery visits", these collective acts of remembrance differ because they are "public, virtual, eternal and direct".[54] Online memorials are much easier to visit than

---

48   See generally P Roberts, "The Living and the Dead: Community in the Virtual Cemetery" (2004) 49 *OMEGA: Journal of Death and Dying* 57.

49   Most of the discussion below will focus in Facebook. However, MySpace is another example, with MyDeathSpace.com as a specialist site for memorialising its users and their obituaries.

50   The protocols for doing so, and the wider issues this raises, are mentioned briefly below. In addition, the deceased's own Facebook 'friends' (which can include family, intimates, and persons tagged as 'friends') can post tributes to the deceased on their own Facebook pages, which then attract responses from that individual's Facebook 'friends' (and can be seen by his/her Facebook 'followers', unless access has been blocked).

51   T Hutchings, "Wiring Death: Dying, Grieving and Remembering on the Internet" in CW Park and DJ Davies, *Emotion, Identity and Death: Mortality Across Disciplines* (Ashgate, 2012), p 51.

52   McManus, for example, observes that many virtual memorials "are framed around traditional funerary practices" and that online visitors can "give virtual flowers, erect virtual monuments and offer condolences" R McManus, *Death in a Global Age* (Palgrave Macmillan, 2013), p 139.

53   What Gibson (n 47), p 343 describes as "socially codified ways" for expressing sympathy to others.

54   R Kern, AE Forman and G Gil-Egui, "RIP: Remain in Perpetuity. Facebook Memorial Pages" (2013) 30 *Telematics and Informatics* 2, p 3. They are also significantly cheaper than physical memorials.

cemeteries, allowing the deceased's family and (online) social network to interact with each other in a way that they would not be able to physically – given the geographically disparate kinship and friendship networks that most individuals are part of today.[55]

Unlike the traditional headstone or grave marker, which can only record a limited amount of information and is constrained by social conventions (most document the deceased's lifespan and their primary relationships, based on traditional constructs of 'family'), virtual memorials are not about lineage; they incorporate the wider family circle, and are a marker of friendships and social networks that the deceased was an active part of in life.[56] Another important factor is that the deceased's identity is not fixed at the time of their death; since virtual memorials are dynamic and interactive sites of memory sharing, they allow "postmortem identities ... to be [continually] crafted and preserved".[57] However, a unique feature of Facebook and other SNS memorials is a marked tendency to communicate directly with the deceased by posting messages to that specific individual in a very public forum.[58] It is not simply a case of maintaining the deceased's presence in everyday life; the living actively engage in ongoing narratives with the dead, who "never really die" but are "perpetually sustained in a digital state of dialogic limbo".[59] In some ways, this practice suggests a new variant of spiritualism and the afterlife; the perception that "the dead live in the virtual cloud, and can hear or read the messages from the living" highlights a "sustained belief in an afterlife ... despite the increase in secularity brought about by modernity and mass

55   SNS "enable expansion – temporally, spatially and socially – of pubic mourning" – J Brubaker, GR Hayes and P Dourish, "Beyond the Grave: Facebook as a Site for the Expansion of Grief and Mourning" (2013) 29 *Information Society* 152, p 152.

56   This idea of a more expansive (and participatory) form of memorialisation is discussed in J Lingel, "The Digital Remains: Social Media and Practices on Online Grief" (2013) 29 *The Information Society: An International Journal* 190. However, virtual memorials can still have exclusionary aspects – see p 228.

57   Brubaker, Hayes and Dourish (n 55), p 153.

58   "Mourners have probably spoken to the dead since time immemorial, but ... in private, whether silently or aloud when no one else was around ... Social media posts, however, often address the deceased ... in the knowledge that others are reading and watching; thus the previously private practice of addressing the dead has become socially acceptable" – T Walter, "New Mourners, Old Mourners: Online Memorial Culture as a Chapter in the History of Mourning" (2015) 15 *New Review of Hypermedia and Multimedia* 10, pp 17–18. Similar observations can be found in T Walter, R Hourizi, W Moncur and S Pitsillides, "Does the Internet Change How We Die and Mourn? Overview and Analysis (2011) 64 *OMEGA: Journal of Death & Dying* 275, p 293 – the authors noting the lack of "embarrassment about speaking to the dead in the presence of an audience, nor about speaking of them in a way that presumes the dead are listening".

59   Kern, Forman and Gil-Egui (n 54), p 3. Degroot describes this as reconnecting with the deceased and preserving a sense of "relational continuity" – JM Degroot, "Maintaining Relational Continuity with the Deceased on Facebook" (2012) 65 *OMEGA: Journal of Death and Dying* 195.

media".[60] However, there is also a sense in which Facebook and other interactive, online memorials blur the distinction between the virtual and real worlds, and prevent users from confronting the reality of death and loss.[61] Physical death may have occurred, but an individual will only truly have died when their online presence is removed.[62]

Unsurprisingly, the internet has fundamentally altered the interface between public and private grief, and there is a significant volume of literature on how SNS in particular impact on death, grieving and memorialisation. Some have questioned whether online interactions actually help people to grieve in a 'healthy' way. McManus, for example, suggests that virtual acts of remembrance potentially undermine the capacity to connect at [an] interpersonal level",[63] while Moss posits that ongoing communications with the deceased (something that the virtual world facilitates) disrupt normal grieving processes because the living refuse to 'let go' of their dead.[64] Online memorials can also be damaging to the deceased's family, and not just because they cannot control the content.[65] These focus on the individual's

60   Kern, Forman and Gil-Egui (n 54), p 9. Walter, however, suggests that online memorials may "foster a partial reversal of secularisation" given the widespread adoption (even by users in secular countries) of religious terminology such as the deceased being 'in 'heaven' or 'an angel'" – Walter (n 58), p 18.

61   For example, Gibson argues that there is a "widening gap and experiential differential between media/technological death culture and 'real life' contexts and temporalities of death and bereavement" – M Gibson, "Death and Mourning in Technologically Mediated Culture" (2007) 16 *Health Sociology Review* 415.

62   Removing the deceased's Facebook page, for example, is effectively a second death, which can be just as traumatic for Facebook 'friends' as that person's biological death. As Lingel notes, the "very permanence and irrevocability of death contribute to the desire for a virtual space in which the deceased 'lives on'" – Lingel (n 56), p 193.

63   McManus (n 52), p 124.

64   M Moss, "Grief on the Web" (2004) 49 *OMEGA: Journal of Death and Dying* 77 (while normal grief does not necessarily terminate continuing bonds with the dead, virtual memorials take this a step further by facilitating bonds that continue indefinitely). The physical separation from the deceased's remains can also suppress normal emotional responses; as Kern and others point out, virtual memorial sites "do not require the mourner to see or even be physically near the deceased, making mourning potentially free of fear or even extreme emotional grief" – Kern, Forman and Gil-Egui (n 54), p 9.

65   In some instances, they may not even be able to access the memorial if, for example, they were not one of the deceased's Facebook 'friends' – see N Wright, "Death and the Internet: The Implications of the Digital Afterlife" (2014) 19 *First Monday*, p 5 http://firstmonday.org/ojs/index.php/fm/article/view/4998/4088 (accessed 30 September 2015). And even if family members have access, the content may be upsetting. For example, virtual memorials (especially public ones) can be subjected to 'trolling', when people post offensive messages – in much the same way as gravestones can be defaced or damaged. However, this can be achieved much more easily (and visibly) online, and breaks the basic social taboo of 'speaking ill of the dead'. See generally see W Phillips, "LOLing at Tragedy: Facebook Trolls, Memorial Pages and Resistance to Grief Online" (2011) 16 *First Monday* http://firstmonday.org/ojs/index.php/fm/article/viewArticle/3168 (accessed 30 September 2015).

virtual identity which, to some extent, is an artificial construct that close relatives might feel inclined to manage in some way;[66] there may also be doubts around the sincerity of multiple outpourings of grief from a wide cohort of social media 'friends',[67] while concerns have also been expressed that Facebook is "too casual a medium for a weighty topic such as grief and grieving".[68] At a more fundamental level, next-of-kin can also feel marginalised by the fact that SNS distort the sense of who is actually 'bereaved' when someone dies.[69] However, others argue that SNS and virtual memorials can be a highly effective forum for grieving and forming vital support networks. As Lingel points out:

> A critical affordance of online grief is the ability to craft individual responses to death in an open venue less constrained by still inchoate social and cultural obligations than a funeral home or cemetery.[70]

As part of this wider narrative, she contrasts the "invocation of familiarity" for habitual Facebook users with the "sense of alienation from the traditional, physical sites of mourning, such as a cemetery", and the "participatory creation of content" by a range of spatially isolated users as effective elements of online grieving and memorialisation.[71] The internet can also provide a forum for expressing and acknowledging disenfranchised grief,[72] in

---

66   See A Marwick and NB Ellison, "'There Isn't Wifi in Heaven!' Negotiating Visibility on Facebook Memorial Pages" (2012) 56 *Journal of Broadcasting & Electronic Media* 378 (suggesting that Facebook allows bereaved friends and relatives to portray the deceased in a specific way and to "engage in impression management" of the dead).

67   See Wright (n 65).

68   Brubaker, Hayes and Dourish (n 55), p 156.

69   Wright (n 65), p 6. Walter (n 58), p 14 notes that the "right and ability to mourn is … extended far beyond close friends and family" with the result that "more people are positioned as co-mourners"; though whether they can all legitimately be classed as 'bereaved' in the conventional sense of the word is open to question.

70   Lingel (n 56), p 191.

71   *Ibid*, p 193. See also SA Dominick, AB Irvine, N Beauchamp, JR Seeley, S Nolan-Hoeksema, KJ Doka and GA Bonanno, "An Internet Tool to Normalise Grief" (2009) 60 *OMEGA: Journal of Death and Dying* 71; B Carroll and K Landry, "Logging On and Letting Out: Using Online Social Networks to Grieve and to Mourn" (2010) 30 *Bulletin of Science and Technology* 341 (looking at young internet users, and suggesting that online social networks enable or empower individuals marginalised by more traditional forms of memorialisation); and K Falconer, M Sachsenweger, K Gibson and H Norman, "Grieving in the Internet Age" (2011) 40 *New Zealand Journal of Psychology* 79 (arguing that online support groups provide a rich and vibrant place to remember loved ones, as well as a place for normalising grief).

72   In other words, grief that is not socially recognised – see generally KJ Doka (ed), *Disenfranchised Grief: Recognizing Hidden Sorrow* (Lexington, 1989). This typically occurs where the relationship between the deceased and a particular individual is no longer socially recognised (for example, an ex-spouse of the deceased) or has no 'legitimate' social status (for example, the relationship between the deceased and a mistress).

a way that would not otherwise be possible.[73] Despite suggestions that virtual memorials have simply filled the void created by the decline of traditional religious values in a secular society,[74] Brubaker, Hayes and Dourish see them as creating new sites of mourning (instead of disrupting traditional practices). According to the authors, Facebook and other SNS provide "a mediated space for grieving and remembrance, and [for] participating in an expanding set of death- and grief-related practices"[75] where the living are "constructing and expressing their own relationship to the deceased".[76]

2.   Regulatory frameworks and contested narratives

As things currently stand, the law does not have an active role to play in this area. The rules around creating digital memorials are governed by company policies and the contractual terms of the user agreement entered into with the service provider. For example, Facebook provides two options when an individual user dies: deleting the deceased's account completely, or converting the account into a memorial page, which is accessible only by the deceased's Facebook 'friends' – with current company policy clearly favouring the latter.[77] A so-called 'memorialization request' can be submitted by a wide range of individuals with some sort of tie to the deceased; at present, the designated categories are 'immediate family' (a spouse, parent, sibling or child), 'extended family' (a grandparent, aunt, uncle or cousin) and 'non-family' (which specifically includes a friend, colleague or classmate).[78] Proof of death is required, in the form of an obituary or some other verification.[79]

---

73   Walter (n 58), p 19.

74   A Sherlock, "Larger Than Life: Digital Resurrection and the Re-Enchantment of Society" (2014) 29 *Information Society* 164.

75   Brubaker, Hayes and Dourish (n 55), p 152.

76   *Ibid*, p 153.

77   See *Facebook: What Will Happen to My Account if I Pass Away?* https://en-gb.facebook.com/help/103897939701143 (accessed 30 September 2015) (though note that creating a new profile in the deceased's name is not an option). For an overview, see K Sherry, "What Happens to Our Facebook Account When We Die?: Property Versus Policy and the Fate of Social-Media Assets Postmortem" (2013) 40 *Pepperdine Law Review* 1 and D McCallig, "Facebook After Death: An Evolving Policy in a Social Network" (2014) 22 *International Journal of Law and Information Technology* 107. Deleting an existing account is a more complex procedure, which requires a birth and a death certificate as well as legal proof that the applicant has authority to administer the deceased's estate (for example, that he/she is executor under the deceased's will). Of course, this assumes that the deceased's log-in details are not available, in which case his/her family could simply delete the account. The procedure for memorialising an account is noted immediately below.

78   See *Facebook: Special Request for Deceased Person's Account* www.facebook.com/help/contact/228813257197480 (accessed 30 September 2015). McCallig notes that, until June 2013, a final generic category listed as 'other' was also available – McCallig (n 77), pp 115–116.

79   See *Facebook: Special Request for Deceased Person's Account* www.facebook.com/help/contact/228813257197480 (accessed 30 September 2015).

Family tensions over access to and the content of the deceased's virtual memorial can be just as prevalent as those over the design and wording of a headstone. Initially, family members may be divided over whether the deceased's Facebook page should be removed or whether it should be memorialised. Those who favour the former may be powerless to prevent the latter from happening;[80] there is no inherent mechanism for resolving intra-familial contests, and (unlike disputes over the deceased's mortal remains where there are clear rankings of entitlement[81]) Facebook policy is silent on whether one potential applicant can veto another's memorialisation request.[82] Assuming that memorialisation is the preferred option, there may be issues around who has access. Despite being open to a potentially large number of people, Facebook memorials (and their equivalents) also have an exclusionary aspect given that they are restricted to those who were already 'friends' with the deceased; there is no power to add someone retrospectively, meaning that some close family members may be unable to access and post on the memorial page.[83] Of course, an excluded individual can still invite tributes to the deceased on their own Facebook page or social networking site, or create a cyber grave in a virtual cemetery; the result may be multiple acts of remembrance to the deceased. Finally, even when close family members all have access to the deceased's Facebook memorial, there are invariably issues around content and contested narratives about the deceased within a very public forum. As Walter points out:

> The more that grief, memories and commemoration are shared, ... the more potential there is for conflict. ... This is partly due to the ability of social media to bring into contact people who might otherwise be isolated from each other.[84]

80   Unless they submit a formal request to have the page removed – though whether this is possible when the deceased's Facebook page has already been memorialised or someone else is actively seeking this, is open to question. Also, a request for removal is just that; it is not something which Facebook is obliged to grant.

81   Sees Ch 3, Pt I and Ch 4.

82   A point that is also highlighted by McCallig (n 77), pp 129–130. It may also be the case that the deceased's family do not wish to memorialise his/her Facebook page but one of the deceased's friends does, and submits a request to this effect. Again, there is no sense from reading Facebook policy that someone within the designated 'family' categories has any power to veto this (or, at a more basic level, is entitled to be notified of the request).

83   Wright (n 65), p 5 and McCallig (n 77), p 116 – the latter noting the distress that this causes for the parents of dead children (especially teenagers) who may not have been added as a 'friend' when the child was alive. Another exclusionary aspect stems from the intergenerational digital divide, with older relatives and friends of the deceased who are not active online users and less familiar with SNS unable to access and post on the virtual memorial. Walter (n 58), p 14 sees this as a "significant exclusion".

84   Walter (n 58), p 19. While the discussion here focuses on disputes within families, there can also be tension within the deceased's wider social network. Sharing personal memories of the dead means that users will also experience friction "as they encounter alternative narratives of the deceased" – Brubaker, Hayes and Dourish (n 55), p 158.

Estranged family members might resume hostilities on the memorial page; tensions between a former spouse and a new partner, or children from different relationships may be reignited in a very public forum as individuals emphasise their own emotional tie with the person who has died and try to 'reclaim' the deceased for themselves.[85] Unless the messages breach the rules around user conduct (for example, where specific posts amount to bullying, intimidation or harassment of another user[86]), there is nothing to prevent this.

Unlike their physical equivalents, Facebook and other virtual memorials do not have definitive rules for resolving potential areas of conflict.[87] Although regulated by company policy and user conditions, they raise broader issues around privacy as well as the "symbolic ownership and management of profiles postmortem".[88] Online memorials will continue to raise complex issues; and just as the "fault lines of both social and technological protocols"[89] will be increasingly exposed, so too will the deficiencies in the applicable regulatory frameworks.[90]

Conclusion

Memorials are not simply about recording someone's existence. They honour the dead, are sites for acts of remembrance and expressions of grief,[91] and allow the bereaved to communicate with the dead and with each other. As highly symbolic representations of the deceased's life and primary relationships, both gravestones and virtual memorials also straddle the boundary between the emotional and the factual, imparting private sentiments in a very public setting.

While the living do not have the right to commemorate their dead in any way they see fit, the law's response to the issues posed by memorialisation has been fairly muted. In churchyards and cemeteries, the design and content of

---

85 This idea was mentioned in Ch 4, p 92.

86 *Facebook: Statement of Rights and Responsibilities* www.facebook.com/legal/terms (accessed 30 September 2015).

87 There has been a number of cases, but most of these have involved disputes between families and the relevant service provider over terms of access – see McCallig (n 77), pp 120–123.

88 Brubaker, Hayes and Dourish (n 55), p 158. Wright correctly highlights that "protecting the privacy of deceased users is not a legally valid concern", since the dead probably do not have a right to privacy – Wright (n 65), p 4. However, the fact that a deceased user need not have agreed in advance to their profile being memorialised raises its own issues.

89 Lingel (n 56), p 194.

90 Current policies have been criticised. For example, Wright (n 65), p 1 cites the "disjointed company policies on what happens to deceased user accounts", while McCallig (n 77), pp 113–117 highlights a number of things about Facebook – including vague policy statements alongside the fact that the help-centre pages and forms are changed frequently and without notice.

91 Both individual and collective.

gravestones are down to internal regulations,[92] and disputes that do end up before courts (usually over the wording on the gravestone) are resolved by established legal principles. Greater challenges are posed by virtual memorials, as new ways of leaving tributes to the dead emerge in the twenty-first century. There is also more potential for conflict amongst the deceased's family and social media networks, not just over the presence of these memorials and who can access them, but the accuracy and content of postings – especially where several different memorials to the deceased are generated online. Effective legal solutions have yet to be developed; in the meantime, company policies and user agreements with individual service providers are the only reference point in an area that is difficult to regulate in any meaningful way.

---

92   Supplemented, for example, by planning regulations, prevailing health and safety laws, and public order offences (such as prohibitions on offensive language).

9 The law and the dead

Time for a re-evaluation?

> Dying is something we, as a species, do a lot, even if we keep evolving better ways to keep from doing so.[1]

In any contemporary society, what constitutes 'appropriate' treatment of the dead is governed by basic legal standards. *The Law and the Dead* has focused on bodily disposal in England and Wales, as well as the related areas of exhumation, commemoration, and the posthumous use of donated bodily material. These are sensitive and increasingly contentious topics, which raise a host of policy issues – not just here, but in other common law jurisdictions with similar legal (and socio-cultural) backgrounds. And while a substantive body of law currently exists in England and Wales around each of these subject-areas, there are obvious defects and deficiencies that highlight the need for reform.

Because the core principles are scattered across a range of sources, the existing framework suffers from a lack of cohesion and clarity. Take the basic burial laws, many of which still date back to nineteenth-century statutes (when a raft of Burial Acts were passed in response to growing urbanisation and public health[2]), some of which are still in force today despite being repealed in part or having only residual relevance. Matters are further 'complicated' by different laws around churchyard and cemetery burial, and the influence of ecclesiastical law on the former[3] – a distinction that also results in very different exhumation requirements for interments in

---

1   E Ferguson, "It's My Funeral: Getting Ready for the End", *The Observer* (London, 13 May 2012) www.theguardian.com/lifeandstyle/2012/may/13/funerals-death-humanist-euan-ferguson (accessed 30 September 2015).

2   The major statutes are collectively referred to as the Burial Acts 1852–1906. However, numerous other measures were passed during this time, with one legal text suggesting that, prior to 1887, there were as many as 120 statutes dealing with some aspect of burial of the dead – see J Brooke Little, *The Law of Burial* (Shaw & Sons, 3rd edn, 1902), p ix, reproducing a statement to this effect in the preface to the first edition (published in 1887). This is noted in I Jones and M Quigley, "Preventing a Lawful and Decent Burial: Resurrecting Dead Offences" (2016) *Legal Studies* (forthcoming).

3   See Ch 2, pp 30–32.

consecrated and unconsecrated land.[4] Meanwhile, cremation laws are fewer in number but are still a curious mix of old and new; parts of the Cremation Act 1902 still apply, with supplementary directives found in the Cremation Act 1952 and the Cremation (England and Wales) Regulations 2008.[5] Even before disposal takes place (and as part of the process itself), corpses attract a range of criminal law offences.[6] Developed on a seemingly *ad hoc* basis over the centuries, some are statutory in origin though most are common law misdemeanours devised by courts to penalise mistreatment of the dead. The result is a definitionally awkward patchwork of offences sustained by antiquated case law, and repeatedly revived to deal with modern exigencies – giving rise to all sorts of interpretative problems.

Similar issues arise in the civil law context. Faced with questions about the legal status of human remains and who has decision-making entitlements over the dead, courts have struggled with the consequences of the centuries-old 'no property' rule.[7] The fact that funeral instructions are not legally binding and are vulnerable to defeat by surviving relatives is one example, despite suggestions in recent cases that such preferences should not be overlooked.[8] Another example is the influence of succession law rules on who controls the deceased's funeral arrangements, though courts in this jurisdiction (and elsewhere) are increasingly willing to deviate from fixed rankings of entitlement and award custody of the deceased's remains to someone else with a more 'meritorious' claim.[9] There is also a sense in which the law's response to these and other issues does not always match our own intuitions about what rights exist over the dead and who can exercise them. For example, there is a widespread assumption that our bodies belong to us, but they do not (at least, not in any strict property law sense); many people assume that their funeral instructions are legally binding when this is not the case; in funeral disputes, the fact that the executor has the final say if the deceased died testate is not generally well known; and personal beliefs about who constitutes the deceased's family do not always match legal realities when it comes to deciding funeral arrangements and other death-related issues.

The latter claim brings us to another important point: the fact that the current legal framework fails to address major socio-cultural shifts that have shaped attitudes towards death and treatment of the dead in the twenty-first century. Recognising new variants of 'kinship' is one aspect of this, with the emergence of blended or reconstituted families; and, despite the fact that the

---

4   See Ch 7. In this context, 'consecrated' refers to ground consecrated according to the rites of the Church of England.

5   SI 2008/2841.

6   See Ch 1, Pt VI and Ch 2, Pt VI.

7   The old common law rule that there is no property in a dead body – see Introduction, pp 2–3.

8   See Ch 5.

9   See Ch 4.

latest Census figures recorded 5.9 million people living in cohabiting relationships in the UK[10] and 554,000 step-families with dependent children,[11] such individuals do not qualify as next-of-kin under intestacy rules and have no legal right to the deceased's remains in funeral disputes.[12] The emergence of an increasingly multi-faith, multi-ethic society also poses challenges. Minority groups have culturally specific beliefs around death and what happens to the dead, which must be recognised and acknowledged within the dominant legal paradigm.[13] To date, however, there are relatively few signs of this happening;[14] for example, spiritual and cultural values have limited impact on the outcome of funeral disputes and the exercise of other decision-making entitlements over the dead. Different ethnic groups may also have very different kinship structures (for example, extended families with greater emphasis on lineal ties), which do not map onto traditional definitions of 'family' in England and Wales.

Looking beyond these factors, *The Law and the Dead* has also highlighted how the existing legal framework has failed to deal with other modern developments: environmental debates around bodily disposal, and the challenges posed by new methods such as resomation and promession;[15] the emergence of a rights-based calculus and the ongoing impact of human rights discourse on bodily disposal and associated issues;[16] the use and legal status of donated bodily material, especially in the field of posthumous reproduction;[17] and the influence of the digital age and, in particular, contested narratives around digital memorials.[18] Other issues, such as regulation of the funeral industry and pre-paid funeral plans, have simply

---

10   Office of National Statistics, *Short Report: Cohabitation in the UK* (2012), located at www.ons.gov.uk/ons/rel/family-demography/families-and-households/2012/cohabi-tation-rpt.html (accessed 30 September 2015) and analysing data from the 2011 Census. The report describes unmarried cohabitation as the fastest growing family type in the UK, with figures doubling since 1996.

11   Office of National Statistics, *Release: Stepfamilies, 2011* (2014), located at www.ons.gov. uk/ons/rel/family-demography/stepfamilies/2011/index.html (accessed 30 September 2015) and analysing date from the 2011 Census. However, the 2011 figures represent a 14 per cent decrease from 2001.

12   See Ch 3, p 63.

13   See Ch 4, pp 104–107. The intersection of common law and customary law may also be an issue here, and was discussed at length in New Zealand at all three levels of the decision in *Takamore v Clarke* [2009] NZHC 901; [2011] NZCA 587; and [2012] NZSC 116.

14   Though one notable exception is the willingness of coroners in England and Wales to sanction less invasive post-mortem investigations for suspicious or sudden deaths, where objections have been lodged by Jewish and Muslim families – see Ch 1, p 17.

15   See Ch 2.

16   Discussed throughout *The Law and the Dead*, but see for example, the respective discussions in Ch 4, pp 107–114 (funeral disputes) and Ch 5, pp 139–144 (funeral instructions).

17   See Ch 6, Pt II.

18   See Ch 8, Pt II.

been mentioned in passing here[19] but may need to be explored in more detail, from both legal and policy perspectives.

Dealing with all of these matters in one composite review would, of course, be impossible. However, drawing on the example of other jurisdictions,[20] a reform agenda that addressed discrete aspects of the existing legal framework in England and Wales would be viable, and is long overdue. The overriding aim would be to modernise and streamline, beginning with the broad area of bodily disposal. New legislation could consolidate existing burial and cremation laws, while introducing new legal rules for resomation and promession; it could revisit the question of who controls the fate of the deceased's remains, setting out a new statutory hierarchy of entitlement based on an amended and more apposite definition of 'family' (which could also allow for cultural nuances across different ethnic groups). As part of this overall assessment, new laws could introduce an effective mechanism for upholding the deceased's own funeral instructions and making these the first point of reference in legal contests involving the fate of the dead. Finally, a simple statutory offence of maliciously, indecently, or improperly interfering with human remains could replace the current raft of criminal law offences.

These are only suggestions, and reform of any of the areas identified here and throughout previous chapters raises its own distinct challenges. At the very least, *The Law and the Dead* has laid the groundwork for a systematic re-evaluation of the laws around death, disposal and related matters in England and Wales (and, perhaps, in other common law jurisdictions facing similar issues). Whether or not this occurs in the near future remains to be seen. In the meantime, to borrow from Mel Blanc's gravestone inscription in the Hollywood Forever Cemetery in Hollywood, California:[21]

"That's All Folks!"

19   See Ch 1, p 21 and Ch 3, p 75.
20   In terms of more recent examples, repeated references have been made to Queensland Law Reform Commission, *A Review of the Law in Relation to the Final Disposal of a Dead Body* (Report No 69, December 2011) and New Zealand Law Commission, *The Legal Framework for Burial and Cremation in New Zealand: A First Principles Review* (Issues Paper 34, October 2013). A wide-ranging review is also taking place in Scotland- Scottish Government, *Consultation on a Proposed Bill Relating to Burial and Cremation and Other Related Matters in Scotland* (January 2015).
21   Mel Blanc, who died in 1989, was the voice of Bugs Bunny, Daffy Duck and other much-loved cartoon characters, and wanted these words (the trademark phrase of Porky Pig) engraved on his tombstone.

Bibliography

Books and Journal Articles

Afrasiabi, PR, "Property Rights in Ancient Skeletal Remains" (1997) 70 *Southern California Law Review* 805

Archer, J, *The Nature of Grief: the Evolution and Psychology of Reactions to Loss* (Routledge, 1999)

Adamson, S and Holloway, M, "Symbols and Symbolism in the Funeral Today: What Do They Tell Us About Contemporary Spirituality" (2013) 3 *Journal for the Study of Spirituality* 140

Atherton, R, "En Ventre sa Frigidaire: Posthumous Children in the Succession Context" (1999) 19 *Legal Studies* 139

Atherton, R, "Claims on the Deceased: The Corpse as Property" (2000) 7 *Journal of Law and Medicine* 361

Atherton, R, "Who Owns Your Body" (2003) 77 *Australian Law Journal* 178

Bainham, A, Sclater, SD and Richards, M, *Body Lore and Laws* (Hart, 2002)

Baker, DM, "Cryonic Preservation of Human Bodies: A Call for Legislative Action" (1994) 98 *Dickinson Law Review* 677

Ball, EBS, "Property and the Human Body: A Proposal for Posthumous Conception" (2008) 15 *Journal of Law and Medicine* 556

NM Banta, "Inherit the Cloud: The Role of Private Contracts in Distributing or Deleting Digital Assets at Death" (2014) 83 *Fordham Law Review* 799

Barish, M, "The Law of Testamentary Disposition: A Legal Barrier to Medical Advance" (1956) 30 *Temple Law Quarterly* 40

Bass, WM and Jefferson, J, *Beyond the Body Farm* (William Morrow, 2007)

Batchelder, PD, "Dust in the Wind? The Bell Tolls for Crematory Mercury" (2008) 2 *Golden Gate University Environmental Law Journal* 118

Beale, H, *Chitty on Contracts, Volume I: General Principles* (Sweet & Maxwell, 31st edn, 2012)

Bennett, B, "Posthumous Reproduction and the Meaning of Autonomy" (1999) 23 *Melbourne University Law Review* 286

Bernard, HY, *The Law of Death and Disposal of the Dead* (Oceana Publications, 1966)

Bernick, A, "Burying an Injustice: Indigenous Human Remains in Museums and the Evolving Obligations to Return Remains to Indigenous Groups" (2014) 1 *Indonesian Journal of International & Comparative Law* 637

Beyer, GW and Cahn, N, "Digital Planning: The Future of Elder Law" (2013) 9 *NAELA Journal* 9 135

Blackstone, W, *Commentaries on the Law of England* (Tucker ed, 1803)

Blake, K and Kushnick, HL, "Ethical Implications of Posthumous Reproduction" in Goldfard, JM (ed) *Third-Party Reproduction* (Springer, 2014), pp 197–207

Boddington, P, "Organ Donation After Death: Should I Decide, or Should my Family?" (1998) 15 *Journal of Applied Philosophy* 69

Bowen, M, "Family Reaction to Death", in Walsh, F and McGoldrick, M (eds), *Living beyond Loss: Death in the Family* (WW Norton and Co, 1991), pp 335–348

Bowlby, J, "The Process of Mourning" (1961) 42 *International Journal of Psycho-Analysis* 331

Brashier, RC, "Disinheritance and the Modern Family" (1994) 45 *Case Western Reserve Law Review* 83

Bray, M, "Personalizing Property: Towards a Property Right in Human Bodies" (1990) 69 *Texas Law Review* 209

Brazier, M, "Retained Organs: Ethics and Humanity" (2002) 22 *Legal Studies* 550

Brecher, B, "Our Obligation to the Dead" (2002) 19 *Journal of Applied Philosophy* 109

Bremenstul, ME, "Victims in Life, Victims in Death: Keeping Burial Rights Out of the Hands of Slayers" (2013) 74 *Louisiana Law Review* 213

Brooke Little, J, *The Law of Burial* (Shaw & Sons, 3rd edn, 1902)

Brooks-Gordon, B, Ebtehaj, F, Herring, J, Johnson, MH and Richards, M (eds), *Death Rites and Rights* (Hart, 2007)

Brubaker, J, Hayes, GR and Dourish, P, "Beyond the Grave: Facebook as a Site for the Expansion of Grief and Mourning" (2013) 29 *Information Society* 152

Bruzzone, P, "Financial Incentives for Organ Donation: A Slippery Slope Toward Organ Commercialism?" (2010) 42 *Transplantation Proceedings*

Callahan, JC, "On Harming the Dead" (1987) 97 *Ethics* 341

Canning, L and Szmigin, I, "Death and Disposal: The Universal, Environmental Dilemma" (2010) 26 *Journal of Marking Management* 1129

Cantor, NL, *After We Die: The Life and Times of the Human Cadaver* (Georgetown University Press, 2010)

Carpenter, B and Tait, G, "The Autopsy Imperative: Medicine, Law and the Coronial Investigation" (2010) 31 *Journal of Medical Humanities* 205

Carpenter, B, Tait, G and Quadrelli, C, "The Body in Grief: Death Investigations, Objections to Autopsy and the Religious and Cultural 'Other'" (2014) 5 *Religions* 165

Carpenter, B, Adkins, G, Barnes, M, Naylor, C and Bequm, N, "Communicating with the Coroner: How Religion, Culture and Family Concerns May Influence Autopsy Decision Making" (2011) 35 *Death Studies* 316

Carpenter, BC, "A Chip off the Old Iceblock: How Cryopreservation Has Changed Estate Law, Why Attempts to Address the Issue Have Fallen Short, and How to Fix It" (2011) 21 *Cornell Journal of Law & Public Policy* 347

Carroll, B and Landry, K, "Logging On and Letting Out: Using Online Social Networks to Grieve and to Mourn" (2010) 30 *Bulletin of Science and Technology* 341

Chambers, D, *A Sociology of Family Life* (Polity, 2012)

Chau, PL and Herring, J, "The Meaning of Death" in Brooks-Gordon, B, Ebtehaj, F, Herring, J, Johnson, MH and Richards, M (eds), *Death Rites and Rights* (Hart, 2007), pp 13–36

Chester, R, "Freezing the Heir Apparent: A Dialogue on Postmortem Conception, Parental Responsibility, and Inheritance" (1996) 33 *Houston Law Review* 967

Chester, R, *From Here to Eternity* (Vandeplas, 2007)

Clark, J and Franzmann, M, "Authority from Grief, Presence and Place in the Making of Roadside Memorials" (2006) 30 *Death Studies* 579

Clayden, A, Hockey, J and Powell, M, "Natural Burial: The De-Materialising of Death" in Hockey, J, Komaromy, C and Woodthorpe, K (eds), *The Matter of Death: Space, Place and Materiality* (Palgrave Macmillan, 2010), pp 148–164

Colleran, E, "My Body, His Property: Prescribing a Framework to Determine Ownership Interests in Directly Donated Human Organs" (2007) 80 *Temple Law Review* 1203

Conway, H, "Burial Instructions and the Governance of Death (2012) 12 *Oxford University Commonwealth Law Journal* 59

Conway, H and Stannard, J, "The Honours of Hades: Death, Emotion and the Law of Burial Disputes" (2011) 34 *University of New South Wales Law Journal* 860

Cook, DS and James, DS, "Necrophilia: Case Report and Consideration of Legal Aspects" (2002) 5 *Medical Law International* 199

Cook, G and Walter, T, "Rewritten Rites: Language and Social Relations in Traditional and Contemporary Funerals" (2005) 16 *Discourse & Society* 365

Cooper, IS and Harper, RM, "Life After Death: The Authority of Estate Fiduciaries to Dispose of Decedents' Reproductive Matter" (2010) 26 *Touro Law Review* 649

Cowling, C, *The Good Funeral Guide* (Continuum, 2010)

Craigie, AW, "Burial of a Tort: The California Supreme Court's Treatment of Tortious Mishandling of Remains in *Christensen v Superior Court*" (1993) 26 *Loyola of Los Angeles Law Review* 909

"Cremation and Burial", 24 *Halsbury's Laws of England* (5th edn, 2010)

Croucher, R "Disposing of the Dead: Objectivity, Subjectivity and Identity" in Freckleton, I and Peterson, K (eds), *Disputes and Dilemmas in Health Law* (Federation Press, 2006), pp 324–342

Cumper, P and Lewis, T, "Last Rites and Human Rights: Funeral Pyres and Religious Freedom in the United Kingdom" (2010) 12 *Ecclesiastical Law Journal* 131

Davies, DJ, *Death, Ritual and Belief: The Rhetoric of Funeral Rites* (Continuum, 2nd edn, 2002)

Davies, DJ, *A Brief History of Death* (Blackwell, 2005)

Davies, DJ and Guest, MJ, "Disposal of Cremated Remains" (1999) 65 *Pharos International* 26

Davies, DJ and Rumble, H, *Natural Burial: Traditional-Secular Spiritualities and Funeral Innovation* (Bloomsbury, 2012)

Deech, R and Smajdor, A, *From IVF to Immortality: Controversy in the Era of Reproductive Technology* (OUP, 2008)

Degroot, JM, "Maintaining Relational Continuity with the Deceased on Facebook" (2012) 65 *OMEGA: Journal of Death and Dying* 195

Delamothe, T, Snow, R and Godlee, F, "Why the Assisted Dying Bill Should Become Law in England and Wales" (2014) 349 *BMJ: British Medical Journal* 4349

Den Hartogh, G, "The Role of the Relatives in Opt-In Systems of Postmortal Organ Procurement" (2012) 15 *Medicine, Health Care and Philosophy* 195

Dickenson, D, *Property in the Body: Feminist Perspectives* (CUP, 2007)

Doka, KJ (ed), *Disenfranchised Grief: Recognizing Hidden Sorrow* (Lexington, 1989)

Dominick, SA, Irvine, AB, Beauchamp, N, Seeley, JR, Nolan-Hoeksema, S, Doka, KJ and Bonanno, GA, "An Internet Tool to Normalise Grief" (2009) 60 *OMEGA: Journal of Death and Dying* 71

Dorries, C, *Coroners' Courts: A Guide to Law and Practice* (OUP, 3rd edn, 2014)

Dowling, A, "Exclusive Rights of Burial and the Law of Real Property" (1998) 18 *Legal Studies* 438

Downie, J, Shea, A and Rajotte, C, "Family Override of Valid Donor Consent to Postmortem Donation: Issues in Law and Practice" (2008) 40 *Transplantation Proceedings* 1255

Dummond, LC, "The Undeserving Heir: Domestic Elder Abuser's Right to Inherit" (2010) 23 *Quinnipiac Probate Law Journal* 214

Dwyer, LA, "Dead Daddies: Issues in Postmortem Reproduction" (2000) 52 *Rutgers Law Review* 881

"Ecclesiastical Law", 34 *Halsbury's Laws of England* (5th edn, 2011)

Elliott, T, "Religious Belief and Choices Regarding the Human Corpse" (2014) 2 *Journal of Medical Law and Ethics* 89

Ellis, TE, "Loved and Lost: Breathing Life Into the Rights of Noncustodial Parents" (2005) 40 *Valparaiso University Law Review* 267

Englefeld, L, *Australian Family Provision Law* (Lawbook Company, Australia, 2011)

Evans, C, *Freedom of Religion under the European Convention on Human Rights* (OUP, 2001)

Everhard, JA, "Whose Body?" (1964) 6 *United States Air Force JAG Bulletin* 17

Evers, L, *I Told You I Was Ill: Dying for a Laugh* (Michael O'Mara Books, 2012)

Falconer, K, Sachsenweger, M, Gibson, K and Norman, H, "Grieving in the Internet Age" (2011) 40 *New Zealand Journal of Psychology* 79

Farrell, AM, Price, D and Quigley, M (eds), *Organ Shortage: Ethics, Law and Pragmatism* (CUP, 2011)

Fegan, R, "Death to Life: Towards My Green Burial" (2007) 10 *Ethics, Place and Environment* 157

Fish, DG, "To Avoid Burial Disputes, New Statutory Form is Available" (2006) 235 *New York Law Journal (Elder Law)* 1

Ford, M, "The Personhood Paradox and the Right to Die" (2005) 13 *Medical Law Review* 80

Foster, C, *Choosing Life, Choosing Death: The Tyranny of Autonomy in Medical Law and Ethics* (Hart, 2009)

Foster, FH, "Towards a Behavior-Based Model of Inheritance?: The Chinese Experiment" (1998) 32 *University of California Davis Law Review* 77

Foster, FH, "The Family Paradigm of Inheritance Law" (2001) 80 *North Carolina Law Review* 199

Foster, FH, "Individualized Justice in Disputes over Dead Bodies" (2008) 61 *Vanderbilt Law Review* 1352

Freckelton, I, "Disputed Family Claims to Bury or Cremate the Dead" (2009) 17 *Journal of Law and Medicine* 178

Friedman, D, "Does Technology Require New Law" (2001) 25 *Harvard Journal of Law & Public Policy* 71

Fuselier, BM, "Trouble with Putting All of Your Eggs in One Basket: Using a Property Rights Model to Resolve Disputes over Cryopreserved Pre-Embryos" (2008) 14 *Texas Journal on Civil Liberties & Civil Rights* 143

Gallagher, S, "Protecting the Dead: Exhumation and the Ministry of Justice" [2008] 5 *Web Journal of Current Legal Issues*

Gallagher, S, "Raising the Dead: Exhumation and the Faculty Jurisdiction: Should We Presume To Exhume?" [2010] 1 *Web Journal of Current Legal Issues*

George, A, "Is 'Property' Necessary? On Owning the Human Body and Its Parts" (2004) 10 *Res Publica* 15

Gibson, M, "Death and Mourning in Technologically Mediated Culture" (2007) 16 *Health Sociology Review* 415.

Gibson, M, "Automatic and Automated Mourning: Messengers of Death and Messages from the Dead" (2015) 29 *Journal of Media & Cultural Studies* 339

Gittings, C and Walter, T, "Rest in Peace? Burial on Private Land" in Madrell, A and Sidaway, JD (eds), *Deathscapes: Spaces for Death, Dying and Bereavement* (Ashgate, 2010), pp 165–177

Giunta, LA, "The Dead on Display: A Call for the International Regulation of Plastination Exhibits" (2010) 49 *Columbia Journal of Transnational Law* 164

Glannon, W, "Persons, Lives and Posthumous Harms" (2001) 32 *Journal of Social Philosophy* 127

Goold, I, Greasley, K, Herring, J, and Skene, L, *Persons, Parts and Property: How Should We Regulate Human Tissue in the 21st Century?* (Hart, 2014)

Goudkamp, J and Peel, E, *Winfield and Jolowicz on Tort* (Sweet & Maxwell, 19th edn, 2014)

Grabenwarter, C, *European Convention on Human Rights: Commentary* (CH Beck, 2014)

Green, J and Green, M, *Dealing with Death: A Handbook of Practices, Procedures and Law* (Jessica Kingsley, 2nd edn, 2008)

Griggs, L and Mackie, K, "Burial Rights: The Contemporary Australian Position" (2007) 7 *Journal of Law and Medicine* 404

Groll, RC and DJ Kerwin, DJ, "The Uniform Anatomical Gift Act: Is the Right to a Decent Burial Obsolete?" (1971) 2 *Loyola University Chicago Law Journal* 275

Grover, D, "Posthumous Harm" (1989) 39 *Philosophical Quarterly* Q 334

Groves, M, "The Disposal of Human Ashes" (2005) 12 *Journal of Law Medicine* 267

Grubb, A, "'I, Me, Mine': Bodies, Parts and Property" (1998) 3 *Medical Law International* 299

Haddleton, RE, "What To Do With The Body? The Trouble With Postmortem Disposition" (2013) 20 *Property & Probate* 55

Haddow, G, "Death, Embodiment and Organ Transplantation" (2005) 27 *Sociology of Health & Illness* 92

Hadwen, W, Tebb, W and Vollum, EP *Premature Burial: How It May Be Prevented* (edited by J Sale, Hesperus, 2012)

Hallam, E and Hockey, J, *Beyond the Body: Death and Social Identity*" (Routledge, 1999)

Hallam, E and Hockey, J, *Death, Memory and Material Culture* (Berg, 2001)

Hans, JD and Yelland, EL, "American Attitudes in Context: Posthumous Uses of Cryopreserved Gametes" (2013) *Journal of Clinical Research & Bioethics* 1

Hans, JD, "Posthumous Gamete Retrieval and Reproduction: Would the Deceased Spouse Consent?" (2014) 119 *Social Science & Medicine* 10

Hans, JD and Dooley, B, "Attitudes Toward Making Babies... With a Deceased Partner's Cryopreserved Gametes" (2014) 38 *Death Studies* 571

Hansen, K, "Choosing To Be Flushed Away: A National Background on Alkaline Hydrolysis and What Texas Should Know about Regulating Liquid Cremation" (2012) 5 *Estate Planning & Community Property Law Journal* 145

Hardcastle, R, *Law and the Human Body: Property Rights, Ownership and Control* (Hart, 2007)

Harris, J, "Law and Regulation of Retained Organs: The Ethical Issues" (2002) 22 *Legal Studies* 527

Harvey, C and Vincent, L, *The Law of Dependants' Relief in Canada* (Carswell, 2nd edn, 2006)

Hawes, C "Property Interests in Body Parts", *Yearworth v North Bristol NHS Trust* (2010) 73 *Modern Law Review* 130

Henaghan, M, "Family Law After Death: Control of the Body of a Child Killed by the Actions for a Parent" (2010) 6 *New Zealand Family Law Journal* 263

Hernández, TK, "The Property of Death" (1999) 60 *University of Pittsburgh Law Review* 971

Herring, J, "Crimes Against the Dead" in Brooks-Gordon, B, Ebtehaj, F, Herring, J, Johnson, MH and Richards, M (eds), *Death Rites and Rights* (Hart, 2007), pp 219–239

Herring, J, *Medical Law and Ethics* (OUP, 5th edn, 2014)

Hill, M, *Ecclesiastical Law* (OUP, 3rd edn, 2007)

Hirst, M, "Preventing the Lawful Burial of a Body" [1996] *Criminal Law Review* 96

Hockey, J, Komaromy, C and Woodthorpe, K (eds), *The Matter of Death: Space, Place and Materiality* (Palgrave Macmillan, 2010)

Holloway, M, *Negotiating Death in Contemporary Health and Social Care* (Policy Press, 2007)

Hooper, A and Omerod, DC (eds), *Blackstone's Criminal Practice* (OUP, 2009)

Hopkins, JP, "Afterlife in the Cloud: Managing a Digital Estate" (2013) 5 *Hastings and Science Technology Law Journal* 210

Horan, JE, "'When Sleep at Last Has Come': Controlling the Disposition of Dead Bodies for Same-Sex Couples" (1999) 2 *Journal of Gender, Race and Justice* 423

Horton, KE, "Who's Watching the Cryptkeeper: The Need for Regulation and Oversight in the Crematory Industry" (2003) 11 *Elder Law Journal* 425

Howarth, G, *Death and Dying: A Sociological Introduction* (Polity, 2007)

Howarth, G, "The Rebirth of Death: Continuing Relationships with the Dead" in Mitchell, M (ed), *Remember Me: Constructing Immortality* (Routledge, 2007), pp 19–34

Howarth, G and Jupp, PC (eds), *Contemporary Issues in the Sociology of Death, Dying and Disposal* (Palgrave Macmillan, 1995)

Hubert, J, "Dry Bones or Living Ancestors? Conflicting Perceptions of Life, Death and the Universe" (1992) 1 *International Journal of Cultural Property* 105

Hume, SG, "Dead Bodies" (1956) 2 *Sydney Law Review* 109

Humphrey, A, Morrell, G, Mills, L, Douglas, G and Woodward, H, *Inheritance and the Family: Attitudes to Will-Making and Intestacy* (National Centre for Social Research, August 2010)

Hurren, ET, *Dying for Victorian Medicine: English Anatomy and Its Trade in the Dead Poor, c.1834–1929* (Palgrave Macmillan, 2014)

Hutchings, T, "Wiring Death: Dying, Grieving and Remembering on the Internet" in Park, CW and Davies, DJ, *Emotion, Identity and Death: Mortality Across Disciplines* (Ashgate, 2012), pp 43–58

Hutchinson, AC, *Is Eating People Wrong? Great Legal Cases and How They Shaped the World* (CUP, 2010)

Jackson, PE, *The Law of Cadavers and of Burial and Burial Places* (Prentice Hall, 2nd edn, 1950)

Jacob, MA, "On Silencing and Slicing: Presumed Consent to Post-Mortem Organ 'Donation' in Diversified Societies" (2003) *Tulsa Journal of Comparative and International Law* 239

Johnston, C, "Advance Decision Making: Rhetoric or Reality" (2014) 34 *Legal Studies* 497

Jones, C, "The Identification of 'Parents' and 'Siblings': New Possibilities under the Reformed Human Fertilisation and Embryology Act", in Wallbank, J, Choudhry, S and Herring, J (eds) *Rights, Gender and Family Law* (Routledge-Cavendish, 2009), pp 219–238

Jones, DG and Whitaker, MA, *Speaking for the Dead: The Human Body in Biology and Medicine* (Ashgate, 2009)

Jones, DG and Whitaker, MA "Engaging with Plastination and the Body Worlds Phenomenon: A Cultural and Intellectual Challenge for Anatomists" (2009) 22 *Clinical Anatomy* 770

Jones, I and Quigley, M, "Preventing a Lawful and Decent Burial: Resurrecting Dead Offences" (2016) *Legal Studies* (forthcoming)

Josias, BL, "Burying the Hatchet in Burial Disputes: Applying Alternative Dispute Resolution Techniques to Disputes Concerning the Interment of Bodies" (2004) 79 *Notre Dame Law Review* 1141

Jupp, PC, "Virtue Ethics and Death: The Final Arrangements" in Flanagan, K and Jupp, PC (eds), *Virtue Ethics and Sociology: Issues of Modernity and Ethics* (Palgrave, 2001), pp 217–235

Katz, KD, "Parenthood from the Grave: Protocols for Retrieving and Utilizing Gametes from the Dead or Dying" (2006) *University of Chicago Legal Forum* 289

Klaiman, MH, "Whose Brain Is It Anyway: The Comparative Law of Post-Mortem Organ Retention" (2005) 26 *Journal of Law and Medicine* 475

Kehoe, PR, "Cemetery Abandonment and Disinterment of Human Remains (1971) 35 *Albany Law Review* 320

Kellaher, L, Prendergast, D and Hockey, J, "In the Shadow of the Traditional Grave" (2005) 10 *Mortality* 237

Kellaher, L, Hockey, J and Prendergast, D, "Wandering Lines and Cul-de Sacs: Trajectories of Ashes in the United Kingdom" in Hockey, J, Komaromy, C and Woodthorpe, K (eds), *The Matter of Death: Space, Place and Materiality* (Palgrave Macmillan, 2010), pp 133–147

Keown, J, *Euthanasia, Ethics and Public Policy: An Argument Against Legalisation* (CUP, 2002)

Kern, R, Forman, AE and Gil-Egui, G, "RIP: Remain in Perpetuity. Facebook Memorial Pages" (2013) 30 *Telematics and Informatics* 2

Kerridge, R and Brierly, AHR, *Parry and Kerridge: The Law of Succession* (Sweet & Maxwell, 12th edn, 2009)

Kester, TM, "Uniform Acts: Can the Dead Hand Control the Dead Body? The Case for Uniform Bodily Remains Law" (2007) 29 *Western New England Law Review* 571

Kindregan, CP and McBrien, M, "Posthumous Reproduction" (2005) *Family Law Quarterly* 579

Knaplund, KS, "Postmortem Conception and a Father's Last Will" (2004) 46 *Arizona Law Review* 91

Knight, C, "A Science Without A Deadline" (2008) 19 *Engineering and Technology* 28

Kopp, SW and Kemp, E, "The Death Care Industry: A Review of Regulatory and Consumer Issues" (2007) 41 *Journal of Consumer Affairs* 150

Kramer, MH, "Do Animals and Dead People Have Legal Rights?" (2001) 14 *Canadian Journal of Law & Jurisprudence* 29

Kr//løkke, C and Adrian, S, "Sperm on Ice: Fatherhood and Life after Death" (2013) 28 *Australian Feminist Studies* 263

Laurie, GT, "*Yearworth v North Bristol NHS Trust*: Property, Principles, Precedents and Paradigms" (2010) 69 *Cambridge Law Journal* 476

Leavitt, J, "The Funeral Director's Liability for Mental Anguish" (1964) 15 *Hastings Law Journal* 464

Lee, J, "*Yearworth v North Bristol NHS Trust* [2009]: Instrumentalism and Fictions in Property Law" in Douglas, S, Hickey, R and Waring, E (eds), *Landmark Cases in Property Law* (Hart, 2015), pp 25–48

Leiboff, M, "Post-Mortem Sperm Harvesting, Conception and the Law: Rationality or Religiosity?" (2006) 6 *Queensland University of Technology Law and Justice Journal* 193

Lingel, J, "The Digital Remains: Social Media and Practices on Online Grief" (2013) 29 *The Information Society: An International Journal* 190

Lorshbough, WH, "The Disposition by Will of One's Body After Death" (1945) 22 *Bar Briefs* 272

Lovatt, C, "Equality Issues from Beyond the Grave" (2000) 19 *Equal Opportunities International* 29

Love, LP, "Mediation of Probate Matters: Leaving a Valuable Legacy" (2001) 1 *Pepperdine Dispute Resolution Law Journal* 255

Lovejoy, B, *Rest in Pieces: The Curious Fate of Famous Corpses* (Simon & Schuster, 2013)

Luce, T, "Coroners and Death Certification Law Reform: The Coroners and Justice Act 2009 and its Aftermath" (2010) 50 *Medicine, Science and the Law* 171

Maclean, A, "Advance Directives, Future Selves and Decision-Making" (2006) 14 *Medical Law and Ethics* 291

Madoff, R, "Unmasking Undue Influence" (1997) 81 *Minnesota Law Review* 571

Madoff, R, *Immortality and the Law: The Rising Power of the American Dead* (Yale University Press, 2010)

Mann, T, *The Magic Mountain* (1924, translated by HT Lowe-Porter)

Marwick, A and Ellison, NB, "'There Isn't Wifi in Heaven!' Negotiating Visibility on Facebook Memorial Pages" (2012) 56 *Journal of Broadcasting & Electronic Media* 378

Mason, JK and Laurie, GT, "Consent or Property? Dealing With The Body and Its Parts in the Shadow of Bristol and Alder Hey" (2001) 64 *Modern Law Review* 710

Mason, JK and Laurie, GT, *Mason & McCall Smith's Law and Medical Ethics* (OUP, 9th edn, 2013)

Matthews, P, "Whose Body? People as Property" (1983) 36 *Current Legal Problems* 193

Matthews, P, *Jervis on the Office and Duties of Coroners* (Sweet & Maxwell, 13th edn, 2014)

Mayersak, T, "Examining the Use of Arbitration and Dealing with Decedent's Wishes in Wills, Trusts and Estates" (2010) 12 *European Journal of Law Reform* 404

McBain, G, "Modernising the Law on the Unlawful Treatment of Dead Bodies" (2014) 7 *Journal of Politics and Law* 89

McCallig, D, "Facebook After Death: An Evolving Policy in a Social Network" (2014) 22 *International Journal of Law and Information Technology* 107

McCandless, J and Sheldon, S, "The Human Fertilisation and Embryology Act (2008) and the Tenacity of the Sexual Family Form" (2010) 73 *Modern Law Review* 175

McEvoy, K and Conway, H, "The Dead, The Law, and the Politics of the Past" (2004) 31 *Journal of Law and Society* 539

McGuinness, S and Brazier, M, "Respecting the Living Means Respecting the Dead Too" (2008) 28 *Oxford Journal of Legal Studies* 297

Mchale, J, "The Human Tissue Act 2004: Innovative Legislation: Fundamentally Flawed or Missed Opportunity?" (2005) 26 *Liverpool Law Review* 169

McLean, S, *Assisted Dying: Reflections on the Need for Law Reform* (Routledge, 2007)

McLean, S, *Autonomy, Consent and the Law* (Routledge-Cavendish, 2010)

McManus, R, *Death in a Global Age* (Palgrave Macmillan, 2013)

Meese, J, Gibbs, M, Carter, M, Arnold, M, Nansen, B and Kohn, T, "Selfies at Funerals: Mourning and Presencing on Social Media Platforms" (2015) 9 *International Journal of Communication* 14.

Mesich-Brant, JL and Grossback, LJ, "Assisting Altruism: Evaluating Legally Binding Consent in Organ Donation Policy" (2005) 30 *Journal of Health Politics, Policy and Law* 687

Mims, C, *When We Die: The Science, Culture and Rituals of Death* (Robinson, 2000)

Mitchell, M (ed), *Remember Me: Constructing Immortality: Beliefs on Immortality, Life and Death* (Routledge, 2007)

Monaghan, A, "Conceptual Niche Management of Grassroots Innovation for Sustainability: The Case of Body Disposal Practices in the UK" (2009) *Technological Forecasting & Social Change* 1026

Monahan, JT and Lawhorn, EA, "Life-Sustaining Treatment and the Law: The Evolution of Informed Consent, Advance Directives and Surrogate Decision-Making" (2009) 19 *Annals of Health Law* 107

Moreman, CM and Lewis, AD (eds), *Digital Death: Mortality and Beyond in the Online Age* (ABC-CLIO, 2014)

Morgan, D and Lee, R, "'In the Name of the Father?' *Ex parte Blood:* Dealing with Novelty and Anomaly" (1997) 60 *Modern Law Review* 840

Morris, B, "You've Got to be Kidneying Me: The Fatal Problem of Severing Rights and Remedies from the Body of Organ Donation Law" (2008) 74 *Brooklyn Law Review* 543

Mortimer, D, "Proprietary Rights in Body Parts" (1993) 19 *Monash University Law Review* 216

Moses, LB "The Problem with Alternatives: The Importance of Property Law in Regulating Excised Human Tissue and In Vitro Human Embryos" in Goold, I, Greasley, K, Herring, J, and Skene, L, *Persons, Parts and Property: How Should We Regulate Human Tissue in the 21st Century?* (Hart, 2014), pp 197–214

Moss, M, "Grief on the Web" (2004) 49 *OMEGA: Journal of Death and Dying* 77

Mullaney, NJ and Handford, P, *Tort Liability for Psychiatric Damage* (Law Book Company of Australia, 2nd edn, 2006)

Munoz, RT and Fox, MD, "Legal Aspects of Brain Death and Organ Donorship" in Novitzky, D and Cooper, DKC (eds), *The Brain-Dead Organ Donor* (Springer, 2013) pp 21–35

Murphy, AM, "Please Don't Bury Me Down in that Cold Cold Ground: The Need for Uniform Laws in the Disposition of Human Remains" (2007) 15 *Elder Law Journal* 400

Nachman, D, "Living Wills: Is It Time to Pull the Plug?" (2010) 18 *Elder Law Journal* 289

Naffine, N, "When Does the Legal Person Die? Jeremy Bentham and the Auto-Icon" (2000) 25 *Australian Journal of Legal Philosophy* 79

Naguit, KE, "Letting the Dead Bury the Dead: Missouri's Right of Sepulcher Addresses the Modern Decedent's Wishes" (2010) 75 *Missouri Law Review* 248

Nedelsky, J, "Property in Potential Life? A Relational Approach to Choosing Legal Categories" (1993) 6 *Canadian Journal of Law and Jurisprudence* 343

Nelkin, D and Andrews, L, "Do The Dead Have Interests? Policy Interests for Research After Life" (1998) 24 *American Journal of Law and Medicine* 261

Neria, Y and Litz, BT, "Bereavement by Traumatic Means: The Complex Synergy of Trauma and Grief" (2004) 9 *Journal of Loss and Trauma* 73

Nwabueze, RN, "Biotechnology and the New Property Regime in Human Bodies and Body Parts" (2002) 24 *Loyola of Los Angeles International & Comparative Law Review* 19

Nwabueze, RN, "Interference with Dead Bodies and Body Parts: A Separate Cause of Action in Tort" (2005) 15 *Tort Law Review* 63

Nwabueze, RN, "Property Interest in a Burial Plot" [2007] *Conveyancer and Property Lawyer* 517

Nwabueze, RN, *Biotechnology and the Challenge of Property: Property Rights in Dead Bodies, Body Parts and Genetic Information* (Ashgate, 2007)

Nwabueze, RN, "Donated Organs, Property Rights and the Remedial Quagmire" (2008) 18 Medical Law Review 201

Nwabueze, RN, "Legal Control of Burial Rights" (2013) 2 *Cambridge Journal of International and Comparative Law* 196

O'Carroll, TL, "Over My Dead Body: Recognizing Property Rights in Corpses" (1996) 29 *Journal of Health and Hospital Law* 238

O'Donovan, K and Gilbar, R, "The Loved Ones: Families, Intimates and Patient Autonomy" (2003) 23 *Legal Studies* 332

Olson, PR, "Flush and Bones: Funeralizing Alkaline Hydrolysis in the United States" (2014) *Science, Technology and Human Values* 1

O'Rourke, T, Spitzberg, BH and Hannawa, AF, "The Good Funeral: Towards an Understanding of Funeral Participation and Satisfaction" (2011) 35 *Death Studies* 729

Oughton, RD, *Tyler's Family Provision* (Tottel, 3rd edn, 1998)

Parsons, B, "Conflict in the Context of Care: An Examination of Role Conflict Between the Bereaved and the Funeral Director in the UK" (2003) 8 *Mortality* 67

Palmer, R, "Death and the Coroner: Some Reflections on Current Practice and Proposed Reforms" (2012) 52 *Medicine, Science and the Law* 63

Partridge, E, "Posthumous Interests and Posthumous Respect" (2001) 91 *Ethics* 243

Pattinson, SD, "Directed Donation and Ownership of Human Organs" (2011) 31 *Legal Studies* 392

Pattinson, SD, *Medical Law and Ethics* (Sweet and Maxwell, 4th edn, 2014)

Pawlowski, M, "Property in Human Body Parts and Products of the Human Body" (2009) 30 *Liverpool Law Review* 35

Perlin, AA, "To Die in Order to Live: The Need for Legislation Governing Post-Mortem Cryonic Suspension" (2007) 36 *Southwestern University Law Review* 33

Perrone, M, "What Happens When We Die: Estate Planning of Digital Assets" (2012) 21 *CommLaw Conspectus* 185

Peterson, KE, "My Father's Eyes and My Mother's Heart: The Due Process Rights of the Next of Kin in Organ Donation" (2006) 40 *Valparaiso University Law Review* 169

Petersson, A, "The Production of a Memorial Place: Materialising Expressions of Grief" in A Maddrell and JD Sidaway (eds), *Deathscapes: Spaces for Death, Dying, Mourning and Remembrance* (Ashgate, 2010), pp 141–158

Pieper, MA, "Frozen Embryos: Persons or Property? *Davis v Davis*" (1990) 23 *Creighton Law Review* 807

Pitcher, G, "The Misfortunes of the Dead" (1984) 21 *American Philosophical Quarterly* 183

Pobjoy, J, "Medically Mediated Reproduction: Posthumous Conception and the Best Interests of the Child" (2008) 15 *Journal of Law and Medicine* 450

Poulson, CJ, Brittain, RP and Marhsall, TK, *The Disposal of the Dead* (English Universities Press, 1953) (updated in 1962, and again in 1975)

Poulter, DJ, "The Case of Dr Shipman" (2003) 24 *American Journal of Forensic Medicine and Pathology* 219

Prendergast, D, Hockey, J and Kellaher, L, "Blowing in the Wind? Identity, Materiality and the Destinations of Human Ashes" (2006) 12 *Journal of the Royal Anthropological Institute* 881

Price, D, *Legal and Ethical Aspects of Organ Transplantation* (CUP, 2000)

Price, D, "The Human Tissue Act 2004" (2005) 68 *Modern Law Review* 798

Price, D, "Property, Harm and the Corpse" in Brooks-Gordon, B, Ebtehaj, F, Herring, J, Johnson, MH and Richards, M (eds), *Death Rites and Rights* (Hart, 2007), pp 199–217

Price, D, *Human Tissue in Transplantation and Research: A Model Legal and Ethical Donation Framework* (CUP, 2010)

Price, TW, "Legal Rights and Duties in Regard to Dead Bodies, Post-Mortems and Dissections" (1951) 68 *South African Law Journal* 403

Quay, PM, "Utilizing the Bodies of Dead" (1984) 28 *St Louis University Law Journal* 889

Quigley, M, "Incentivising Organ Donation" in Farrell, AM, Price, D and Quigley, M (eds), *Organ Shortage: Ethics, Law and Pragmatism* (CUP, 2011), pp 89–103

Quigley, M, "Property in Human Biomaterials: Separating Persons and Things?" (2012) 32 *Oxford Journal of Legal Studies* 659

Rao, R, "Property, Privacy and the Human Body" (2000) 80 *Boston University Law Review* 359

Rees, D, *Death and Bereavement: The Psychological, Religious and Cultural Interfaces* (Whurr, 2nd edn, 2001)

Render, M "The Law of the Body" (2013) 62 *Emory Law Journal* 549

Renteln, AD, "The Rights of the Dead: Autopsies and Corpse Mismanagement in Multicultural Societies" (2001) 100 *South Atlantic Quarterly* 1005

Richardson, R, *Death, Dissection and the Destitute: The Politics of the Corpse in Pre-Victorian Britain* (University of Chicago Press, 2001)

Roach, M, *Stiff: The Curious Life of Human Cadavers* (Penguin, 2003)

Roberts, P, "The Living and the Dead: Community in the Virtual Cemetery" (2004) 49 *OMEGA: Journal of Death and Dying* 57

Robertson, JA, "Posthumous Reproduction" (1993) 69 *Indiana Law Journal* 1027

Rodriguez-Dod, EC, "Ashes to Ashes: Comparative Law Regarding Survivors' Disputes Concerning Cremation and Cremated Remains" (2008) 17 *Transnational Law and Contemporary Legal Problems* 311

Rugg, J, "Defining the Place of Death: What Makes a Cemetery a Cemetery?" (2000) 5 *Mortality* 259

Rumble, H, Troyer, J, Walter, T and Woodthorpe, K, "Disposal or Dispersal: Environmentalism and Final Treatment of the British Dead" (2014) 19 *Mortality* 243

Sandberg, R, "Human Rights and Human Remains: The Impact of *Dödsbo v Sweden*" (2006) 8 *Ecclesiastical Law Journal* 453

Sanders, G, "Themed Death: Novelty in the Funeral Industry" (2008) 10 *Consumers, Commodities and Consumption*

Sanders, G, "The Dismal Trade as Culture Industry" (2010) 38 *Poetics* 47

Santino, J (ed), *Spontaneous Shrines and the Public Memorialization of Death* (Palgrave Macmillan, 2006)

Schafer, C, "Corpses, Conflict and Insignificance? A Critical Analysis of Post-Morten Practices" (2012) 17 *Mortality* 305

Scarre, G, "Privacy and the Dead" (2012) 19 *Philosophy in the Contemporary World* 1

Scarre, G, "Speaking of the Dead" (2012) *Mortality* 36

Seale, C, *Constructing Death: The Sociology of Dying and Bereavement* (CUP, 1998)

Selket, K, Glover, M and Palmer, S, "Normalising Post-Mortem: Whose Cultural Imperative? An Indigenous View on New Zealand Post-Mortem Policy" (2012) *New Zealand Journal of Social Sciences Online*

Shaw, D, "Cryoethics: Seeking Life After Death (2009) 23 *Bioethics* 515

Shaw, D, "We Should Not Let Families Stop Organ Donation From Their Dead Relatives" (2012) *BMJ: British Medical Journal* 345.

Shaw, D and Elger, B, "Persuading Bereaved Families to Permit Organ Donation" (2014) 40 *Intensive Care Medicine* 96

Shepherd, L and O'Carroll, R, "When do Next-of-Kin Opt-In? Anticipated Regret, Affective Attitudes and Donating Deceased Family Members' Organs" (2013) 19 *Journal of Health Psychology* 1508

Sher, BD, "Funeral Prearrangement: Mitigating the Undertaker's Bargaining Advantage" (1963) 15 *Stanford Law Review* 414

Sherlock, A, "Larger Than Life: Digital Resurrection and the Re-Enchantment of Society" (2014) 29 *Information Society* 164

Sherry, K, "What Happens to Our Facebook Account When We Die?: Property Versus Policy and the Fate of Social-Media Assets Postmortem" (2013) 40 *Pepperdine Law Review* 1

Skegg, PD, "The 'No Property' Rule and Rights Relating to Dead Bodies" [1997] *Tort Law Review* 222

Skene, L, "Arguments Against People Legally 'Owning' Their Bodies, Body Parts and Tissue" (2002) 2 *Macquarie Law Journal* 165

Skene, L, "Proprietary Interests in Human Bodily Material: *Yearworth*, Recent Australian Cases on Stored Sperm and Their Implications" (2012) 20 *Medical Law Review* 277

Skene, L and Masters, B, "What Legal Rights Do You Have Over Your Body After Your Death?" (2002) 81 *Australian Law Reform Commission Reform Journal* 38

Smale, DA, *Davies' Law of Burial, Cremation and Exhumation* (Shaw & Sons, 7th edn, 2002)

Smith, ATH, "Stealing the Body and its Parts" [1976] *Criminal Law Review* 623

Smith, SW and Deazley, R, *The Legal, Medical and Cultural Regulation of the Body: Transformation and Transgression* (Ashgate, 2009)

Smolensky, KR, "Rights of the Dead" (2009) 37 *Hofstra Law Review* 763

Sparkes, P, "Exclusive Burial Rights" (1991) 2 *Ecclesiastical Law Journal* 133

Sperling, D, *Posthumous Interests: Legal and Ethical Perspectives* (CUP, 2008)

Stawicki, SP, Aggrawal, A, Dean, AJ, Bahner, DA, Steinberg, SM, Stehly, CD and Hoey, BA "Postmortem Use of Advanced Imaging Techniques: Is Autopsy Going Digital" (2008) 12 *OPUS* 17

Strahilevitz, LJ, "The Right to Destroy" (2005) 114 *Yale Law Journal* 781

Strong, C, "Ethical and Legal Aspects of Sperm Retrieval After Death or Persistent Vegetative State" (1999) 27 *Journal of Law, Medicine & Ethics* 347

Strong, C, Gingrich, JR and Kutteh, WH, "Ethics of Postmortem Sperm Retrieval: Ethics of Sperm Retrieval After Death or Persistent Vegetative State" (2000) 15 *Human Reproduction* 739

Strong, C, "Gamete Retrieval After Death or Irreversible Unconsciousness: What Counts as Informed Consent?" (2006) 15 *Cambridge Quarterly of Healthcare Ethics* 161

Stueve, T, *Mortuary Law* (Cincinnati Foundation for Mortuary Education, 6th edn, 1984)

Sykas, A, "Waste Not, Want Not: Can the Public Policy Doctrine Prohibit the Destruction of Property by Testamentary Direction?" (2001) 25 *Vermont Law Review* 911

Taylor, JH, *Death and the Afterlife in Ancient Egypt* (University of Chicago Press, 2001)

Taylor, JS, "The Myth of Posthumous Harm" (2005) 42 *American Philosophical Quarterly* 311

Taylor, JS, *Death, Posthumous Harm and Bioethics* (Routledge, 2012)

Tritt, L, "Sperms and Estates: An Unadulterated Functionally Based Approach to Parent-Child Property Succession" (2009) 62 *Southern Methodist University Law Review* 367

Troyer, J, "Abuse of a Corpse: A Brief History and Re-Theorization of Necrophilia Laws in the USA" (2008) 13 *Mortality* 132

Valentine, C and Woodthorpe, K, "From the Cradle to the Grave: Funeral Welfare from an International Perspective" (2013) 48 *Social Policy & Administration* 515

Vines, P, "Objections to Post-mortem Examination: Multiculturalism, Psychology and Legal Decision-Making" (2000) 7 *Journal of Law and Medicine* 422

van Dijk, P, van Hoof, F, van Rijn, A and Zwaak, L (eds), *Theory and Practice of the European Convention on Human Rights* (Intersertia, 4th edn, 2006)

Vines, P, "Consequences of Intestacy for Indigenous People: The Passing of Property and Burial Rights" (2004) 8 *Australian Indigenous Legal Reporter* 1

Vines, P and Croucher, RF, "Law and Religion: Religion and Death in the Common Law" in Radan, P, Meyerson D and Croucher, RF, *Law and Religion: God, the State and the Common Law* (Routledge, 2012), pp 295–320

Walker, RL, Juengst, ET, Whipple, W and Davis, AM, "Genomic Research With the Newly Dead: A Crossroads for Ethics and Policy" (2014) 42 *Journal of Law, Medicine and Ethics* 220

Wall, J, "The Legal Status of Body Parts: A Framework" (2011) 31 *Oxford Journal of Legal Studies* 659

Walter, T, "Plastination for Display: A New Way to Dispose of the Dead" (2004) 10 *Journal of the Royal Anthropological Institute* 603

Walter, T, "New Mourners, Old Mourners: Online Memorial Culture as a Chapter in the History of Mourning" (2015) 15 *New Review of Hypermedia and Multimedia* 10

Walter, T and Gittings, C, "What Will the Neighbours Say? Reactions to Field and Garden Burial" in Hockey, J, Komaromy, C and Woodthorpe, K (eds), *The Matter of Death: Space, Place and Materiality* (Palgrave Macmillan, 2010), pp 165–177

Walter, T, Hourizi, R, Moncur, W and Pitsillides, S, "Does the Internet Change How We Die and Mourn? Overview and Analysis" (2011) 64 *OMEGA: Journal of Death & Dying* 275

Wear, T, "Wills: Direction in Will to Destroy Estate Property Violates Public Policy" (1976) 41 *Missouri Law Review* 309

Welbourn, H, "A Principlist Approach to Presumed Consent for Organ Donation" (2014) 9 *Clinical Ethics* 10

West, K, *A Guide to Natural Burial* (Sweet & Maxwell, 2010)

White, RCA and Ovey, C, *Jacobs, White and Ovey: The European Convention on Human Rights* (OUP, 5th edn , 2010)

White, S, "An End to D-I-Y Cremation?" (1993) 33 *Medicine, Science and the Law* 151

White, S, "Rights to (Buried?) Cremated Remains" (1998) 4 *Pharos International* 37

White, S, "The Law Relating to Dead Bodies" (2000) 4 *Medical Law International* 145

White, S, "A Burial Ahead of its Time? The Crookenden Burial Case and the Sanctioning of Cremation in England & Wales" (2002) 7 *Mortality* 171

White, S, "Funeral Pyres in a Legal Limbo" (2010) *Pharos International* 30

White, S, "The Public Health (Aquification) (England and Wales) Regulations?" (2011) 77 *Pharos International* 10

White, S, "Cremation Act 1902, s 5 (The 'Distance' or 'Radius' Clause): The Balloon and String Theory of Statutory Interpretation" (2013) 79 *Pharos International* 79

White, S, "The Legal Status of Corpses and Cremains: When and Where Can You Steal a Dead Body?" in Buckham, S, Jupp, PC and Rugg, J (eds), *Death in Scotland 1855–1955: Beliefs, Attitudes and Practices* (Peter Lang, 2016) (forthcoming)

Wienrich, S and Speyer, J, *The Natural Death Handbook* (Rider Books, 4th edn, 2003)

Wilkinson, TM, "Individual and Family Consent to Organ and Tissue Donation: Is the Current Position Coherent?" (2005) 31 *Journal of Medical Ethics* 587

Wilkinson, TM, *Ethics and the Acquisition of Organs* (OUP, 2011)

Williams, DD, "Over My Dead Body: The Legal Nightmare and Medical Phenomenon of Posthumous Conception Through Postmortem Sperm Retrieval" (2011) 34 *Campbell Law Review* 181

Wisnewski, JJ, "When the Dead Do Not Consent: A Defense of Non-Consensual Organ Use" (2008) 22 *Public Affairs Quarterly* 289

Wojcik, E, "Discrimination After Death" (2000) 53 *Oklahoma Law Review* 389

Woodthorpe, K "Private Grief in Public Spaces: Interpreting Memorialisation in the Contemporary Cemetery" in Hockey, J, Komaromy, C and Woodthorpe, K (eds), *The Matter of Death: Space, Place and Materiality* (Palgrave Macmillan, 2010), pp 117–132

Woodthorpe, K, Rumble, H and Valentine, C, "Putting 'The Grave' into Social Policy: State Support for Funerals in Contemporary UK Society" (2013) 42 *Journal of Social Policy* 605

Young, H, "The Right to Posthumous Bodily Integrity and Implications of Whose Right It Is" (2013) 14 *Marquette Elder's Advisor* 197

Young, H, "Presuming Consent to Posthumous Reproduction" (2014) 27 *Journal of Law & Health* 68

Zbarsky, I and Hutchinson, S, *Lenin's Embalmers* (Harvill Press, 1999)

Zablotsky, P, "'Curst Be He That Moves My Bones': The Surprisingly Controlling Role of Religion in Equitable Disinterment Decisions" (2007) 83 *North Dakota Law Review* 361

Zwicker, MZ and Sweatman, MJ, "Who Has the Right to Choose the Deceased's Final Resting Place" (2002) 22 *Estates, Trusts & Pensions Journal* 43

Government Reports and Official Publications

Chief Coroner and Ministry of Justice, *The Chief Coroner's Guide to the Coroners and Justice Act 2009* (September 2013), located at https://www.judiciary.gov.uk/wp-content/uploads/JCO/Documents/coroners/guidance/chief-coroners-guide-to-act-sept2013.pdf (accessed 30 September 2015)

DEFRA, *Mercury Emissions from Crematoria* (2003)

Home Office Consultation Paper, *Burial Law and Policy in the 21st Century: The Need for a Sensitive and Sustainable Approach* (2004)

Law Commission for England and Wales, *Intestacy and Family Provision Claims on Death* (Law Com No 331, 2011)

Ministry of Justice, *Burial Law and Policy in the 21st Century: The Need for a Sensitive and Sustainable Approach* (2004*): Government Response to the Consolation Carried Out by the Home Office DCA* (2007)

New Zealand Law Commission, *The Legal Framework for Burial and Cremation in New Zealand: A First Principles Review* (Issues Paper 34, October 2013)

Office of Fair Trading, "Funerals: A Report of the OFT Inquiry into the Funerals Industry" (2001)

Office of National Statistics, *Short Report: Cohabitation in the UK* (2012), located at www.ons.gov.uk/ons/rel/family-demography/families-and-households/2012/cohabitation-rpt.html (accessed 30 September 2015)

Office of National Statistics, *Release: Stepfamilies, 2011* (2014), located at (www.ons.gov.uk/ons/rel/family-demography/stepfamilies/2011/index.html (accessed 30 September 2015)

Queensland Law Reform Commission, *A Review of the Law in Relation to the Final Disposal of a Dead Body* (Report No 69, December 2011)

Redfearn, M, *The Royal Liverpool Children's Hospital Inquiry Report* (London: The Stationary Office, 2003)

Newspaper, Magazines and Subscription Publications, and Media Reports

By author

Barkham, P, "Should Fake Flowers Be Banned From Cemeteries?", *The Guardian* (London, 12 January 2011) www.theguardian.com/lifeandstyle/2011/jan/12/cemetery-ban-for-fake-flowers (accessed 30 September 2015)

Barton, F, "Sisters Keep Mother's Body in the Fridge for Ten Years: And Visit Her Every Weekend, *Daily Mail* (London, 6 September 2007) www.dailymail.co.uk/news/article-480276/Sisters-mothers-body-fridge-years—visit-weekend.html (accessed 30 September 2015)

Beckford, M, "Aristocrat to be Exhumed in Bird Flu Fight", *The Telegraph* (London, 28 February 2007) www.telegraph.co.uk/news/uknews/1544094/Aristocrat-to-be-exhumed-in-bird-flu-fight.html (accessed 30 September 2015)

Bloomfield, A, "Family Given Permission to Extract Eggs from Ovaries of Dead Daughter in World First", *The Telegraph* (London, 8 August 2011) www.telegraph.co.uk/news/health/news/8689479/Family-given-permission-to-extract-eggs-from-ovaries-of-dead-daughter-in-world-first.html (accessed 30 September 2015)

Bove, AA and Langa, M, "Ted Williams: Is He Headed for the Dugout or the Deep Freeze? Property Rights in a Dead Body Resurrected", *Massachusetts Lawyers Weekly* (19 August 2002)

Brown, J, "As a Final Act, Elizabeth Taylor Is Late For Her Own Funeral", *The Independent* (London, 26 March 2011) www.independent.co.uk/news/world/americas/as-a-final-act-elizabeth-taylor-is-late-for-her-own-funeral-2253475.html (accessed 30 September 2015)

Chapman, J, "It's Like Poundland! Council Bans Grave Trinkets and Chimes", *Daily Express*, (London, 2 February 2011) www.express.co.uk/news/uk/226622/It-s-like-Poundland-Council-bans-grave-trinkets-and-chimes (accessed 30 September 2015)

Christafis, A, "Freezer Failure Ends Couple's Hopes of Life After Death", *The Guardian* (London, 17 March 2006)

Clinton, J, "We Need Human Body Farms, Says Real-Life Dr Bones and Forensic Expert Anna Williams", *Sunday Express* (London, 22 September 2013) www.express.co.uk/news/science-technology/431287/We-need-human-body-farms-says-real-life-Dr-Bones-and-forensic-expert-Anna-Williams (accessed 30 September 2015)

Collinson, P, "Are Cemeteries the New Safe Investment?", *The Guardian* (London, 16 October 2010) www.theguardian.com/money/2010/oct/16/cemeteries-burial-investment (accessed 30 September 2015)

de Bruxelles, S, "Artist Kept Tramp's Corpse in a Cupboard", *The Times* (London, 12 October 2002)

Ferguson, E, "It's My Funeral: Getting Ready for the End", *The Observer* (London, 13 May 2012) www.theguardian.com/lifeandstyle/2012/may/13/funerals-death-humanist-euan-ferguson (accessed 30 September 2015)

Foster, D, "Too Poor to Die: How Funeral Poverty is Surging in the UK", *The Guardian* (London, 9 June 2015)www.theguardian.com/commentisfree/2015/jun/09/poor-die-funeral-poverty-costs-uk (accessed 30 September 2015)

Gafson, I, "The Facts Don't Lie: We Haven't Cracked Egg Freezing. Not Even Close", *The Telegraph* (London, 17 October 2014) www.telegraph.co.uk/women/womens-life/11169420/Facebook-egg-freezing-The-facts-dont-lie-We-havent-cracked-egg-freezing.-Not-even-close.html (accessed 30 September 2015)

Herzog, K, "A Greener Afterlife: Is Human Composting the Future for Funerals?", *The Guardian* (London, 10 March 2015) www.theguardian.com/environment/2015/mar/10/a-greener-afterlife-is-human-composting-the-future-for-funerals (accessed 30 September 2015)

Jones, R, "Cost of Dying Outstrips Inflation", *The Guardian* (London, 5 October 2015) www.theguardian.com/money/2015/oct/05/cost-of-dying-outstrips-inflation-funeral-3500 (accessed 15 October 2015)

Kamanev, M, "Aquamation: A Greener Alternative to Cremation?", *Time Science* (online) (28 September 2010) http://content.time.com/time/health/article/0,8599,2022206,00.html (accessed 30 September 2015)

Kelly, J, "Happy Funerals: A Celebration of Life?", *BBC News Magazine* (14 June 2015) www.bbc.co.uk/news/magazine-31940529 (accessed 30 September 2015)

Jones, S, "Police Say Sikh Funeral Pyre May Have Broken Cremation Laws", *The Guardian* (London, 13 July 2006) www.guardian.co.uk/uk/2006/jul/13/religion.world (accessed 30 September 2015)

Lynn, G, "Exhumation After Wrong Bodies Buried in Hospital Mix-Up", *BBC News Online* (4 July 2011) www.bbc.co.uk/news/uk-england-london-13993666 (accessed 30 September 2015)

Malnick, E, "Bodies 'Buried in Former Car Parks Due to Graves Shortage'", *The Telegraph* (London, 27 September 2013) www.telegraph.co.uk/news/religion/10337249/Bodies-buried-in-former-car-parks-due-to-graves-shortage.html (accessed 30 September 2015)

Mangan, L, "One House For Sale, Two Bodies in the Garden – But No Discount", *The Guardian* (London, 13th January 2014) www.theguardian.com/lifeandstyle/shortcuts/2014/jan/12/house-for-sale-bodies-garden-no-discount (accessed 30 September 2015)

Marsden, S, "Billionaire Heir Hans Rausing Escapes Jail for Preventing His Wife Eva's Burial", *The Telegraph* (London, 1 August 2012) www.telegraph.co.uk/news/uknews/crime/9443573/Billionaire-heir-Hans-Rausing-escapes-jail-for-preventing-his-wife-Evas-burial.html (accessed 30 September 2015)

Miller, J, "Britons Spend More Time on Tech Than Asleep, Study Suggests", *BBC News Online* (7 August 2014) www.bbc.co.uk/news/technology-28677674 (accessed 30 September 2015)

O'Rourke, M and Williams, G, "Burial of a Child's Remains: Resolving Parental Disputes", *Family Law Week* (28 February 2013) www.familylawweek.co.uk/site.aspx?i=ed112114 (accessed 30 September 2015)

Patterson, T, "BodyWorlds Impresario 'Used Corpses of Executed Prisoners for Exhibition'", *The Telegraph* (London, 25 January 2004)

Penny, L, "Mourning in the Digital Age: Selfies at Funerals and Memorial Hashtags", *The New Statesman* (London, 14 April 2014) www.newstatesman.com/culture/2014/04/selfies-funerals-and-memorial-hashtags-mourning-digital-age (accessed 30 September 2015)

Phillips, W, "LOLing at Tragedy: Facebook Trolls, Memorial Pages and Resistance to Grief Online" (2011) 16 *First Monday* http://firstmonday.org/ojs/index.php/fm/article/viewArticle/3168 (accessed 30 September 2015)

Pleasance, C, "Bereaved Mother Ordered to Tear Down Stillborn Daughter's Shrine Decorated with Windmills, a Pink Fence and Pink Gravel – Because It's Not 'In Keeping' with Graveyard", *Daily Mail* (London, 29 January 2015) www.dailymail.co.uk/news/article-2931795/Bereaved-mother-ordered-tear-stillborn-daughter-s-shrine-decorated-windmills-pink-fence-pink-gravel-s-not-kee ping-graveyard.html (accessed June 2015)

Rayner, G, "Grave of Catholic Shadrack Smith Will Not Be Exhumed, Family Told in Row Over Muslim Burial Plot Next Door", *The Telegraph* (London, 11 February 2015) www.telegraph.co.uk/news/religion/11406710/Grave-of-Catholic-Shadrack-Smith-will-not-be-exhumed-family-told-in-row-over-Muslim-burial-plot-next-door.html (accessed 30 September 2015)

Rohrer, F, "Should Graveyard Wind Chimes and Plastic Displays be Banned?", *BBC News Magazine* (9 February 2011) www.bbc.co.uk/news/magazine-12396991 (accessed 30 September 2015)

Smith Squire, A, "Couple's Grief as Vicar Bans Teddies From The Grave of Their Stillborn Baby In Case They Upset Other Churchgoers", *Daily Mail* (London, 27 September 2013) www.dailymail.co.uk/news/article-2436092/Couples-grief-vicar-bans-teddies-grave-stillborn-baby-case-upset-churchgoers.html (accessed 30 September 2015)

Swerling, G, "Live Streaming Takes Funerals into Digital Age", *The Times* (London, 27 December 2014) www.thetimes.co.uk/tto/technology/article4307581.ece (accessed 30 September 2015)

Townsend, L, "Digging Up The Dead", *BBC News Magazine* (26 June 2012) www.bbc.co.uk/news/magazine-18505222 (accessed 30 September 2015)

Usburne, D, "Missing Van Gogh Feared Cremated With Its Owner", *The Independent* (London, 27 July 1999) www.independent.co.uk/news/world/missing-van-gogh-feared-cremated-with-its-owner-1108973.html (accessed 30 September 2015)

Wheeler, C, "RIP … Until You Move Home: Brits Taking Dead Relatives With Them", *Daily Express* (London, 1 August 2015) www.express.co.uk/news/uk/595436/Relatives-moving-home-deceased (accessed 30 September 2015)

White, J and Wilkes, D, "Latte me to rest! Coffee-Lover Who Died After Battle with Cancer is Buried in Costa-Themed Coffin…With Her Order Written on the Side", *Daily Mail* (London, 17 February 2014) www.dailymail.co.uk/news/article-2561169/Making-mocha-ry-tradition-Coffee-lover-buried-Costa-themed-coffin.html (accessed 30 September 2015)

Williams, O, "How Could They Kill a Healthy Guide Dog?", *Daily Mail* (London, 4 April 2013) www.dailymail.co.uk/news/article-2303841 (accessed 30 September 2015)

Wright, N, "Death and the Internet: The Implications of the Digital Afterlife" (2014) 19 *First Monday*, p 5 http://firstmonday.org/ojs/index.php/fm/article/view/4998/4088 (accessed 30 September 2015)

Zukerman, W, "Dissolving Your Earthly Remains Will Protect the Earth", *New Scientist* (online) (19th August 2010) www.newscientist.com/article/dn19333-dissolving-your-earthly-remains-will-protect-the-earth.html (accessed 30 September 2015)

No author listed

American Bar Association Journal (8 April 2009), "Texas Judge Tells Medical Examiner to Help Mom Get Dead Son's Sperm"

BBC News Online (31 May 2000), "Keeping Body Frozen Ruled Illegal", http://news.bbc.co.uk/1/hi/world/europe/771811.stm (accessed 30 September 2015)

BBC News Online (10 January 2005), "Crematoria Warned Over Mercury", http://news.bbc.co.uk/1/hi/health/4160895.stm (accessed 30 September 2015)

BBC News Online, (21 January 2005), "Body Burning Son Spared Jail Term", http://news.bbc.co.uk/1/hi/england/west_yorkshire/4196447.stm (accessed 30 September 2015)

BBC News Online (16 March 2006), "Frenchman Cremates Frozen Parents", http://news.bbc.co.uk/1/hi/world/europe/4814540.stm (accessed 30 September 2015)

BBC News Online (21 July 2011), "Top Nazi Rudolf Hess Exhumed From 'Pilgrimage' Grave", www.bbc.co.uk/news/world-europe-14232768 (accessed 30 September 2015).

BBC News Online (9 January 2015), "Police Called to Soldier's Funeral After Family Feud, www.bbc.co.uk/news/uk-scotland-tayside-central-30750749 (accessed 30 September 2015)

Chicago Tribune (Chicago, 3 August 2005), "Nudist to be Buried Clothed", http://articles.chicagotribune.com/2005-08-03/news/0508030440_1_arrests-buried-robert-norton (accessed 30 September 2015)

Daily Mail (London, 4 May 2011), "Grandmother's Last Wish to Have Gnomes Lining Funeral Route Scuppered – By 'ELF and Safety killjoys", www.dailymail.co.uk/news/article-1383075/Grandmothers-wish-gnomes-lining-funeral-route-scuppered-health-safety.html#ixzz1RK8wwS4y (accessed 30 September 2015)

Daily Mail (London, 17 October 2013), "It Doesn't Add Up! Family's Anger After Council Orders Them to Remove 'Offensive' Sudoku Headstone Tribute to Late Father Mathematician", www.dailymail.co.uk/news/article-2465290/Familys-anger-council-orders-removal-sudoku-headstone-mathematician-Allan-Robinson.html (accessed 30 September 2015)

International Business Times (18 July 2014), "Nelson Mandela Day: How Mandela Family Feud Overshadowed Madiba's Final Year and Death", www.ibtimes.co.uk/nelson-mandela-day-how-mandela-family-feud-overshadowed-madibas-final-year-death-1457117 (accessed 30 September 2015)

International Business Times (30 January 2015), "Facebook: One Out Of Every Five People On Earth Have An Active Account", www.ibtimes.com/facebook-one-out-every-five-people-earth-have-active-account-1801240 (accessed 30 September 2015)

National Review (9 July 2002), "Frozen Future".

The Belfast Telegraph (Belfast, 4 June 2014), "Mass Grave of 796 Babies Found in Septic Tank at Catholic Orphanage in Tuam, Galway", www.belfasttelegraph.co.uk/news/local-national/republic-of-ireland/mass-grave-of-796-babies-found-in-septic-tank-at-catholic-orphanage-in-tuam-galway-30327483.html (accessed 30 September 2015)

The Courier-Mail (Brisbane, 12 August 2010), "Queenslanders Can Now Have a Watery Grave with 'Aquamation' Centre Opening on the Gold Coast", www.couriermail.com.au/news/queenslanders-can-now-have-a-watery-grave-with-aquamation-centre-opening-on-the-gold-coast/story-e6frep26-122590443 5756 (accessed 30 September 2015)

The Guardian (London 8 May 2015), "Mother in Legal Battle to Bear Dead Daughter's Child", www.theguardian.com/uk-news/2015/may/08/mother-in-legal-battle-to-bear-dead-daughters-child (accessed 30 September 2015)

The Sentinel (Stoke, 6 July 2013), "Church Bans Headstone of Ozzy Osbourne Roadie", www.stokesentinel.co.uk/Video-Church-bans-headstone-Ozzy-Osbourne-roadie/story-19485728-detail/story.html (accessed 30 September 2015)

The Telegraph (London, 10 May 2010), "Cemetery Bans Artificial Flowers Over Fear of Being Sued", www.telegraph.co.uk/news/uknews/7705476/Cemetery-bans-artificial-flowers-over-fear-of-being-sued.html (accessed 30 September 2015)

The Telegraph (London, 25 January 2011), "Crematorium Heats Swimming Pool", www.telegraph.co.uk/news/newstopics/howaboutthat/8279648/Crematorium-heats-swimming-pool.html (accessed 30 September 2015)

The Telegraph (London, 3 June 2011), "Gravestone Needs Killer's Consent", www.telegraph.co.uk/news/uknews/law-and-order/8552935/Gravestone-needs-killers-consent.html (accessed 30 September 2015)

The Telegraph (London, 26 February 2015), "Frankenstein-Style Human Head Transplant 'Could Happen in Two Years'", www.telegraph.co.uk/news/science/science-news/11436319/Frankenstein-style-human-head-transplant-could-happen-in-two-years.html (accessed 30 September 2015).

The Telegraph (London, 2 August 2015), "Church Warns Against Treating Bereaved Relatives as 'Portable Remains'", www.telegraph.co.uk/news/religion/11779209/Church-warns-against-treating-bereaved-relatives-as-portable-remains.html (accessed 30 September 2015)

The Times of Israel (Israel, 6 May 2014), "Swastika Covered Up Austrian Tombstone", www.timesofisrael.com/swastika-covered-up-on-austrian-tombstone/ (accessed 30 September 2015)

The Telegraph (London, 6 November 2015), "Siblings in High Court Battle Over Whether Mother Should Be Buried in Jewish or Christian Ceremony", www.telegraph.co.uk/news/uknews/law-and-order/11980060/Siblings-in-High-court-battle-over-whether-mother-should-be-buried-in-Jewish-or-Christian-ceremony.html (accessed 1 December 2015)

Miscellaneous

Cremation Society of Great Britain, *History of Modern Cremation in Great Britain From 1874: The First Hundred Years* (1974), located at www.cremation.org.uk/ (accessed 30 September 2015)

IBISWorld, *Funeral Activities in the UK: Market Research Report* (January 2015)

Human Tissue Authority, Code of Practice 1: Consent (Updated July 2014, www.hta.gov.uk/code-practice-1-consent) (accessed 30 September 2015)

Human Tissue Act, Code of Practice 2: Donation of Solid Organs for Transplantation (Updated July 2014, www.hta.gov.uk/code-practice-2-donation-solid-organs-transplantation) (accessed 30 September 2015)

Scottish Government, *Consultation on a Proposed Bill Relating to Burial and Cremation and Other Related Matters in Scotland* (January 2015), [35]–[40].

Sun Life, *Cost of Dying: 2015* (13 October 2015) www.sunlifedirect.co.uk/press-office/cost-of-dying-2015/ (accessed 15 October 2015)

Woodthorpe, K, *Affording a Funeral: Social Fund Funeral Payments* (Axa Sun Life Direct Report, 2012)

Index